Science Policy Reports

The series Science Policy Reports presents the endorsed results of important studies in basic and applied areas of science and technology. They include, to give just a few examples: panel reports exploring the practical and economic feasibility of a new technology; R&D studies of development opportunties for particular materials, devices or other inventions; reports by responsible bodies on technology standardization in developing branches of industry.

Sponsored typically by large organizations - government agencies, watchdogs, funding bodies, standards institutes, international consortia - the studies selected for Science Policy Reports will disseminate carefully compiled information, detailed data and in-depth analysis to a wide audience. They will bring out implications of scientific discoveries and technologies in societal, cultural, environmental, political and/or commercial contexts and will enable interested parties to take advantage of new opportunities and exploit on-going development processes to the full.

More information about this series at http://www.springer.com/series/8882

Paul Janmey • Daniel Fletcher
Sharon Gerecht • Ross Levine
Parag Mallick • Owen McCarty
Lance Munn • Cynthia Reinhart-King
Editors

Physical Sciences and Engineering Advances in Life Sciences and Oncology

A WTEC Global Assessment

Springer

Editors
Paul Janmey
Institute for Medicine and Engineering
University of Pennsylvania
Philadelphia, PA, USA

Sharon Gerecht
JH Physical Sciences-Oncology Center, Institute for NanoBioTechnology, Department of Chemical and Biomolecular Engineering
John Hopkins University
Baltimore, MD, USA

Parag Mallick
Cancer Institute
School of Medicine
Stanford University
Stanford, CA, USA

Lance Munn
Radiation Oncology Edwin L. Steele Lab for Tumor Biology
Harvard Medical School
Charlestown, MA, USA

Daniel Fletcher
Bioengineering
University of California
Berkeley, CA, USA

Ross Levine
Human Oncology & Pathogenesis Program
Memorial Sloan-Kettering Cancer Center
New York, NY, USA

Owen McCarty
Department of Biomedical Engineering
Oregon Health & Science University
Portland, OR, USA

Cynthia Reinhart-King
Department of Biomedical Engineering
Cornell Univesity
Ithaca, NY, USA

ISSN 2213-1965 ISSN 2213-1973 (electronic)
Science Policy Reports
ISBN 978-3-319-17929-2 ISBN 978-3-319-17930-8 (eBook)
DOI 10.1007/978-3-319-17930-8

Library of Congress Control Number: 2015956024

Springer Cham Heidelberg New York Dordrecht London

Printed on acid-free paper

Springer International Publishing AG Switzerland is part of Springer Science+Business Media (www.springer.com)

World Technology Evaluation Center, Inc. (WTEC)

Acknowledgments

We at WTEC wish to thank all the panelists for their valuable insights and their dedicated work in conducting this international assessment of physical sciences and engineering advances in life sciences and oncology (APHELION) research and development and to thank all the site visit hosts for so generously sharing their time, expertise, and facilities with us. For their sponsorship of this important study, our sincere thanks go to the National Cancer Institute at the National Institutes of Health (NIH), the National Science Foundation, and the National Institute of Biomedical Imaging and Bioengineering at the NIH.

WTEC, Lancaster, PA, USA R.D. Shelton

WTEC Mission

WTEC provides assessments of international research and development in selected technologies under awards from the National Science Foundation (NSF), the Office of Naval Research, and other agencies. Formerly part of Loyola University Maryland, WTEC is now a separate nonprofit research institute. Sponsors interested in international technology assessments and related studies can provide support for the program through NSF or directly through separate grants or GSA task orders to WTEC.

WTEC's mission is to inform US scientists, engineers, and policy makers of global trends in science and technology. WTEC assessments cover basic research, advanced development, and applications. Panels of typically six technical experts conduct WTEC assessments. Panelists are leading authorities in their field, technically active, and knowledgeable about US and foreign research programs. As part of the assessment process, panels visit and carry out extensive discussions with foreign scientists and engineers in their labs.

The WTEC staff helps select topics, recruits expert panelists, arranges study visits to foreign laboratories, organizes workshop presentations, and, finally, edits and publishes the final reports. Dr. R.D. Shelton is the WTEC point of contact: 717 299–7130, shelton@wtec.org.

Foreword

This study covers some of the applications of physical sciences and engineering to oncology and to biology generally. It provides an international perspective of the subject, through extensive visits to some of the leading labs abroad by a team of American peer reviewers. The results provide a unique snapshot of the status and trends in this field and help identify opportunities for further progress.

The main purpose of this foreword is to acknowledge the contributions of those who helped make the study a success. Most of all, we are deeply indebted to our foreign hosts who generously shared their research results with us and provided many constructive discussions. How many? Well, we visited 63 sites in Europe, Asia, and Israel and met with more than 200 host scientists. This does not count many others who presented at several workshops organized in Asia. The details are in Appendices B, C, and D. That is a lot of hospitality and a lot of travel by our indefatigable panelists.

The panel was ably led by Paul Janmey, who cheerfully managed to get everyone to provide their findings on time, despite their having a few other things to do. We much appreciate his hard work and that of the other panelists: Dan Fletcher, Sharon Gerecht, Ross Levine, Parag Mallick, Owen McCarty, Lance Munn, and Cynthia Reinhart-King. Our NCI sponsors were also closely engaged in the study, including attending many of the site visits abroad. Tito Fojo and Denis Wirtz provided sage advice when it was most needed. We continue to be impressed by the quality and insight of these scientists and the high regard with which they are held by members of the worldwide oncology community. It is their expertise that makes this report credible. Short bios are in Appendix A.

It takes resources to conduct a study like this. The study was planned by Jerry S.H. Lee, Larry Nagahara, and Nastaran Kuhn at the National Cancer Institute, which provided most of the funding. Program officers at the National Science Foundation (Kesh Narayanan, Mike Roco, Semahat Demir, Kaiming Ye, and Clark Cooper) and at the National Institute for Biomedical Imaging and Biotechnology (Chris Kelley) also contributed to the study.

Prof. Janmey has prepared an executive summary that reflects consensus conclusions by the whole panel. In his Introduction chapter, he provides the background and scope of the study. In particular, he details the process used, including sites visited abroad, and provides an overview of the chapters to follow, which use an analytical organization of the topics covered geographically in the site reports.

Two earlier versions of this report have been posted at:

- http://www.wtec.org/aphelion/AphelionEuropeanReport09.23.12.pdf
- http://www.wtec.org/aphelion/AphelionFinalReport-web.pdf

WTEC, Lancaster, PA, USA R.D. Shelton
February 11, 2015

Executive Summary

The mission of publically funded investments in health-related scientific work, including the billions of dollars invested in cancer research, is to determine the origins of diseases and thereby develop diagnostic methods and countermeasures to prevent, reverse, or control their devastating consequences. Cancer research in the United States and throughout the world has been dominated by programs in cell biology, genetics, biochemistry, and animal models of cancer, which have led to important scientific advances and improvements in the diagnosis and treatment of many cancers. However, in contrast to the enormous advances in the prevention, treatment, and cure of infectious disease, cardiovascular disease, and other major causes of death throughout the world, the diagnosis of cancer is as devastating a reality today as it was decades ago. The more we learn about cancer biology, the more it has become apparent that the relationship between a specific gene mutation and disease is often staggeringly complex, the interaction of cancer cells with their local environment is an essential but largely obscure aspect of the disease, and traditional methods of cancer biology research might not be sufficient to produce the results required for effective clinical improvements. In particular, concepts and methods of physical science in cancer biology research have until very recently been largely focused on engagement of physicists and engineers in design of instrumentation for diagnosis and therapy, rather than directly on issues related to how malignant cells arise, develop, and spread to produce clinical symptoms.

To address these issues, the National Cancer Institute (NCI) held a series of three Physical Sciences in Oncology Workshops and Think Tanks between February and October 2008 (http://physics.cancer.gov/workshops/). The aim of these meetings was to explore the opportunities to advance cancer research by integrating physicists and physical science approaches with the more traditional research effort in cancer biology. The ideas and discussions at these meetings helped guide an initiative within the NCI to establish an Office of Physical Sciences Oncology (OPSO). OPSO facilitates the development and implementation of physical science-based initiatives supporting cancer research for NCI and integrates such efforts in other divisions of the National Institutes of Health (NIH) and throughout the research community.

The focus of this effort is to go beyond involving physicists and engineers in the development of new instrumentation or methods and to engage the methods and concepts of physics to foster discoveries and new fields of study related to cancer research. Broadly speaking, the goal is to study cancer as a physical process. The National Science Foundation (NSF) has traditionally been involved in funding research in engineering and the physical sciences. Such efforts can have an important impact in biomedical research while maintaining a fundamentally basic research perspective. Recently, NSF and NCI have collaborated on a funding opportunity titled Physical and Engineering Sciences in Oncology (PESO) Awards that dovetails with the efforts of OPSO. The main mission of OPSO is to fund physical sciences-oncology research in various centers throughout the United States. As a result, 12 leading institutions were selected in 2009 to build a collaborative network of Physical Science-Oncology Centers (PS-OCs) (http://physics.cancer.gov/centers/).

PS-OCs are now reaching the fruition of their initial assignments. To compare the missions of this and other initiatives with related research efforts abroad, NCI, in cosponsorship with NSF, commissioned the World Technology Evaluation Center Inc. (WTEC) to undertake an international Assessment of Physical Sciences and Engineering Advances in Life Sciences and Oncology (APHELION).

On 18 January 2012, the sponsors/chair meeting of the WTEC APHELION study was held at NSF headquarters. The main goals of the sponsors' meeting were to provide an overview of the plans for the study, to solicit interest and participation of other US Government agencies, and to coordinate the study design with WTEC.

On 1 February 2012, the kickoff meeting of the WTEC APHELION study was held at NIH's Bethesda, MD, campus where the scientific panel and advisors met with the sponsors.

The initial phase of APHELION determined the status and trends of applying physical sciences and engineering principles to oncology research and development in leading laboratories and organizations in Europe via an on-site peer review process. The study group visited laboratories in France, Italy, Israel, Germany, the Netherlands, Spain, Sweden, and Switzerland, typically meeting with representatives of multiple institutions at each stop. Assessments of the activities in the physics/biomedicine interface at European sites are provided in Appendix B. This project was completed in 2012, and details of this stage of the study are available at http://www.wtec.org/aphelion.

On 12 June 2012, panel members presented a workshop at the Cloisters building lecture room on the NIH campus to report the findings of the study group's visit to European laboratories and to discuss these findings with the sponsors and the public. Details and documents presented at this workshop are available at www.tvworldwide.com/events/nih/120612/.

The second phase of APHELION was initiated in June 2013 with visits to laboratories in Asia working at the interface of physics and biomedical sciences. These visits involved sites in Singapore, China, Taiwan, Hong Kong, and Japan. A third phase of the project included visits from a subset of the committee to laboratories in England and Scotland in October 2013. Reports on the activities at sites visited in Asia and the United Kingdom are in Appendices C and D, respectively. Site visit reports were also prepared for two sites in Brazil, and they are found in Appendix E. A final workshop to summarize findings in Asia and the United Kingdom and to discuss the findings with reference to research efforts in Europe took place at Fishers Lane Conference Center, Rockville, MD, on 21 November 2013. Details and documents presented at this workshop are available at www.tvworldwide.com/events/nih/131121/.

Scientific panel members:

- Daniel A. Fletcher, Ph.D., D.Phil. Professor of Bioengineering and Biophysics at the University of California, Berkeley
- Sharon Gerecht, Ph.D., Associate Professor of Chemical and Biomolecular Engineering at Johns Hopkins University
- Paul Janmey, Ph.D. (study chair) Professor of Physiology, Physics, and Bioengineering at the Institute of Medicine and Engineering at the University of Pennsylvania
- Ross Levine, M.D., Laurence Joseph Dineen Chair in Leukemia Research, Human Oncology and Pathogenesis Program, Memorial Sloan-Kettering Cancer Center
- Parag Mallick, Ph.D., Assistant Professor of Radiology, Bio-X Program, at the Canary Center for Cancer Early Detection, Stanford University
- Owen McCarty, Ph.D., Associate Professor of Biomedical Engineering at the Oregon Health and Science University
- Lance L. Munn, Ph.D., Associate Professor of Radiation Oncology at the Massachusetts General Hospital/Harvard Medical School
- Cynthia A. Reinhart-King, Ph.D., Associate Professor of Biomedical Engineering at Cornell University

Expert advisors to the study panel:

- Antonio Tito Fojo, M.D., Ph.D., Head, Experimental Therapeutics Section Medical Oncology Branch and Affiliates at the National Institutes of Health
- Denis Wirtz, Ph.D., Theophilus H. Smoot Professor, Department of Chemical and Biomolecular Engineering at Johns Hopkins University

Short biographies of the panel members and advisors are provided in Appendix A.

The goals of the APHELION study are:

1. Compare the US research and development activities related to the interface between physics and oncology, or more generally between physical science and biomedicine, with similar work being done in Europe and Asia.
2. Identify the gaps and barriers for research groups and clinicians in the United States by working with leading European and Asian institutions.
3. Identify major innovations that are emerging abroad.
4. Identify opportunities for cooperation and collaboration with research groups and industry in Europe and Asia.
5. Guide US research investments at the physics/oncology interface.

The initial meeting of the working group and sponsors allowed for extensive discussion of the more important topics. We identified areas of research and technology development with the greatest potential to advance understanding and treatment of cancer and other diseases. As a result of this exchange, six topic areas were identified. Each study group member took responsibility for one topic, analyzed information collected during the site visits, and integrated their findings with the current state of understanding.

The following topics form the basis for each of the chapters in this book:

1. Information and complexity: New methods for dealing with the enormous data sets generated by modern imaging methods and integrating methods developed by the physics community to understand complex, nonlinear systems and emergent properties that cannot be predicted by traditional biological models.
2. Microenvironment: The influence of chemical composition, spatial patterning, nutrient supply, oxygen stress, and other features of the tissue and extracellular environment on the growth and homeostasis of normal tissues and tumors.
3. Cell and tissue mechanics: How the forces generated by cells and the viscoelasticity of the cell and extracellular matrix affect cell growth, survival, differentiation, and movement.
4. Transport: How the movement of cancer cells, nutrients, growth factors, drugs, and fluids affects cell survival and tissue mechanics. How the removal of metabolic waste products and cell debris is controlled and how they are altered in the tumor environment.
5. Dynamics: How the rates and patterns of cell shape change, migration, and division can be measured, understood, and integrated with biochemical and genetic information.
6. Devices and new diagnostic principles: New technologies based on physical principles, especially those in which the physical properties of tissues are exploited for cancer diagnosis or treatment.

Outcomes and Summary of Findings

The purpose of our visits was focused on learning about new scientific advances and plans for future studies. We also examined each institution's facilities, traditions, advantages, and challenges related to performing interdisciplinary or multidisciplinary work. There was a clear perception among the investigators at every site that the interface of physical and biomedical sciences is a growth area with potential for both scientific discovery and medical applications. In the context of cancer research, there is also a clearly evident trend to engage physicists in roles beyond those of traditional "medical physics" focused on radiation physics or diagnostic instrumentation. New research programs throughout the world increasingly engage scientists to consider cancer as a physical process that, despite its complexity and heterogeneity, nonetheless has limits imposed by physical laws that can be addressed by thermodynamics, information theory, mechanics, hydrodynamics, and other fields in which physicists and engineers can act as partners with biologists.

It is impossible to generate a comprehensive analysis of the relative strengths of research efforts throughout Europe and Asia from the limited sites that were visited and the personnel constraints of this project, but several consensus views emerged from the study group. It is also evident that, especially in multidisciplinary projects, national boundaries are blurred and nearly all large groups include partners from other countries and very often collaborators in North America or other important research centers in India, Australasia, South America, and other sites to which visits could not be arranged.

At several sites, new research programs were explicitly motivated by the initiatives taken by OPSO and, in some cases, have involved researchers in the United States as advisors. More frequently, the initiatives started by the NCI have guided funding and scientific policy agencies in other countries to facilitate related efforts tailored to the expertise and traditions of existing research institutions. In other cases, for example, at the Institut Curie in France, integrated physics/biology programs have a long tradition and have become part of the established curriculum and research programs. At institutions in Heidelberg and Munich, Germany, there was even a sense that a critical mass of researchers at this interface might already have been reached. In most institutions we visited, interdisciplinary studies at the physics/biomedicine interface were highly attractive to graduate students and young faculty and were often increasingly supported by granting agencies. The funding mechanisms that support these efforts vary widely, ranging from support of individual researchers or groups by focused interdisciplinary grants (common in many countries) to massive investments in infrastructure, instrumentation, and new hiring in rapidly developing academic systems in Singapore and China Many of our hosts told us that interdisciplinarity cannot be optimized without a firm basic grounding in a specific physical or biological science in the education of students and young researchers.

Throughout Europe and Asia there is evidence that the vision of NCI and NSF to engage physics more deeply in cancer research coincided with initiatives based on similar beliefs that engagement of not only physicists but the concepts and methods of physics research could benefit cancer research. One example of this type of initiative is the document, "Progress in the Domain of Physics Applications in Life Science with an Invention for Substantial Reduction of Premature Cancer Deaths: The Need for a Paradigm Change in Oncology Research" (www.crosettofoundation.org/uploads/371.pdf) which received nearly 1,000 signatures from 29 to 31 January 2010. The document argues for the need to engage new ways of thinking in cancer research, including using physical science to combat cancer. This study surveyed the World Health Organization data to conclude that "despite annual cancer costs of $741 billion/year ($750/citizen), the 38 most industrialized nations had only a 5% reduction in cancer deaths over the past 50 year (heart disease was reduced by 64%)."

Such considerations have led to many new conferences and funding initiatives. For example, in 2012, the Cancer ITMOs and Health Technologies ITMOs of the French National Alliance for Life and Health Sciences, in partnership with the French National Cancer Institute, initiated a call for research projects in physics, mathematics, or engineering sciences related to cancer (https://www.eva2.inserm.fr/EVA/jsp/AppelsOffres/CANCER/index_F.jsp). New laboratories of excellence have also been funded in France, including CELTISPHYBIO, initiated in 2012 at the Institut Curie, to establish a center for physics in cell biology. In Sweden, the Science for Life Laboratory (http://www.scilifelab.se/), which integrates research across multiple intuitions to enable collaborations between technical universities, medicals schools, and basic science research, is one of the largest scientific investments in Swedish history. New funding programs for interdisciplinary projects at the physical science/biomedicine interface funded by the German Science Foundation and the Max-Planck-Society are almost too numerous to list. Overall, despite the many funding constraints for science throughout the world, this area of research appears to be robust and in some cases even growing. An expanded list of recent conferences focused on the interface of physics and biomedical research is provided in Appendix F.

Investments in new research efforts that combine physical and biological science are especially strong in Asia, in particular in Singapore and parts of China. A significant part of the research programs established by Singapore's Agency for Science, Technology and Research in 2002 have helped foster collaboration between biological and physical scientists and have helped provide momentum for more recent large programs such as the Center for Biomedical Imaging and the Mechanobiology Institute at the National University of Singapore, where collaboration of physical and biological scientists is integral to the future of these new facilities. State-of-the-art imaging by both light and electron microscopies, adapted to problems in cancer biology and other fields of biomedicine, appears to be especially active in Asia, with world-class imaging facilities also developed in Japan, Hong Kong, and elsewhere. An integrated approach involving a wide range of expertise in experimental as well as theoretical work to study specific problems in biology has long been developed in Japan with groundbreaking results; some examples are detailed in Appendix C. Taiwan also has active and highly productive collaborations among scientists at the biology/physics interface, with research groups in physics making important advances in improved methods for diagnostics, imaging, and design of new materials for biomedical research. In Asia as well as Europe, the integration of researchers with clinicians, as well as access of clinical specimens and data for laboratory research, depends on traditions of training clinical investigators or the existence of M.D./Ph.D. training programs, which vary widely from one country to another.

In summary, the vision of the NCI to bring fresh ideas and expertise from physicists and engineers to the study of cancer biology is now actively pursued in major research and clinical centers throughout the world. Some new or growing programs have been directly influenced by the initiatives of the PS-OC, whereas in other centers such work has a long and independent history that provides opportunities for collaboration and new perspectives. The potential of physics and engineering approaches to contribute to cancer biology is in many places no longer a novel idea, but an established practice that is increasingly becoming a mainstream interest of young researchers. The following chapters in this volume provide some examples of the scientific directions where this field is now heading.

Paul Janmey

Preface

The National Cancer Institute (NCI) Office of Physical Sciences Oncology (OPSO) has initiated a large-scale interdisciplinary research effort at the interface of the physical sciences and oncology through 12 centers located throughout the United States. The NCI Physical Sciences-Oncology Center (PS-OC) program brings together experts from physics/engineering and cancer biology/oncology to enable cross-disciplinary research that merges these fields and defines a new physics of cancer. Physicists strive to explain nature by precise mathematical equations, which could bring a new perspective to cancer research.

From a molecular perspective, cancer is not a specific disease. Cancer arises as a result of a succession of randomly occurring mutations. Tumors are inherently molecularly diverse. This complexity might give the wrong impression that cancer is not accessible to physics, which strives to describe nature by precise quantitative laws. Nevertheless, statistical physics has proven to be able to find the laws behind the stochastic processes underlying thermodynamics, and nonlinear dynamics has even uncovered the principles that govern chaotic behavior in nature. Molecular background and pathogenesis of solid tumors may vary, but the pattern of tumor progression—uncontrolled proliferation, invasive growth, and metastasis—is the same. Defining and unifying physical laws that are rooted in soft matter physics are required to understand these three functions. The concept of functional modules developed in biological physics will greatly facilitate understanding the laws that govern tumor progression. In tumor cells, the modules that are responsible for division, tumor growth, and metastasis may not have identical molecular architecture, but the same physics is essential for their functions. All cells in a tissue can be motile and are viscous on long time scales, behaving very much like liquid droplets. Consequentially, tissue boundaries are comparable to fluid boundaries. Tissues can be described as a new form of fluid matter, which is a significant topic in the novel research area of active soft matter.

The most common chemotherapy agents act by killing cells that divide rapidly. Newer anticancer drugs act directly against cancer-specific proteins or inhibit tumor angiogenesis. In all these cases the goal is to reduce the tumor. Yet, the primary tumor can often be removed by surgery and radiation. It is the residual tumor cells and their ability to transgress boundaries that have to be hindered. Inducing changes in physical and material properties of tumor cells that disrupt the functional modules required for metastasis will provide a broad treatment option.

The physics of cancer is substantially more than providing new techniques for oncology. Soft matter physics as a basis for the physics of cancer has been strong in Europe. Institutions such as the Institute Curie in Paris have traditionally demonstrated that a solid connection between physics and medicine is feasible. As well, the German strength in cell biophysics has provided a good foundation. The NCI PS-OC program, which is unfortunately not yet paralleled in Europe, will jump-start the physics of cancer throughout the United States and will serve to guide similar initiatives worldwide.

Soft Matter Physics Division,
Institute for Experimental Physics I,
Faculty of Physics and Earth Science, University of Leipzig,
Leipzig, Germany

Josef A. Käs

Contents

List of Figures

Chapter 1
Introduction

Paul Janmey

Background

Physics and medicine have a long history of mutual support. The words "physics" and "physician" stem from the same word in Greek (physika) meaning "natural things," and historically some famous physicists have also been physicians. For example, Thomas Young (1777–1823) is perhaps best known for studies in mechanics—a commonly used measure of material stiffness, the Young's modulus, is named after him—but he was also a physician who was the first to diagnose astigmatism. Young understood that the vision condition originated from inappropriate shaping of the lens. Two centuries later, astigmatism is still diagnosed and treated using physical principles, even though this medical condition arises from complex genetic expression programs in the cells of the eye that deviate from the controls required for normal eye function.

The idea that physical effects help determine biological structure and function has a long if often neglected history in cell biology and physiology. The classical work of D'Arcy Thompson explicitly emphasized the importance of incorporating the laws of physics into biological models (Thompson 1942), and many experimental studies have revealed the important effects of force application, substrate stiffness, or surface topography on cell growth in culture and tissue function *in vivo*. From the relations established by Kramers (1940) and Bell (1978) that defined the effects of force on the dissociation rates of bonds at the molecular and cellular levels, respectively, to Wolff's law, which predicts how bones develop and are structured in response to imposed loads at the whole organism level (Wolff 1892), the evidence that physical effects are important, quantifiable, and controllable in biology

P. Janmey (✉)
Institute for Medicine and Engineering, University of Pennsylvania,
Philadelphia, PA 19104, USA
e-mail: janmey@mail.med.upenn.edu

P. Janmey et al. (eds.), *Physical Sciences and Engineering Advances in Life Sciences and Oncology*, Science Policy Reports,
DOI 10.1007/978-3-319-17930-8_1

and medicine is compelling. New technologies and interest in mechanical effects enabled groundbreaking studies in the late 1990s that unambiguously showed how direct application of forces to cell adhesion sites or changes in the elastic modulus of the substrate altered cell function and structure. These advances have shown how specific, controllable, and in some cases reversible effects of mechanical stimuli on cell function can act in concert with, or in some cases override or prevent, chemical stimulation.

The influences of physical effects on cancer are not unreasonable because tumor growth and metastasis are, from a macroscopic or cellular perspective, physical processes. Cells need to move through tissues and withstand the stresses of the bloodstream. Pressures build up in tumors and impinge on surrounding organs, and materials need to move into and out of tumors to facilitate their growth, spread, or metastasis. Diagnostic principles also often take advantage of physics. Clinical diagnoses are still routinely based on the size and shape of a cell or its nucleus, how a tissue feels when palpated, how it blocks radiation, and how it yields to a knife. The physical properties of cells and tissue that prove useful in diagnosis might also be informative about the differences in function and response of normal and cancerous cells.

In the context of cancer, physics has had a long involvement. The essential interactions among the genetic, biochemical, and physical properties in cancer biology were long ago recognized at leading research institutes in Europe. Notably, the Institut Curie, established in 1909 by University of Paris and the Institut Pasteur as the Radium Institute to study the biological and medical effects of radiation, is one of the world's leading cancer treatment and research centers. In 1995, it fused its physics and biology research enterprises into one unified research center. Many other European research centers have integrated biological and physical studies directed at cancer and other diseases both to develop new diagnostic and treatment methods and to understand the basic cell biology of cancer.

Pure and Applied Research

Scientific discovery is, by its nature, not possible to predict or dictate, but policy and funding decisions can do a lot to create environments that support or suppress it. It is important to find the correct balance between pure and applied research needed to achieve a specific goal. Many of the best institutions throughout the world working at the interface of physical science and medicine integrate basic and applied work, both in terms of funding mechanisms and interactions among different institutions and laboratories. In many settings, pure research is strongly supported and the results have often led to spectacularly useful applications. For example, a report in 2001 concluded that 30 % of the gross national product of the United States stems from discoveries in quantum mechanics made nearly a century ago, and mostly in Europe (Tegmark and Wheeler 2001). It seems unlikely that the originators of quantum mechanics had in mind any of its (peaceful) practical applications that are now routine. Much of this technology as well as other results of physics research, such as X-rays and proton beams, are used routinely in clinical medicine. These physical methods and the engineering it took to make them practical are all the more impressive in that they are used so routinely in western medicine that the physics behind them is almost invisible. X-ray and proton beams are widely used in treating tumors. Their effectiveness depends not only on the physics that generates these radiation streams, but also on the biophysics of the cells and tissues with which they collide. Patients whose diagnosis and treatment are aided by PET imaging to detect possible metastasis are helped because the tumor can be seen as it accumulates tracer molecules that emit a positron (a piece of antimatter!) leading to emission of gamma rays that sensitive detectors can quantify.

Complexity and Emergence

Fields of physics research besides mechanics and electromagnetism have potential for fundamental contributions to cancer research. As more genomic information accrues to describe differences between normal and cancerous tissue, it is becoming increasingly clear that the number of mutations in tumors and the differences between mutations in similar tumors in different patients is far more complex than the simple dogma of one gene/one phenotype. It is possible that the deterministic paradigms that have traditionally proved useful for many aspects of cell biology research are inadequate for understanding cancer development, and that the formalism and methods developed in physics and engineering to study complex systems and emergent properties will provide new insights into cancer etiology. Similarly, the vast amount of data derived from modern sequencing and imaging methods presents great challenges even for data storage, let alone analysis. The physics and engineering communities have a long tradition of dealing with such challenges. As one example, the recent success of the Higgs boson search depended not only on the ability to generate enough energy so that collisions between protons could produce this particle, but also to the ability to track and analyze the trillions of events that resulted from such collisions.

Overview

In the following seven chapters, experts in each of the topics of the WTEC study examine how different aspects of physics are being used to study problems in cancer biology, with an emphasis on recent results from European laboratories. In Chap. 2, **Parag Mallick** considers the potential for thinking about cancer as an emergent phenomenon and on the utility of its analysis as a complex system. In Chap. 3, **Sharon Gerecht** discusses how cancer cells react with the chemical, spatial, and physical features of their microenvironments. The mechanical properties of tissues and tumors are discussed in greater depth by **Cynthia Reinhart-King** in Chap. 4, where the effects of stiffness and forces on cancer cell biology are examined. Transport and fluid flow in tumors and the resulting physical effects are discussed by **Lance Munn** in Chap. 5. **Owen McCarty** covers the dynamics of cancer cells in Chap. 6 and looks at how aberrant movement and alteration of the cytoskeleton can arise and affect tumor growth and metastasis. Chapter 7 by **Dan Fletcher** covers recent advances in physics and methods that consider the physical properties of cells in development of diagnostic and treatment methods for cancer. Finally, in Chap. 8, **Ross Levine** looks at research advances from a medical perspective to identify new ideas and results with potential for impact in clinical applications.

Studies at the interface between physics and oncology that have been initiated by the National Cancer Institute and its counterparts in Europe, Asia, and elsewhere have generated a great deal of activity and interest in the research community. The next few years will very likely reveal new advances and surprises, and a hope of practical application to one of the most important unsolved medical challenges.

References

Bell, G.I.: Models for the specific adhesion of cells to cells. Science **200**, 618–627 (1978)

Kramers, H.A.: Brownian motion in a field of force and the diffusion model of chemical reactions. Physica **7**, 230–284 (1940)

Tegmark, M., Wheeler, J.A.: 100 years of quantum mysteries. Sci. Am. **284**, 68–75 (2001)

Thompson, D.W.: On Growth and Form. Cambridge University Press, Cambridge (1942)

Wolff, J.: Das Gesetz der Transformation der Knochen. Verlag von August Hirschwald, Berlin (1892)

Chapter 2
Complexity and Information: Cancer as a Multi-Scale Complex Adaptive System

Parag Mallick

Life is a relationship among molecules and not a property of any molecule.

—Linus Pauling

Cancer is no more of a disease of cells than a traffic jam is a disease of cars. A lifetime of study of the internal combustion engine would not help anyone to understand our traffic problems. The causes of congestion can be many. A traffic jam is due to failure of the normal relationship between driven cars and their environment and can occur whether they themselves are running normally or not.

—D.W. Smithers, Lancet, March 1962 (Smithers 1962)

Introduction

Our current understanding of biology and cancer is an implicit model of cellular and organismic regulation with its roots in early biochemical genetics inquiries. The concept that a gene is responsible for a particular protein and can be responsible for a disease was first proposed in 1908 by Archibald Garrod, an English physician (Garrod 1908). Garrod was interested in heritable diseases containing "inborn errors of metabolism." He suggested (correctly) that alkaptonuria results from a single recessive gene, which causes a deficiency in the enzyme that normally breaks down alkapton. It is now known that alkaptonuria is caused by a defect in homogentisate 1,2-dioxygenase which impairs the degradation of tyrosine (La Du et al. 1958; Zatkova 2011). Beadle and Tatum's subsequent work demonstrated that single gene mutations could incapacitate specific enzymes, so that neurospora with these

P. Mallick (✉)
Cancer Institute, School of Medicine, Stanford University,
Stanford, CA 94305, USA
e-mail: paragm@stanford.edu

P. Janmey et al. (eds.), *Physical Sciences and Engineering Advances in Life Sciences and Oncology*, Science Policy Reports,
DOI 10.1007/978-3-319-17930-8_2

mutations had significantly altered physiology—they required an external supply of nutrients to generate something that endogenous enzymes normally produced (Beadle and Tatum 1941).

These results led them to the single-gene/single-enzyme hypothesis, which states that each gene is responsible for directing the construction of a single, specific enzyme. Many researchers, including Meyerhof (Meyerhof 1945; Meyerhof and Junowicz-Kocholaty 1943; Meyerhof and Oesper 1947), have contributed to advancing the concept of "enzymatic pathways" through the elucidation of glycolysis. Taken together, these studies suggested that aberrant physiology (i.e., disease) could readily occur through the alteration of one or several genes that had immediate implications in "pathways."

Through the worldview proposed by early biochemical geneticists, the relationship between genotype and phenotype was straightforward. Furthermore, the gene-centric approach was a robust, self-consistent model for biology and was able to readily explain a number of diseases and biological phenomena. A natural consequence of this single-gene/single-enzyme view of biology is that the major focus of cancer investigation has been identifying genes and gene products whose alteration leads to carcinogenesis or to changes in the "phenotype" of cancer cells. In this worldview—the phenotype is a sum of its parts, or "genotype."

The approach of determining the origins of phenotype by deconstructing the system into its component parts may have roots in much earlier thinking, like René Descartes, who posited that complex situations can be analyzed by reducing them to manageable pieces, examining each individually, and reassembling the whole from the behavior of the pieces. At the time, during the scientific revolution of the sixteenth and seventeenth centuries, this mathematical, positivistic perception was novel relative to the prevalent descriptive-metaphysical perceptions they replaced. Descartes' reductionism preceded biology and could not have anticipated many of the challenges that arise in complex systems. Reductionist investigations continue to drive present-day biology and lead to the simple assumption that higher levels in a biological hierarchy can easily be understood from the behavior of the lower levels.

Mechanistic biology also had its roots in the seventeenth century and was influenced by the same factors that led to reductionism. The successes in Newtonian physics and in clockworks led to the belief that everything, including organisms, was based on simple clockwork-like, easily understood, deterministic principles. Mechanistic biology was most formally expressed by Jacques Loeb in 1912. His book reflected the common view of the time. All biological behavior, he concluded, was predetermined, forced, and identical between all individuals of a particular species; organisms were thus merely complex machines. Although environmental transduction mechanisms were completely unknown, Loeb assumed they must be rigid, invariant, physico-chemical mechanisms like the cogs in a clock.

In much the same way that Newtonian physics explains a lot, but not all of the behavior of objects in motion, the early views of biological regulation fail to fully explain or predict the biology. The largest hole in early models is a failure to account for the impact of context. By applying formalisms from systems and complexity theory we arrive at a very different view of the disease. We find that biology, in general, and cancer in particular can be viewed very naturally as a complex adaptive system (Deisboeck and Kresh 2006; Schwab and Pienta 1996). By altering our perception of cancer we may gain a deeper understanding of the disease, uncovering new ways to prevent it, diagnose it, and treat it.

Though systems-thinking can be traced back to early pre-Socratics of the sixth century B.C.E., it is clearly articulated in an Aristotelian world view, which focuses on the holistic as summarized in his statement "the whole is more than the sum of its parts." In modern times, systems approaches were significantly advanced in the late 1960s and 1970s by researchers such as Bertalanffy (1973; Von Bertalanffy 1972) and Laszlo (1972).

At a basic level, a system can be defined as a set of interacting, interdependent components. Systems theory provides a vocabulary and approach for modeling the behavior of any group of objects that work in concert to produce some result. Simple systems display superposition, scaling, and homogeneity, thus allowing one to readily explain behaviors driven purely by the components and not interactions amongst those components. However, interdependence is a critical feature of systems. Mathematically, if there were no interdependence and the result of a set of variables contained no cross-terms, by definition, the whole would be the sum of its parts.

A system is considered complex if it displays emergence and self-organization. In other words, if the behavior of the whole is difficult to predict from the behavior of its parts the system is complex (e.g., water formation). A complex system is adaptive if the agents as well as the system are adaptive. Systems (simple, complex, or adaptive) may be composed of other systems. Importantly, complicated and complex are not the same. There are many systems with numerous interacting parts (e.g., your laptop) whose behavior is not complex.

Typically, when studying complex systems we ask a set of questions:

- What are the components?
- What are the connections between components?
- What are the states of the components and the system as a whole?
- How do those states evolve and transition?
- What impacts the evolution of those states?
- What are the emergent behaviors?
- How does the system itself evolve?

Historically, much of systems biology has focused on the first two questions. However, there is a much wider set of questions affiliated with complex systems studies. Furthermore, complex adaptive systems display a variety of sophisticated properties, including:

Nonlinear behavior: The component parts do not act in linear ways. The superposition of the actions of the parts is not the output of the system. Small perturbations may lead to large effects (e.g., transitions in bi-stable systems).

Emergent behaviors: Properties are not obvious from the properties of the individual parts.

Self-organization: Order appears from the chaotic interactions of individuals and the rules they obey.

Adaptation (evolution): The environment becomes encoded in the rules governing the structure and/or behavior of the parts by a process of selection in which those that are better become more numerous than those that are not as fit.

Layers of description (nesting): A complex system may be composed of other complex systems. Additionally, a rule may apply at some higher levels of description but not at lower layers. Sometimes systems exhibit fractal scale-independent behavior and can be represented by the same models at different scales.

We use these properties as an organizing principle for this chapter showcasing diverse studies that provide examples of how these properties are widely prevalent throughout biology.

Research

Genome-Scale Models of Cellular Regulation: Nonlinearity

The torrential flood of data generated by -omics technologies has given us a fine-grained, detailed view of the world of genes, biomolecules, and cells—drowning us with data of immense complexity that we are just barely beginning to understand. Unfortunately, there is a deep chasm separating our knowledge of the molecular components of a cell and observations of cellular and organismic physiology—how these components interact and function together to enable cells to sense and respond to their environment, and to determine actions such as proliferation, migration, and apoptosis.

We do not understand on a fundamental level how information is transferred and processed in a biological system. Through mysterious processes, cells are able to take signals from their environment, process those signals, and then act. Unfolding this mystery of information transfer in biological systems is a critical challenge to modern biology. To unfold this mystery, physical sciences researchers attempt to develop models of information transfer and communication. Importantly, these communication systems have been shaped by millions of years of evolution and are additionally shaped by evolutionary forces within tumors. This extensive history makes it extremely difficult to develop effective and accurate models of cellular behavior.

Models of cellular regulation range from qualitative to quantitative, or from the conceptual to the mathematical. Biologists typically formulate their hypotheses (or models) in intuitive and conceptual ways, often through comparison amongst well-known systems. These biological models can be transformed into more quantitative models. In physics, mathematics is employed to describe physical phenomena. Similar approaches are required in biology to develop mechanistic and kinetic models of cellular phenomena.

Much of modern systems biology has focused on elucidating the components of a cell (which transcripts are present and in what abundance) and their connections to each other (which proteins interact or are co-regulated). These studies have led to increasingly complicated models of cellular regulation. These models often contain thousands or tens of thousands of components and are fundamentally rooted in the concept of "pathway." There are currently hundreds of molecular interaction and pathway databases. In theory, these resources should enable building or validating models of how cells use their component machinery to achieve homeostasis and response; however, there is a significant lack of compact, principle-based models illustrating the ways in which biology self-regulates.

Much like an architectural model is a replica of a building, models of cellular regulation are meant to be *in silico* replicas of the system. A biologic model conjoins a set of assumptions and declarations to reproduce or illustrate the behavior of a system and, importantly, to offer predictions for testing the model's validity. A clear definition of the system is the required first step in modeling. For example, the system of earth, its moon, and the sun is complicated. It is potentially very complicated if one includes details such as the composition of the earth and its atmosphere, as well as details about each crater on the moon. However, if the aim of the modeling is to plot the trajectories of earth around the sun and the moon around the earth, then it is sufficient to model the earth, sun, and moon as point masses and use Newton's universal law of gravitation to calculate the trajectories from aphelion to perihelion. Sometimes such compressions are not possible, i.e., there are no details that can be abstracted away. All of the details available are necessary to accurately describe the behavior of the system.

One of the key initiatives of our study was to examine work involving very compact models of regulatory mechanisms. This work has attempted to uncover fundamental properties of biological systems, asking why they are designed (or have evolved) to operate the way that they do and how it is that they are able to display non-linearity—a critical feature of biological regulation.

Dr. Jens Timmer of the Freiburg Institute, Germany, (site report, Appendix B) is attempting to uncover the general principles governing regulatory processes (Bachmann et al. 2012; Becker et al. 2010, 2012). In one study, he looked at a bacterial signaling network to investigate the impact that diverse topologies might have on its function (Kollmann et al. 2005). In particular, he asserted that a network should have the following properties: (1) It should be robust to noise, stable under cell-to-cell fluctuations of protein concentration by factor of 10, have the ability to sense and respond to relative changes of attractant concentrations as small as 2 % over a dynamic concentration range of five orders of magnitude and precise adaptation; and (2) It should be able to return to the same level of pathway activity under conditions of continuous stimulation. Given these design constraints, Timmer evaluated a range of topologies for the impact they might have on regulatory behavior determining what the necessary complexity might be and the source of the non-linear regulation. He also has expanded this work to other systems, including cytokines. These principles of network design are likely to help interpret a wide variety of systems with larger numbers of components. Notably, even this compact model, which did not contain the rest of the regulatory circuitry, was able to match experimental data. Additional work in the role of noise in biological regulation is actively ongoing in the El-Samad lab (Stewart-Orstein et al. 2012). Dr. van Oudenaarden discusses the issue of noise in biological systems extensively in two of his recent publications (Munsky et al. 2012; Balázsi et al. 2011).

Work at Kyoto University, Japan, directly explored nonlinear processes in biological systems in their Laboratory of Bioimaging and Cell Signaling directed by Dr. Michiyuki Matsuda. The lab is focused on visualizing the growth signal transduction cascades in living cells. In general, there is a significant focus on dynamic living systems based on multi-dimensional quantitative imaging and mathematical modeling. The program involves investigators from a number of departments (graduate schools of Medicine, Biostudies, and Informatics; Institute for Frontier Medical Sciences; Institute for Virus Research; and Imaging Platform for Spatio-Temporal Information). They developed a unique FRET-based pipeline that measures key parameters associated with protein signaling (Aoki et al. 2008, 2011). They were able to measure most, if not all, of the kinetic parameters required for kinetic simulation of the MAPK/ERK signaling pathway, using HeLa cells as a model system. The model requires four classes of parameters: protein concentrations, association/dissociation rates, nuclear import/export rates, and phosphorylation/dephosphorylation rates. They experimentally determined approximately 30 parameters. Through a combination of this rigorous experimentation coupled with kinetic modeling they were able to show complex non-linear dynamics in signaling systems. They also showed how molecular crowding could further complicate signaling (Aoki et al. 2013).

At Hong Kong's Institute of Computational and Theoretical Studies, Dr. Lei-han Tang is investigating nonlinearity and noise in gene expression. In recent work, their group developed a model of transcription that includes three processes: transcriptional bursting in the nucleus, mRNA transport, and mRNA decay in the cytoplasm (Xiong et al. 2010). They generally observed that the extent of burst attenuation is governed by the rate of transport. The slower the mRNA transport, the smaller the noise in the cytoplasmic mRNA number. In the case of the Michaelis-Menten transport, the saturation effect of transport mediators or nuclear pores further reduces mRNA copy number fluctuations in the cytoplasm allowing for a dampening of the noise. In the context of gene expression in eukaryotic cells, their results indicate that transcriptional bursting can be substantially attenuated by the transport of mRNA from nucleus to cytoplasm. Saturation of the transport mediators or nuclear pores contributes further to the noise reduction.

A common thread in biological models has been the role of cooperativity in DNA structure as a control element. Dr. José Vilar from the University of the Basque Country, Spain, (site report, Appendix B) highlighted two examples of DNA proximity leading to nonlinear regulatory effects (Fig. 2.1). In one example, Dr. Vilar presented early work on the lac repressor (Vilar et al. 2003; Vilar and Leibler 2003), which binds to a primary operator O1 and prevents the RNA polymerase from transcribing the genes (Fig. 2.2). If it is not bound, transcription proceeds at a given rate. In addition to O1, there are two sites outside the control region, the

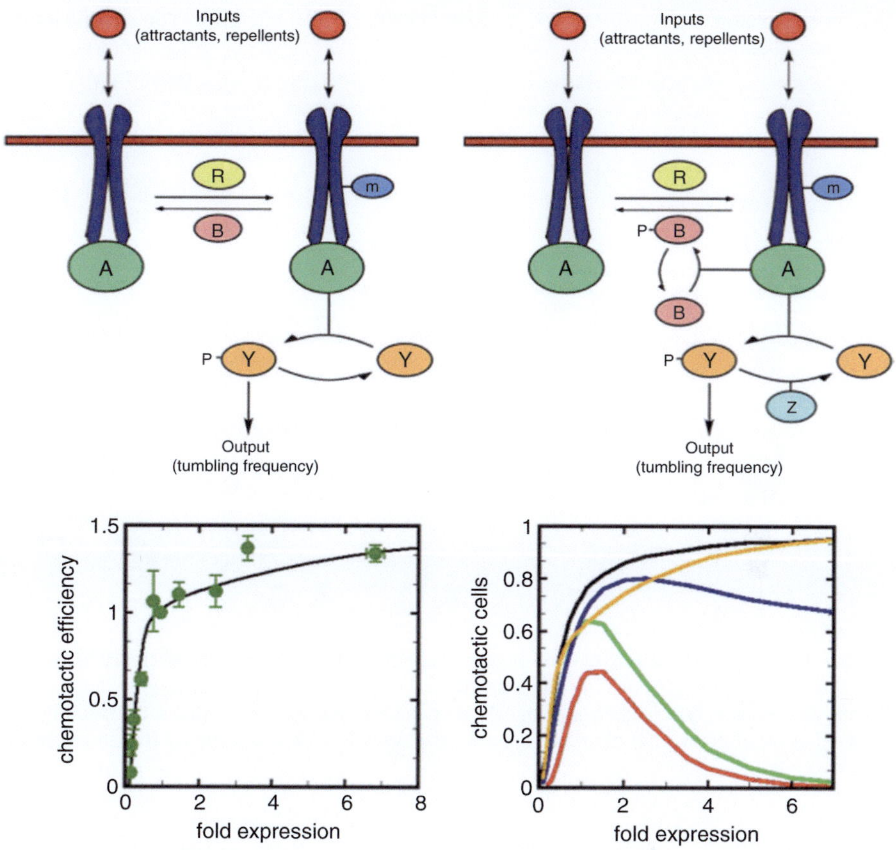

Fig. 2.1 Two possible topologies of a regulatory network (From Kollmann et al. 2005)

Fig. 2.2 The structure of the *lac* operon (From Vilar and Leibler 2003)

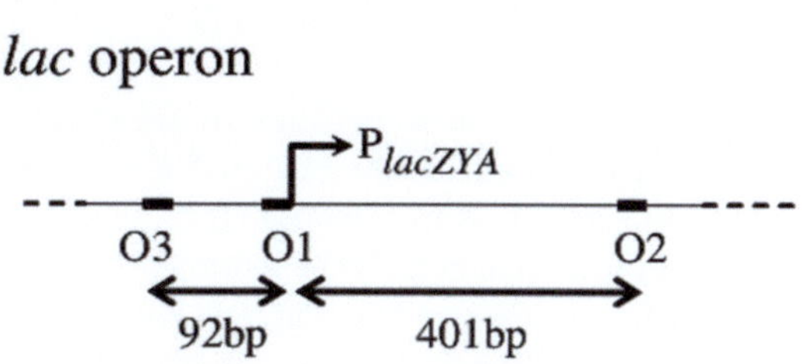

so-called auxiliary operators O2 and O3, which closely resemble O1 and can also bind with the repressor. However, they are much weaker than O1 (10 and 300 times weaker). Moreover, elimination of either one of them leaves the repression level practically unchanged. However, the role of O2 and O3 are actually quite significant: simultaneous elimination of both of these operators reduced the repression level about 100 times. Deeper investigations and computational modeling were able to detail how DNA looping could explain this result (Saiz and Vilar 2007). Dr. Vilar

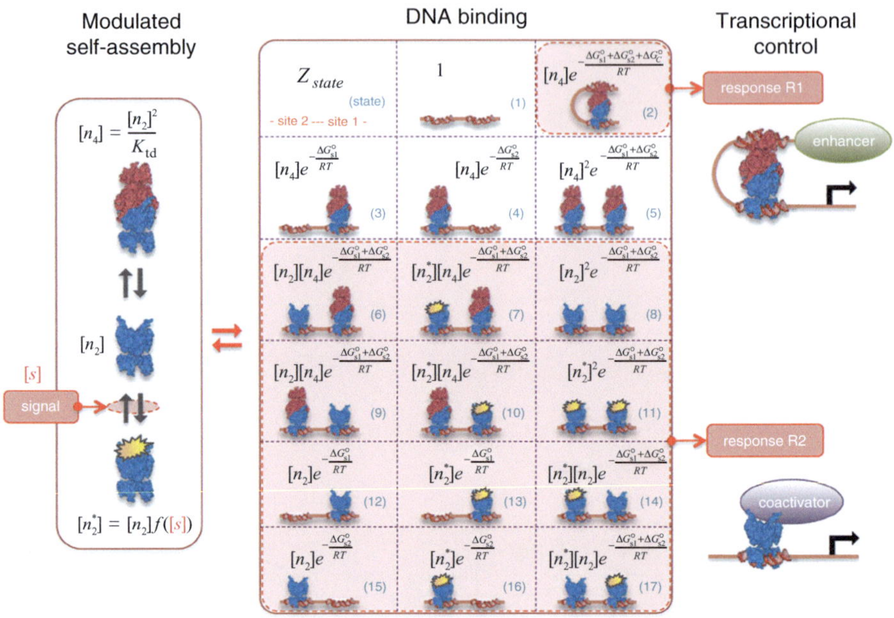

Fig. 2.3 Quantitative modeling of control gene expression by modulated self-assembly of the retinoid X receptor (RXR) (From Vilar and Saiz 2011)
The model considers how intracellular signals are processed through modulated self-assembly into populations of different RXR oligomeric species that upon DNA binding engage in transcriptional control

then proceeded to demonstrate how general processes of self-assembly could lead to these sorts of emergent behaviors in regulation by mechanisms previously thought to be distinct, such as in retinoid X receptor regulation (Fig. 2.3; Vilar and Saiz 2011).

Dr. Vilar's gene-scale findings are recapitulated at the chromosomal scale by Dr. Marc Marti-Renom of the National Center for Genomic Analysis in Barcelona, Spain (site report, Appendix B) (Bau and Marti-Renom 2012; Marti-Renom and Mirny 2011; Sanyal et al. 2011; Umbarger et al. 2011). This work parallels the work of Drs. Michor (De and Michor 2011) and Mirny (Fudenberg et al. 2011) supported by the National Cancer Institute Physical Sciences-Oncology Center (NCI PS-OC) program in the United States (Fig. 2.4).

Techniques from structural biology can be used to reconstruct chromosomal structure and demonstrate how long-range interactions may play a role in regulation. Dr. Helmut Schiessel at the Instituut-Lorentz, Netherlands, (site report, Appendix B) also identified many examples of genome structure playing a role in cellular regulation (Prinsen and Schiessel 2010). His group showed how the wrapping and unwrapping of the nucleosome allowed regulatory DNA binding sites to become exposed (Fig. 2.5).

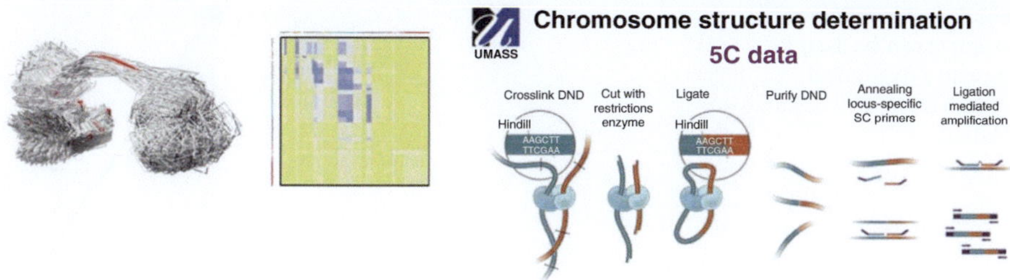

Fig. 2.4 Two computational approaches for determining the 3D structure of genomic domains and genomes (From Marti-Renom and Mirny 2011)

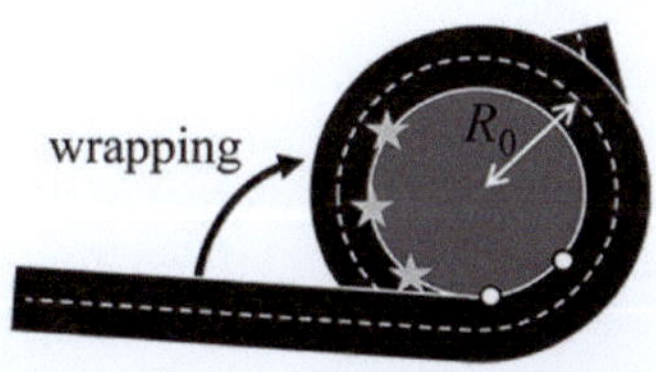

Fig. 2.5 A partially unwrapped nucleosome with exposed nucleosomal binding sites (*stars*). The nucleosome can lower its energy by closing those binding sites at the cost of bending the DNA (From Prinsen and Schiessel 2010)

Dynamical Systems, Cell States, and State Transitions

Dynamical systems models attempt to describe the temporal evolution of a system. To create a dynamical system we need to identify the "something" that will evolve and the rules describing that evolution.

In order to identify the "something," we need to come up with a set of variables that give a compact description of the system at any particular time. The variables do not have to fully describe a real-life system. However, the more complete the model, the more likely it will be able to accurately predict the system's behavior. It is assumed that by knowing the values of these variables at a particular time, we can accurately predict the state of the system at a future time. To model a real-life system, the modeler must decide what variables will form the complete description for the mathematical model. The variables used to describe the state of the dynamical system are called the state variables. The "state space" is the set of all possible states of the dynamical system; each state of the system corresponds to a unique point in the state space. The axes of the space are defined by the state variables. The state space may be of infinite size.

The second step in creating a dynamical system is to specify the rule for the time evolution of the dynamical system. This rule must be defined such that one can use current values of the state variables in combination with the rule to infer all future states. If the time evolution depends on a variable not included in the state space,

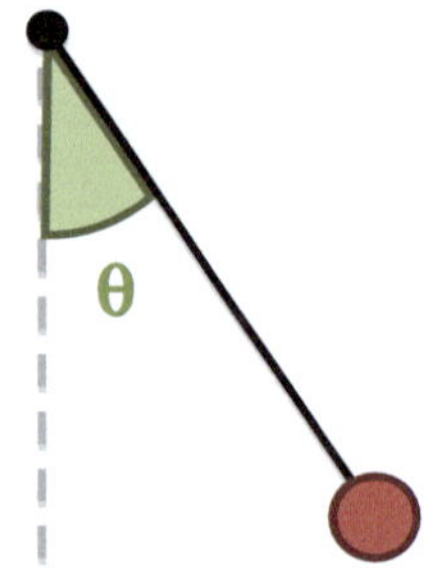

Fig. 2.6 A pendulum example can be used to explain a dynamical system (Courtesy of Parag Mallick)

then the rule combined with the state space does not specify a dynamical system. One must either change the rule or augment the state space by the necessary variables to form a dynamical system.

To make this more concrete, consider the example of a pendulum. The angle θ completely specifies the position of the pendulum (Fig. 2.6). However, we cannot use θ as the only state variable. If the above picture of the pendulum were a snapshot of a pendulum, we would not have enough information to know where the pendulum will move next.

Determining the future behavior of the pendulum requires knowing not only its position, but also its angular velocity. Therefore, the state space is the set of all possible pairs (angle, velocity). For this idealized pendulum, the angle θ and the angular velocity ω completely determine the state of the system.

Both θ and ω will evolve over time, and their value at one time determines all their future values. The dynamical system is 2D, and since θ and ω evolve continuously, it is a continuous dynamical system.

The dynamical systems approach is highly appropriate in biology. In biology we frequently refer to cells as having specific phenotypes, which may be analogized as states. For example, a cell may inhabit states such as dividing, apoptosing, and migrating. Accordingly, cells may have particular likelihoods of inhabiting particular states and of transitioning between states (e.g., metastatic potential).

Though significant effort has been made to determine diverse cellular phenotypes and to understand how endogenous and exogenous perturbations (e.g., mutations) lead to transitions in those phenotypes, our current approaches typically generate state spaces of very high dimension wherein each gene or protein in a cell might be considered as components of state.

During our survey abroad, we saw several exciting examples of dynamical systems approaches. For example, in Leipzig, Germany, the study team met Dr. Adalinde Uhrmacher from the University of Rostock (site report, Appendix B). Dr. Uhrmacher, a computer scientist, emphasized the importance of both modeling and simulation. Her group has designed a general purpose plug-in-based modeling and simulation framework that has already been applied to develop different modeling and simulation tools for cell biology (Ewald et al. 2010). Currently, the framework includes more than 700 plug-ins and more than 100 plugin types (e.g., different modeling formalisms, execution algorithms, steady state analyzers). It also provides intelligent support to configure suitable experiments on demand.

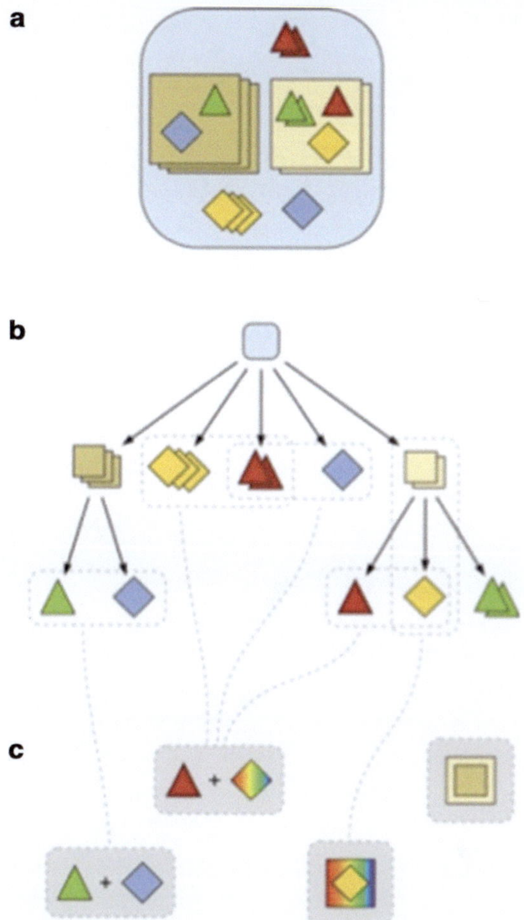

Fig. 2.7 Nested model structure (From Maus et al. 2011)
The hierarchical modeling concept. Different-shaped nodes correspond to different species while attributes are color-coded. Stacking of identical nodes represents the amount of a certain species. (**a**) A graphical representation of a hierarchical model structure via nested nodes. (**b**) The same model structure alternatively depicted as a directed tree graph. Note that in addition to atomic species (*triangles* and *diamonds*), species containing a sub-solution (*squares*) might be attributed so that each species at each level might have its own state. (**c**) Examples of matching different reactant patterns within the hierarchical model structure. The rainbow shadings in the second pattern (*diamond*) and third pattern (*rectangle*) illustrate variable instead of defined colors, i.e., attributes

Dr. Uhrmacher's general approach relies upon ML-Rules—a multilevel rule-based modeling method (Maus et al. 2011, Fig. 2.7), and a spatial variant—ML-Space (Bittig et al. 2011, Fig. 2.8). ML-Rules allows users to compactly describe and combine compartmentalized dynamics, including inter- and intra-cellular dynamics and processes at the cellular level such as proliferation of cells, apoptosis, and cell differentiation (Mazemondet et al. 2011). ML-Rules assumes

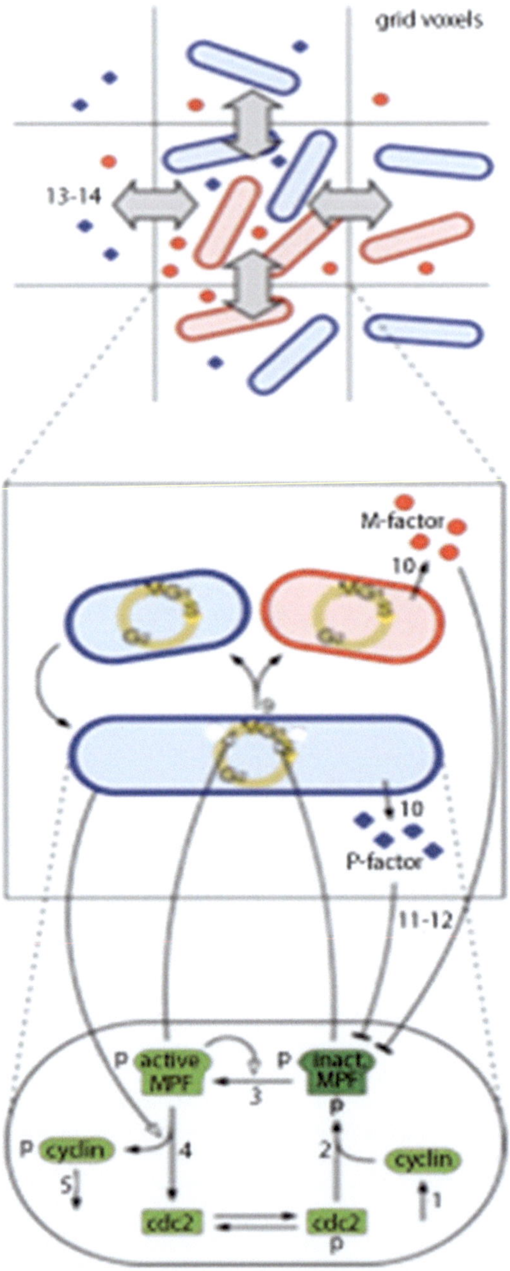

Fig. 2.8 Schematic description of the example model (From Bittig et al. 2011)
The model comprises three distinct hierarchical levels. At the *bottom level*, interacting proteins describe the intracellular dynamics of a fission yeast cell (reactions 1–5). The intermediate level describes dynamics of entire cell states, i.e., cell growth (6), cell cycle phase transitions (7–9), and division including mating type switching (9). In addition, cells may secrete pheromone molecules

well-mixed solutions within the compartments. It does not capture phenomena that are induced by the molecular crowding within cells. Therefore, the language for ML-Space has been developed with decidedly spatial semantics. Here, species can be defined as individual particles that react due to collisions, or as a population of species residing in a small area. It inherits from ML-Rules the compact description and the ability to describe processes at different organizational levels however they adhere to spatial physical constraints. ML-Space has been used to investigate lipid rafts as compartments, with a focus on their movements and the activity of receptors in rafts.

Dr. Hauke Busch of the Freiburg Institute for Advanced Studies, Germany (site report, Appendix B) presented his work, which is built upon the work done in the United States by Drs. Huang and Ingber (2006, Fig. 2.9) to characterize cell states and fate decisions. In this work (Busch et al. 2008) Dr. Busch asserted that the long-term phenotypic response of a cell can be expressed in terms of its slowest evolving functional elements. Postulating that a cell reaches a decision on a timescale of hours, its phenotype should be controlled by the slow protein turnover rates. To validate this finding they looked at the impact of hepatocyte growth factor (HGF) stimulation on keratinocyte cell migration.

Similar approaches are being investigated by Dr. Ping Ao at Shanghai Jiao Tong University. He hypothesizes (Ao et al. 2008; Wang et al. 2012) that in order to maintain the normal physiological function and developmental process for tissue specific function shaped by evolution, a minimal set of fundamental functional modules or pathways (e.g., cell cycle, cell death, inflammation, metabolism, cell adhesion, and angiogenesis) are needed. Each module can accomplish a relatively autonomous function, and cross-talk between modules allows one function to influence another. At the molecular-cellular level, it is hypothesized that the functional modules are deeply hierarchical and may be specified by important molecular and cellular agents, such as oncogenes, suppressor genes, and related growth factors, hormones, cytokines, etc. The interactions among these agents form an autonomous, nonlinear, stochastic, and collective dynamical network. The endogenous network may generate many locally stable states with obvious or non-obvious biological function. Normal state and cancer state are assumed to be stable states of the endogenous molecular- cellular network. The endogenous network may stay in each stable state for a considerably long time, and in certain conditions, stable states can switch between each other. In this hypothesis, the genesis and progression of cancer can be regarded as transition of the endogenous molecular-cellular network from normal stable state to cancer state (Fig. 2.10).

Fig. 2.8 (continued) (P-factor and M-factor) to the extracellular medium (10). Various inter-level causalities between the intermediate and the *bottom level* influence processes both in an upward (7–9) and downward causation manner (4, 11–12). The *top level* discretizes the environment of cells into multiple fictive compartments in order to study spatial dynamics of pheromone diffusion and displacement of cells (13–14). Note that, although spatial dynamics referring to compartments and particle diffusion between cells can be modeled, excluded volume effects cannot be described in ML-Rules therefore one has to move to ML-Space

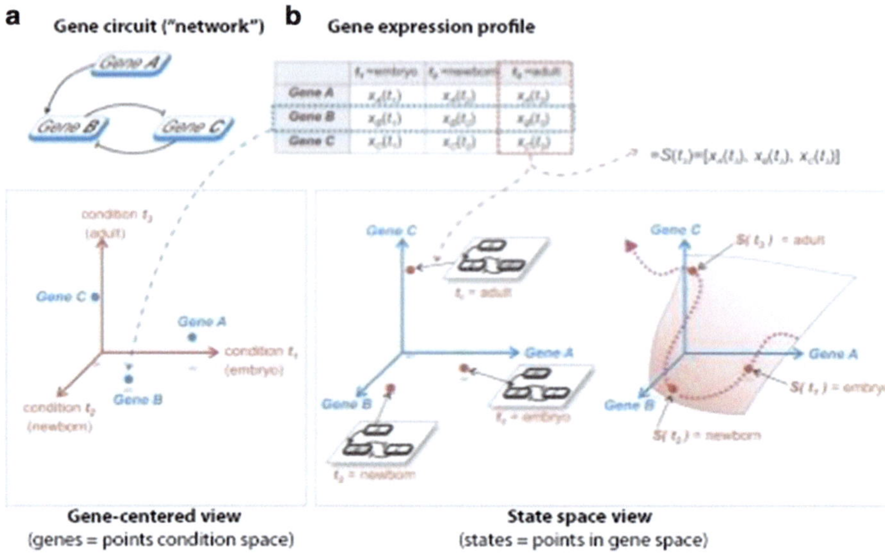

Fig. 2.9 The long-term phenotypic response of a cell can be described as a state space (From Huang and Ingber 2006–2007)

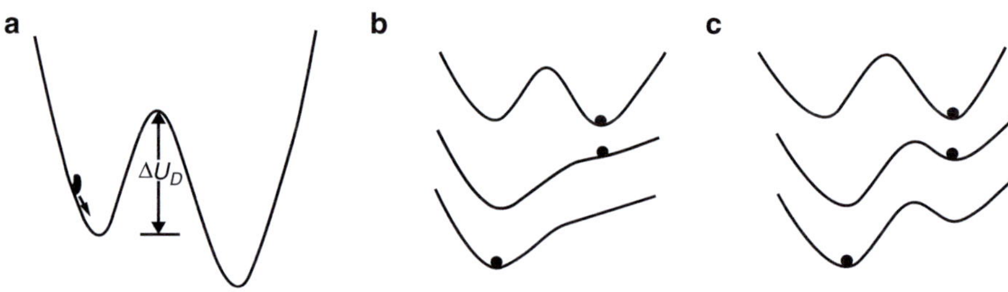

Fig. 2.10 Schematic diagram of the adaptive landscape of the phage lambda genetic switch, where the dynamic state of the biological system is represented as a *black dot* (From Wang et al. 2012) As the system progresses through different configurations, the dynamic state of the system may change or remain in a local minimum of the adaptive landscape. For full details of the endogenous molecular-cellular network hypothesis and its applications, see Wang et al. (2012)

State transitions and fate determination is hugely important in understanding therapeutic efficacy. At The Center for Quantitative Systems Biology in Hong Kong, Dr. Jue Shi is exploring cell-type variation in anticancer drug response (Chen et al. 2013). By combining experimental measurements of cellular alterations induced by drug treatment with kinetic modeling of pathway dynamics, the ultimate goal is to understand how different anticancer drugs perturb cellular behaviors and what variables as well as interaction modules in cellular pathways are the determining factors in engendering distinct drug response phenotypes in different cell types. In particular, they identified a novel, bimodal switch of p53 dynamics modulated by

DNA-damage strength that is crucial for cell-fate control. After low DNA damage, p53 underwent periodic pulsing and cells entered cell-cycle arrest. After high DNA damage, p53 underwent a strong monotonic increase and cells activated apoptosis. Their findings not only uncover a new mode of regulation for p53 dynamics and cell fate, but also suggest that p53 oscillation may function as a suppressor, maintaining a low level of p53 induction and pro-apoptotic activities so as to render cell-cycle arrest that allows damage repair.

Related studies are under way by Dr. Mariko Okada-Hatakeyama at the RIKEN Laboratory for Cellular Systems Modeling, Japan. The lab is investigating how activation of ErbB receptors by epidermal growth factor (EGF) or heregulin (HRG) determines cell fate decisions using a mix of experimental and computational approaches (Nakakuki et al. 2010). Although signals propagate through shared pathways, HRG and EGF generate distinct, all-or-none responses of the phosphorylated transcription factor c-Fos. In the cytosol, EGF induces transient and HRG induces sustained extracellular-signal-regulated kinase (ERK) activation. In the nucleus, however, ERK activity and *c-fos* mRNA expression are transient for both ligands. Knockdown of dual-specificity phosphatases extends HRG-stimulated nuclear ERK activation, but not *c-fos* mRNA expression, implying the existence of an HRG-induced repressor of *c-fos* transcription. Further experiments confirmed that this repressor is mainly induced by HRG, but not EGF, and requires new protein synthesis. Dr. Okada-Hatakeyama shows how a spatially distributed, signaling-transcription cascade robustly discriminates between transient and sustained ERK activities at the c-Fos system level. The proposed control mechanisms are general and operate in different cell types, stimulated by various ligands (Fig. 2.11).

An experimental approach to fate characterization was taken by Dr. Matthias Lutolf at the École Polytechnique Fédérale de Lausanne, Switzerland, (site report, Appendix B) who is attempting to control cell fates through microenvironment. He has developed 2D microwells (Gobaa et al. 2011) molded in hydrogel (Fig. 2.12 and see also Chap. 3). A major focus of his group's research is on the neural stem cell niche, in which he has shown that notch, jagged, and dll4 are involved in self-renewal of stem cells in his devices. Operationally, he is experimentally defining cellular state spaces for diverse cell types.

Self-Organization and Emergence

As noted above, complex systems are characterized to display emergent behaviors and self-organization. These properties are prevalent throughout chemistry and biology. For example, spontaneous collective motion can be observed in flocks of birds and schools of fish. One of the greatest mysteries in cancer biology arises from the observation that small length-scale perturbations (e.g., gene mutations) can lead to significant large length-scale effects (e.g., death).

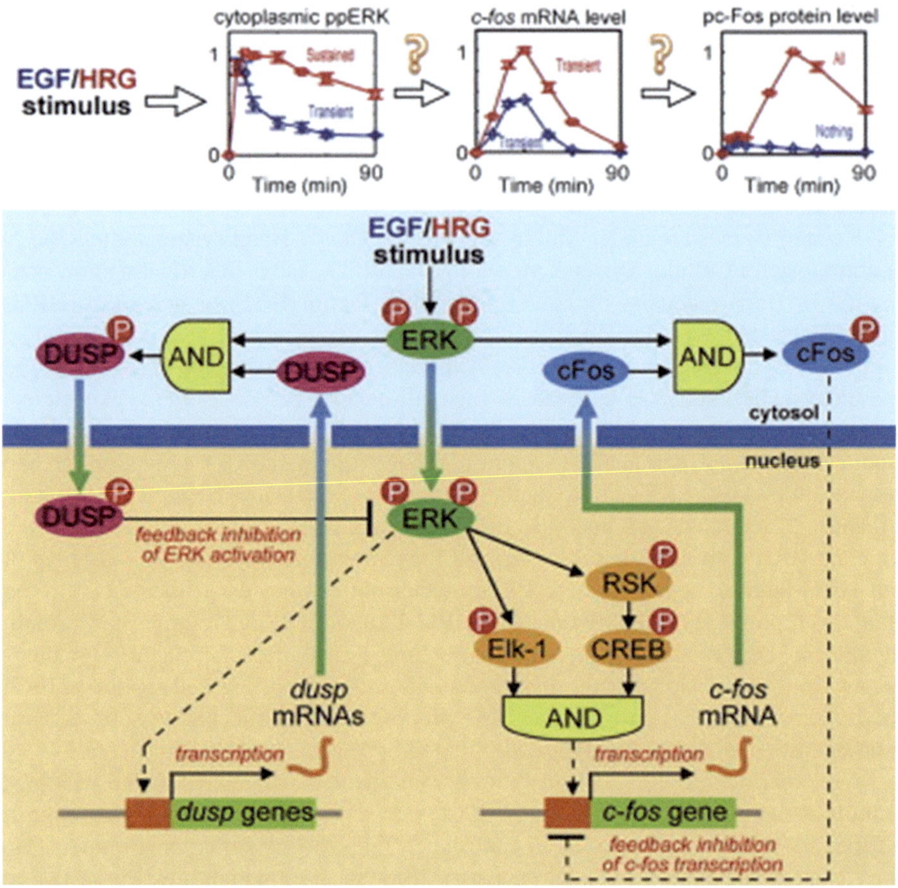

Fig. 2.11 Activation of ErbB receptors by epidermal growth factor (EGF) or heregulin (HRG) determines distinct cell-fate decisions, although signals propagate through shared pathways (From Nakakuki et al. 2010)
In the cytosol, EGF induces transient and HRG induces sustained ERK activation. In the nucleus, however, ERK activity and c-fos mRNA expression are transient for both ligands. Knockdown of dual-specificity phosphatases extends HRG-stimulated nuclear ERK activation, but not c-fos mRNA expression, implying the existence of a HRG-induced repressor of c-fos transcription. Further experiments confirmed that this repressor is mainly induced by HRG, but not EGF, and requires new protein synthesis

In looking at self-organization and emergence, we typically ask questions such as:

- How do collections of entities behave differently than either entity alone?
- What properties emerge from aggregate behavior?
- What information is communicated to aid in that self-organization?
- How is that information transduced?

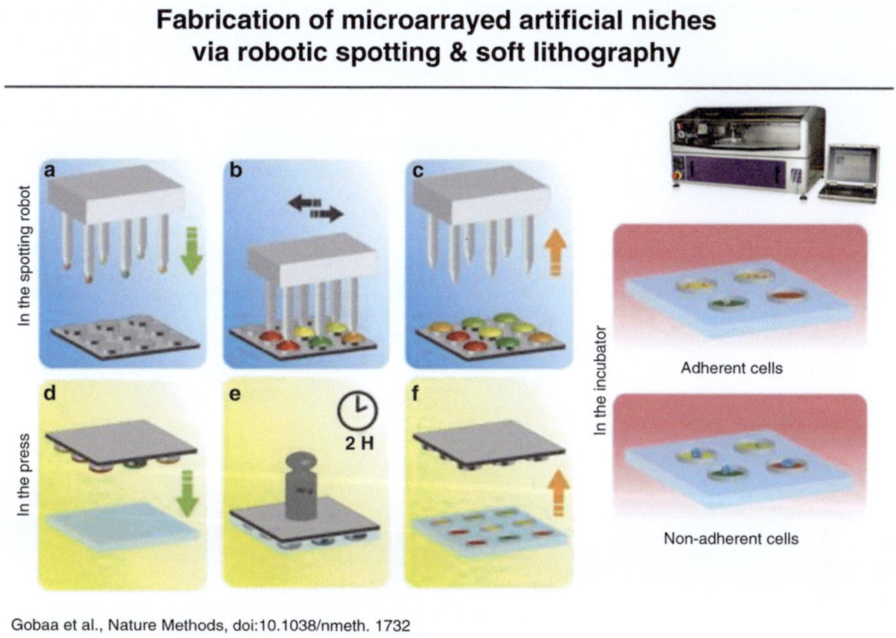

Fig. 2.12 A 2D microwell molded in hydrogel (From Gobaa et al. 2011)

Several groups are now actively pursuing this area at multiple scales.

At the molecular scale, Dr. Andreas Bausch's group at the Technical University of Munich, Germany (site report, Appendix B) has been actively engaged in studying the dynamics of actin assembly (Kohler et al. 2011; Schaller and Bausch 2012; Schaller et al. 2010). They have shown the emergence of collective motion in an actin/myosin motility assay. Motility assays, in which protein filaments are densely placed on a planar substrate, can show collective motion for high densities of motors and attached filaments. Notably, this motion is density dependent. At low density, fibrils have near random motion. However, above a threshold density, the filaments self-organize to form diverse moving structures such as swirls and interconnected bands (Fig. 2.13). These polar nematic structures are long-lived and can span length scales that are orders of magnitudes larger than their constituents. Recent work in Japan (Sumino et al. 2012) showed a similar pattern for microtubules.

At a cellular scale, the group led by Dr. Xavier Trepat at the University of Barcelona, Spain (site report, Appendix B) has focused on defining how cell and tissue dynamics are integrated to drive function. In particular, his group is one of the leaders in the emerging field of plithotaxis—the mechanism of innately collective cell guidance (Trepat and Fredberg 2011). To study this process, Trepat and colleagues have created a novel technique, monolayer stress microscopy, to characterize the local state of stress within a monolayer (Tambe et al. 2011). The technology allows the measurement of stresses within and between cells comprising a mono-

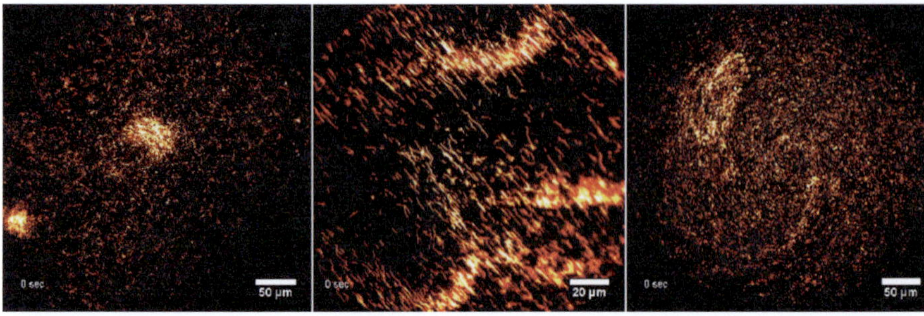

Fig. 2.13 An actin/myosin motility assay that shows that the motion is density dependent (From Schaller et al. 2010)

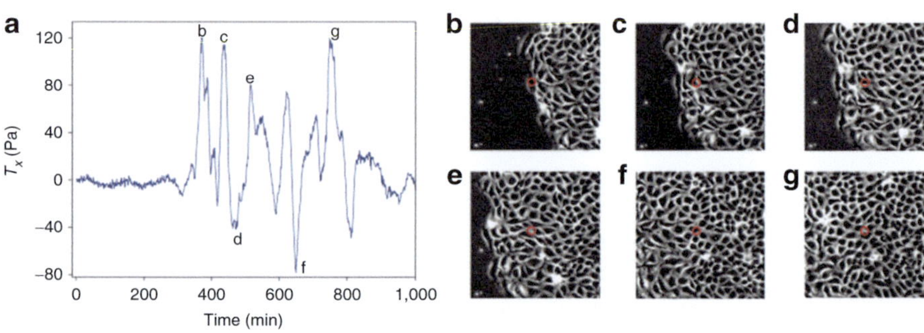

Fig. 2.14 High-resolution maps of stress components within an advancing monolayer sheet of cells (From Trepat and Fredberg 2011)

layer for the first time. Monolayer stress microscopy can generate high-resolution maps of stress components within an advancing monolayer sheet of cells (Trepat and Fredberg 2011). This work (Fig. 2.14) demonstrated that as cells in a monolayer expand, they do so "center-out," thus generating sinusoidal force patterns (Trepat et al. 2009). Though cells might migrate and grow in a number of different ways (e.g., front plane driven, uniformly, etc.), Trepat's technique was able to uncover a novel pulsing mechanism. Similar emergent multiscale phenomena are also being pursued by Dr. Yasuhiro Inoue at Kyoto University, Japan (Okuda et al. 2013a, b).

Tumor Evolution and Heterogeneity

Tumors, as collections of cells, are complex systems. They can be viewed from an evolutionary perspective as collections of cells that accumulate genetic and epigenetic changes, which are then evaluated relative to the selective pressures prevalent within an environment. Beneficial heritable changes can cause rapid expansion of the mutant clone since they enable their carriers to outcompete cells that have not accumulated similar improvements. Mutations advantageous to the cancer cell are normally detrimental to the organism, ultimately causing death of both the patient and the tumor. Evolution generally selects for increased proliferation, survival, and evolvability on the cellular level, which leads to organ-scale consequences of progression, invasion, and resistance.

Investigations of evolution have been ongoing for hundreds of years—pre-dating Darwin. Physical sciences approaches combined with recent advances in genomic technologies have led to a renewed emphasis on cancer evolution. Work done in the United States by Dr. Rong Fan has shown that it is now possible to look at evolutionary processes with single-cell resolution (Fan et al. 2011). Furthermore, recent work from the United Kingdom has shown the extensive heterogeneity prevalent in cancers (Gerlinger et al. 2012). Evolutionary studies have analyzed the full spectrum of cancer from initiation through acquisition and penetrance of resistance.

With support from the NCI PS-OC program in the United States, Dr. Franziska Michor (Dana-Farber Cancer Institute; Chmielecki et al. 2011; Pao and Chmielecki 2010), and Dr. Robert A. Gatenby (H. Lee Moffitt Cancer Center & Research Institute; Gillies et al. 2012) have been leading evolutionary studies with a variety of stochastic models that rely on quantifying selective advantage. Other efforts, such as those of Dr. Carlo C. Maley (Greaves and Maley 2012), have focused on using evolutionary ecology approaches.

One approach employed by Dr. Oskar Hallatschek, from the Max Planck Institute for Dynamics and Self-Organization, Göttingen, Germany (site report, Appendix B) described how evolution begins with colony growth and proceeds to whole tumor growth. He is interested in range expansions—the movement of populations to different areas where they evolve separately. In his philosophy, there is typically a competition between Darwinian selection and genetic drift to drive evolutionary change. Genetic drift can have significant effects on small populations that may even lead to speciation. This is contrasted with large populations, where genetic drift is considered weak. Notably, in cancer, this general principle is violated when large populations undergo range expansions. The descendents of individuals first settling in a new territory are most likely to dominate the gene pool as the expansion progresses. Random sampling effects among these pioneers results in genetic drift that can have profound consequences on the diversity of the expanding population.

In this project, Hallatschek used simple microbial systems (Hallatschek et al. 2007) to study the nature of these number fluctuations (genetic drift) in range expansions of large populations.

This finding was first validated in bacteria (Fig. 2.15) with Dr. David Nelson, but has since been adapted to colon cancer and clonal expansion in neoplastic tissues (Martens et al. 2011). In these cases, the work allows for mutations to come in that confer a certain growth rate advantage. This model may be good for understanding the growth of intestinal epithelial cells out of the crypt (Fig. 2.16).

The lab of Dr. Jian-Dong Huang at Hong Kong Baptist University is taking advantage of evolution and fitness in a novel treatment strategy (Yu et al. 2012). Specifically, they take a synthetic biology approach to create a novel tumor-targeting

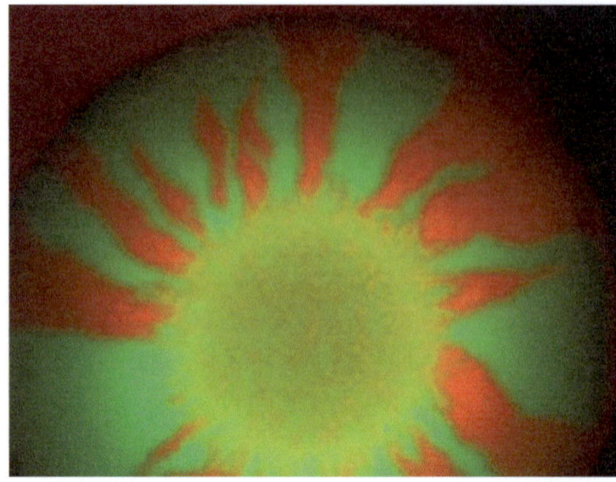

Fig. 2.15 Spatial distribution of evolving cell populations (From Hallatschek et al. 2007)

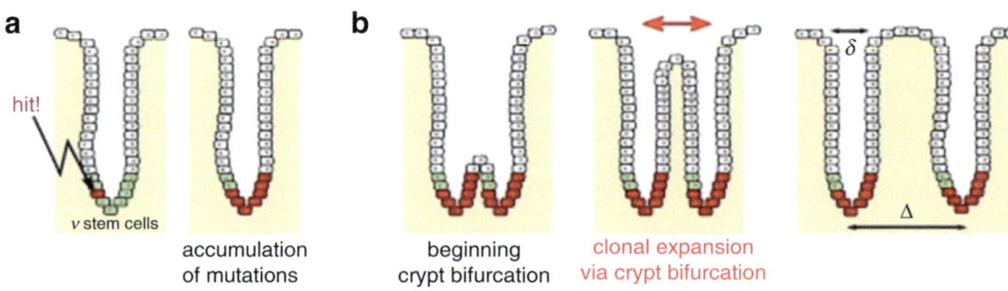

Fig. 2.16 Colon cancer model for understanding the growth of intestinal epithelial cells out of the crypt (From Martens et al. 2011)

bacteria. Built upon a salmonella backbone, they attempted to develop a bacterium that effectively targeted the tumor microenvironment, and could deliver a toxic payload. To accomplish this targeting they put an essential gene under the control of an oxygen sensitive promoter. Consequently, in presence of oxygen, the gene is not expressed and the bacteria die. Particularly notable, though exploiting evolution in their mechanism, the bacteria themselves were created by design and not by evolution.

We saw several notable examples of evolutionary studies in the work of Drs. Stefano Zapperi and Alberto d'Onofrio at the European Institute of Oncology, Italy (site report, Appendix B). Dr. Zapperi uses approaches very similar to those of Dr. Michor (branching birth-death processes). He has used these approaches to investigate the implications of cancer stem cells within a population (La Porta et al. 2012). Dr. Zapperi introduced a recent study on a novel approach to investigating tumor growth from a cancer-stem-cell perspective in melanoma. It is commonly believed that cell senescence—the loss of replicative capacity of cells—acts as a barrier for tumor growth. Dr. Zapperi and colleagues are investigating this phenomenon.

In their study, Dr. Zapperi and colleagues followed the evolution of senescence markers in melanoma cells and found that while most cancer cells eventually turn senescent, it is irrelevant for the long-term growth rate of a tumor. To demonstrate

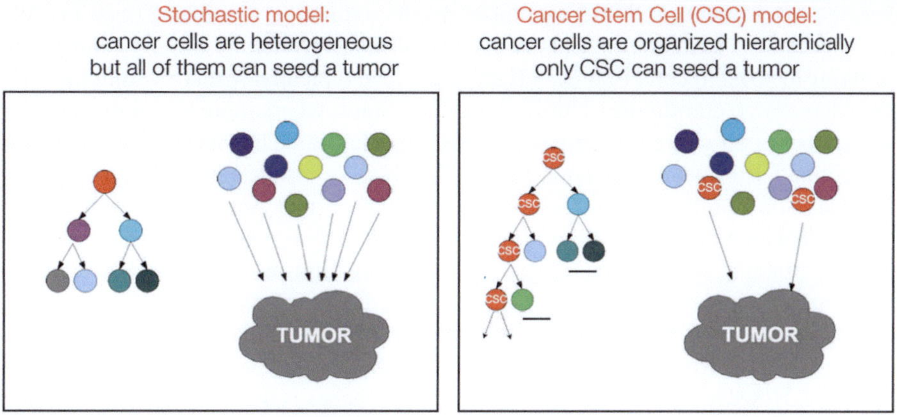

Fig. 2.17 Mathematical population dynamics model (From La Porta et al. 2012)

this phenomenon, they constructed a mathematical population dynamics model (Fig. 2.17, right) incorporating cancer stem cells, which is able to reproduce quantitatively the experimental data. Their results support the existence of cancer stem cells in melanoma and explain why it is difficult to fight cancer by inducing senescence in cancer cells. Only a fraction of the cells are susceptible to senescence, but those cells are irrelevant for tumor growth. A successful therapeutic strategy should instead target cancer stem cells, which are, however, likely to be strongly resistant to drug-induced senescence. This result is quite important and highlights the need of evolutionary modeling of tumor growth as well as the possible insights that come from formal modeling approaches. Notably, additional light on the existence of cancer stem cells has been provided by Drs. Clevers, Blanpain, and Parada (Schepers et al. 2012; Driessens et al. 2012; Chen et al. 2012).

Dr. d'Onofrio used similar approaches, but focused on the area of noise to develop strategies for optimizing anti-angiogenic therapies (Bertolini et al. 2011; d'Onofrio and Gandolfi 2010).

Discussion

We identified numerous examples of cancer behaving as a complex adaptive system throughout our study. Experimentally, this was observed at a variety of length scales from the single-protein to the tumor. Notably, this has engendered an impressive and diverse collection of modeling approaches. There is significant research being conducted abroad in all aspects of cancer as an information transfer system and of the evolutionary processes, state-evolution functions, and emergent properties. Among the major bottlenecks were a frequent need for close integration between experimentalists and modelers. In addition, to appropriately ask and answer a question about complex systems in biology, it was often necessary to design and perform specific experiments and integrate those results with larger published data sets. We also observed a greater emphasis on the use of model systems (ranging from single proteins to yeast to cell culture) in Europe than we typically observe within the U.S. cancer research community. Unlike the United States where there is significant hesitation about non-clinically mimetic biosystems, that same hesitation did not appear to dominate European or Asian research. Notably, we also found significantly more emphasis on compact biomodel systems in Europe and Asia than in the United States. In Japan and Hong Kong, in particular, there was a significant emphasis on cellular-imaging approaches for getting incredibly high-resolution, views of cellular dynamical processes. There also was a significant emphasis on exploring specific types of effects and extracting principles that might be scaled up. This approach differed widely from the typical U.S. approach, which favors large-scale, global analyses. However, the advantage of global approaches is their ability to interrogate complex systems as a whole, rather than as a subset, such as in recent work to build total cell models (Karr et al. 2012). A major contributor to successful research endeavors was funding environment. Successful projects depended upon close collaboration amongst groups of researchers, particularly including a mix of experimentalists, bioinformaticians, and modelers. Consequently, multi-investigator funding mechanisms have been critical for pushing innovation at the frontier of information transfer and complex systems analysis of biology. Through programs, such as those engendered by the NCI's Office of Physical Sciences Oncology, and foundation awards at the interface between the physical and life sciences, investigators in the United States have been fortunate to have access to interdisciplinary funding opportunities. Generally, we conclude that the areas of information transfer, evolution, and complex adaptive systems research are rapidly progressing, and critically important for impacting cancer and more generally understanding biology.

References

Ao, P., Galas, D., Hood, L., Zhu, X.: Cancer as robust intrinsic state of endogenous molecular-cellular network shaped by evolution. Med. Hypotheses **70**, 678–684 (2008)

Aoki, K., Kiyokawa, E., Nakamura, T., Matsuda, M.: Visualization of growth signal transduction cascades in living cells with genetically encoded probes based on Förster resonance energy transfer. Philos. Trans. R. Soc. B **363**, 2143–2151 (2008). doi:10.1098/rstb.2008.2267

Aoki, K., Takahashi, K., Kunida, K., Yasuda, S., Matsuda, M.: Processive phosphorylation of ERK MAP kinase in mammalian cells. Proc. Natl. Acad. Sci. U. S. A. **108**(31), 12675–12680 (2011). doi:10.1073/pnas.1104030108

Aoki, K., Yamada, M., Kaizu, K., Matsuda, M.: A quantitative model of ERK MAP kinase phosphorylation in crowded media. Sci. Rep. **3**, 1541 (2013). doi:10.1038/srep01541, 8 pp. [Epub 26 Mar 2013]

Bachmann, J., Raue, A., Schilling, M., Becker, V., Timmer, J., Klingmuller, U.: Predictive mathematical models of cancer signalling pathways. J. Int. Med. **271**, 155–165 (2012)

Balázsi, G., van Oudenaarden, A., Collins, J.J.: Cellular decision making and biological noise: from microbes to mammals. Cell **144**, 910–925 (2011)

Bau, D., Marti-Renom, M.A.: Genome structure determination via 3C-based data integration by the Integrative Modeling Platform. Methods **58**, 300–306 (2012) [Epub ahead of print, April 13]

Beadle, G.W., Tatum, E.L.: Genetic control of biochemical reactions in neurospora. Proc. Natl. Acad. Sci. U. S. A. **27**, 499–506 (1941)

Becker, V., Schilling, M., Bachmann, J., Baumann, U., Raue, A., Maiwald, T., Timmer, J., Klingmuller, U.: Covering a broad dynamic range: information processing at the erythropoietin receptor. Science **328**, 1404–1408 (2010)

Becker, V., Timmer, J., Klingmuller, U.: Receptor dynamics in signaling. Adv. Exp. Med. Biol. **736**, 313–323 (2012)

Bertalanffy, L.V.: General System Theory: Foundations, Development, Applications. Penguin, Harmondsworth (1973)

Bertolini, F., Marighetti, P., Martin-Padura, I., Mancuso, P., Hu-Lowe, D.D., Shaked, Y., D'Onofrio, A.: Anti-VEGF and beyond: shaping a new generation of anti-angiogenic therapies for cancer. Drug Discov. Today **16**, 1052–1060 (2011)

Bittig, A.T., Haack, F., Maus, C., Uhrmacher, A.M.: Adapting rule-based model descriptions for simulating in continuous and hybrid space. In: Proceedings of the 9th International Conference on Computational Methods in Systems Biology, pp. 161–170. ACM, Paris (2011)

Busch, H., Camacho-Trullio, D., Rogon, Z., Breuhahn, K., Angel, P., Eils, R., Szabowski, A.: Gene network dynamics controlling keratinocyte migration. Mol. Syst. Biol. **4**, 199 (2008)

Chen, J., Li, Y., Yu, T.-S., McKay, R., Burns, D.K., Kernie, S.G., Parada, L.F.: A restricted cell population propagates glioblastoma growth after chemotherapy. Nature **488**, 522–526 (2012). doi:10.1038/nature11287

Chen, X., Chen, J., Gan, S., Guan, H., Zhou, Y., Ouyang, Q., Shi, J.: DNA damage strength modulates a bimodal switch of p53 dynamics for cell-fate control. BMC Biol. **11**, 73 (2013). doi:10.1186/1741-7007-11-73

Chmielecki, J., Foo, J., Oxnard, G.R., Hutchinson, K., Ohashi, K., Somwar, R., Wang, L., Amato, K.R., Arcila, M., Sos, M.L., Socci, N.D., Viale, A., de Stanchina, E., Ginsberg, M.S., Thomas, R.K., Kris, M.G., Inoue, A., Ladanyi, M., Miller, V.A., Michor, F., Pao, W.: Optimization of dosing for EGFR-mutant non-small cell lung cancer with evolutionary cancer modeling. Sci. Transl. Med. **3**(90), 90ra59 (2011). doi:10.1126/scitranslmed.3002356

d'Onofrio, A., Gandolfi, A.: Resistance to antitumor chemotherapy due to bounded-noise-induced transitions. Phys. Rev. E **82**, 061901 (2010)

De, S., Michor, F.: DNA replication timing and long-range DNA interactions predict mutational landscapes of cancer genomes. Nat. Biotechnol. **29**, 1103–1108 (2011)

Deisboeck, T.S., Kresh, J.Y.: Complex Systems Science in Biomedicine. Springer, New York (2006)

Driessens, G., Beck, B., Caauwe, A., Simons, B.D., Blanpain, C.: Defining the mode of tumour growth by clonal analysis. Nature **488**, 527–530 (2012). doi:10.1038/nature11344

Ewald, R., Himmelspach, J., Jeschke, M., Leye, S., Uhrmacher, A.M.: Flexible experimentation in the modeling and simulation framework JAMES II–implications for computational systems biology. Brief. Bioinform. **11**, 290–300 (2010)

Fan, H.C., Wang, J., Potanina, A., Quake, S.R.: Whole-genome molecular haplotyping of single cells. Nat. Biotechnol. **29**, 51–57 (2011)

Fudenberg, G., Getz, G., Meyerson, M., Mirny, L.A.: High order chromatin architecture shapes the landscape of chromosomal alterations in cancer. Nat. Biotechnol. **29**, 1109–1113 (2011)

Garrod, A.: The Croonian lectures on inborn errors of metabolism. Lancet **172**, 1–7 (1908)

Gerlinger, M., Rowan, A.J., Horswell, S., Larkin, J., Endesfelder, D., Gronroos, E., Martinez, P., Matthews, N., Stewart, A., Tarpey, P., Varela, I., Phillimore, B., Begum, S., McDonald, N.Q., Butler, A., Jones, D., Raine, K., Latimer, C., Santos, C.R., Nohadani, M., Eklund, A.C., Spencer-Dene, B., Clark, G., Pickering, L., Stamp, G., Gore, M., Szallasi, Z., Downward, J., Futreal, P.A., Swanton, C.: Intratumor heterogeneity and branched evolution revealed by multiregion sequencing. N. Engl. J. Med. **366**, 883–892 (2012)

Gillies, R.J., Verduzco, D., Gatenby, R.A.: Evolutionary dynamics of carcinogenesis and why targeted therapy does not work. Nat. Rev. Cancer **12**, 487–493 (2012)

Gobaa, S., Hoehnel, S., Roccio, M., Negro, A., Kobel, S., Lutolf, M.P.: Artificial niche microarrays for probing single stem cell fate in high troughput. Nat. Methods **8**, 949–955 (2011)

Greaves, M., Maley, C.C.: Clonal evolution in cancer. Nature **481**, 306–313 (2012)
Hallatschek, O., Hersen, P., Ramanathan, S., Nelson, D.R.: Genetic drift at expanding frontiers promotes gene segregation. Proc. Natl. Acad. Sci. U. S. A. **104**, 19926–19930 (2007)
Huang, S., Ingber, D.E.: A non-genetic basis for cancer progression and metastasis: self-organizing attractors in cell regulatory networks. Breast Dis. **26**, 27–54 (2006–2007)
Karr, J.R., Sanghvi, J.C., Macklin, D.N., Gutschow, M.V., Jacobs, J.M., Bolival, B., Assad-Garcia, N., Glass, J.I., Covert, M.W.: A whole-cell computational model predicts phenotype from genotype. Cell **150**, 389–401 (2012)
Kohler, S., Schaller, V., Bausch, A.R.: Collective dynamics of active cytoskeletal networks. PLoS One **6**, e23798 (2011)
Kollmann, M., Lovdok, L., Bartholome, K., Timmer, J., Sourjik, V.: Design principles of a bacterial signalling network. Nature **438**, 504–507 (2005)
La Du, B.N., Zannoni, V.G., Laster, L., Seegmiller, J.E.: The nature of the defect in tyrosine metabolism in alcaptonuria. J. Biol. Chem. **230**, 251–260 (1958)
La Porta, C.A., Zapperi, S., Sethna, J.P.: Senescent cells in growing tumors: population dynamics and cancer stem cells. PLoS Comput. Biol. **8**, e1002316 (2012)
Laszlo, E.: The Systems View of the World; The Natural Philosophy of the New Developments in the Sciences. G. Braziller, New York (1972)
Martens, E.A., Kostadinov, R., Maley, C.C., Hallatschek, O.: Spatial structure increases the waiting time for cancer. New J. Phys. **13**, 115014 (2011)
Marti-Renom, M.A., Mirny, L.A.: Bridging the resolution gap in structural modeling of 3D genome organization. PLoS Comput. Biol. **7**, e1002125 (2011)
Maus, C., Rybacki, S., Uhrmacher, A.M.: Rule-based multi-level modeling of cell biological systems. BMC Syst. Biol. **5**, 166 (2011)
Mazemondet, O., Hubner, R., Frahm, J., Koczan, D., Bader, B.M., Weiss, D.G., Uhrmacher, A.M., Frech, M.J., Rolfs, A., Luo, J.: Quantitative and kinetic profile of Wnt/beta-catenin signaling components during human neural progenitor cell differentiation. Cell. Mol. Biol. Lett. **16**, 515–538 (2011)
Meyerhof, O.: The origin of the reaction of Harden and Young in cell-free alcoholic fermentation. J. Biol. Chem. **157**, 105–120 (1945)
Meyerhof, O., Junowicz-Kocholaty, R.: The equilibria of isomerase and aldolase, and the problem of the phosphorylation of glyceraldehyde phosphate. J. Biol. Chem. **149**, 71 (1943)
Meyerhof, O., Oesper, P.: The mechanism of the oxidative reaction in fermentation. J. Biol. Chem. **170**, 1 (1947)
Munsky, B., Neuert, G., van Oudenaarden, A.: Using gene expression noise to understand gene regulation. Science **336**, 183–187 (2012)
Nakakuki, T., Birtwistle, M.R., Saeki, Y., Yumoto, N., Ide, K., Nagashima, T., Brusch, L., Ogunnaike, B.A., Okada Hatakeyama, M., Kholodenko, B.M.: Ligand-specific c-Fos expression emerges from the spatiotemporal control of ErbB network dynamics. Cell **141**(5), 884–896 (2010). doi:10.1016/j.cell.2010.03.054
Okuda, S., Inoue, Y., Eiraku, M., Sasai, Y., Adachi, T.: Reversible network reconnection model for simulating large deformation in dynamic tissue morphogenesis. Biomech. Model. Mechanobiol. **12**(4), 627–644 (2013a). doi:10.1007/s10237-012-0430-7
Okuda, S., Inoue, Y., Eiraku, M., Sasai, Y., Adachi, T.: Modeling cell proliferation for simulating three-dimensional tissue morphogenesis based on a reversible network reconnection framework. Biomech. Model. Mechanobiol. **12**(5), 987–996 (2013b). doi:10.1007/s10237-012-0458-8
Pao, W., Chmielecki, J.: Rational, biologically based treatment of EGFR-mutant non-small-cell lung cancer. Nat. Rev. Cancer **10**, 760–774 (2010)
Prinsen, P., Schiessel, H.: Nucleosome stability and accessibility of its DNA to proteins. Biochimie **92**, 1722–1728 (2010)
Saiz, L., Vilar, J.M.: Multilevel deconstruction of the *In vivo* behavior of looped DNA-protein complexes. PLoS One **2**, e355 (2007)

Sanyal, A., Bau, D., Marti-Renom, M.A., Dekker, J.: Chromatin globules: a common motif of higher order chromosome structure? Curr. Opin. Cell Biol. **23**, 325–331 (2011)

Schaller, V., Bausch, A.R.: Materials science: a fresh twist for self-assembly. Nature **481**, 268–269 (2012)

Schaller, V., Weber, C., Semmrich, C., Frey, E., Bausch, A.R.: Polar patterns of driven filaments. Nature **467**, 73–77 (2010)

Schepers, A.G., Snippert, H.J., Stange, D.E., van den Born, M., van Es, J.H., van de Wetering, M., Clevers, H.: Lineage tracing reveals Lgr5+ stem cell activity in mouse intestinal adenomas. Science **337**, 730–735 (2012)

Schwab, E.D., Pienta, K.J.: Cancer as a complex adaptive system. Med. Hypotheses **47**, 235–241 (1996)

Smithers, D.W.: An attack on cytologism. Lancet **1**, 493–499 (1962)

Stewart-Orstein, J., Weissman, J.S., El-Samad, H.: Cellular noise regulons underlie fluctuations in *Saccharomyces cerevisiae*. Mol. Cell **45**, 483–493 (2012)

Sumino, Y., Nagai, K.H., Shitaka, Y., Tanaka, D., Yoshikawa, K., Chate, H., Oiwa, K.: Large-scale vortex lattice emerging from collectively moving microtubules. Nature **483**, 448–452 (2012)

Tambe, D.T., Hardin, C.C., Angelini, T.E., Rajendran, K., Park, C.Y., Serra-Picamal, X., Zhou, E.H., Zaman, M.H., Butler, J.P., Weitz, D.A., Fredburg, J.J., Trepat, X.: Collective cell guidance by cooperative intercellular forces. Nat. Mater. **10**, 469–475 (2011)

Trepat, X., Fredberg, J.J.: Plithotaxis and emergent dynamics in collective cellular migration. Trends Cell Biol. **21**, 638–646 (2011)

Trepat, X., Wasserman, M.R., Angelini, T.E., Millet, E., Weitz, D.A., Butler, J.P., Fredberg, J.J.: Physical forces during collective cell migration. Nat. Phys. **5**, 426–430 (2009)

Umbarger, M.A., Toro, E., Wright, M.A., Porreca, G.J., Bau, D., Hong, S.H., Fero, M.J., Zhu, L.J., Marti-Renom, M.A., McAdams, H.H., Shapiro, L., Dekker, J., Church, D.M.: The three-dimensional architecture of a bacterial genome and its alteration by genetic perturbation. Mol. Cell **44**, 252–264 (2011)

Vilar, J.M., Leibler, S.: DNA looping and physical constraints on transcription regulation. J. Mol. Biol. **331**, 981–989 (2003)

Vilar, J.M., Saiz, L.: Control of gene expression by modulated self-assembly. Nucleic Acids Res. **39**, 6854–6863 (2011)

Vilar, J.M., Guet, C.C., Leibler, S.: Modeling network dynamics: the lac operon, a case study. J. Cell Biol. **161**, 471–476 (2003)

Von Bertalanffy, L.: The history and status of general systems theory. In: Klir, G. (ed.) Trends in General Systems Theory, pp. 407–426. Wiley, New York (1972)

Wang, G.-W., Zhu, X.-M., Hood, L., Ao, P.: From Phage lambda to human cancer: endogenous molecular-cellular network hypothesis. Quant. Biol. **1**(1), 32–49 (2012). doi:10.1007/s40484-013-0007-1

Xiong, L., Ma, Y., Tang, L.: Attenuation of transcriptional bursting in mRNA transport. Phys. Biol. **7**(1), 016005 (2010). doi:10.1088/1478-3975/7/1/016005

Yu, B., Yang, M., Shi, L., Yao, Y., Jiang, Q., Li, X., Tang, L., Zheng, B.J., Yuen, K., Smith, D.K., Song, E., Huang, J.D.: Explicit hypoxia targeting with tumor suppression by creating an "obligate" anaerobic *Salmonella Typhimurium* strain. Sci. Rep. **2**, 436 (2012). doi:10.1038/srep00436, 10 pp. [EPub 1 Jun 2012]

Zatkova, A.: An update on molecular genetics of Alkaptonuria (AKU). J. Inherit. Metab. Dis. **34**, 1127–1136 (2011)

Chapter 3
Mimicking the Microenvironment

Sharon Gerecht

Introduction

Cancer growth and vascularization is regulated by complex biochemical, biomechanical, and biophysical cues from the surrounding niche (Eccles 2004; Entschladen et al. 2004). Specifically, the extracellular matrix (ECM) is a non-cellular entity, comprised of a variety of macromolecules that act as a scaffold to support overlying cells and the tissues they comprise (Jarvelainen et al. 2009). Locally resident cells, such as fibroblasts, secrete the components that make up the ECM. The main constituents of ECM include collagens, elastin, proteoglycans, and glycoproteins (Jarvelainen et al. 2009). While previously considered an inert, filler substance, the ECM actively influences numerous cellular activities, including cell adhesion, proliferation, differentiation, self-renewal, survival, and migration. It also provides mechanical support to overlying cells (Jarvelainen et al. 2009). The ECM contributes to these diverse cellular functions by providing attachment sites for cells, sequestering bioactive molecules (which are released once proteolytic degradation takes place), and providing mechanical support to overlying cells (Jarvelainen et al. 2009). Disturbances in any of these properties induce changes in cell phenotype and function. These changes in ECM protein expression and mechanical stiffness may promote and contribute to tumorigenic progression. A well-known example is breast cancer: Increased mammographic density, recently attributed to alterations in stroma and ECM deposition (Alowami et al. 2003; Guo et al. 2001; Li et al. 2005), is associated with an increased risk for developing breast cancer (Boyd et al. 1998; Boyd et al. 2001; McCormack and dos Santos Silva 2006). Lu and colleagues pro-

S. Gerecht (✉)
JH Physical Sciences-Oncology Center, Institute for NanoBioTechnology,
Department of Chemical and Biomolecular Engineering, John Hopkins University,
Baltimore, MD 21218, USA
e-mail: gerecht@jhu.edu

P. Janmey et al. (eds.), *Physical Sciences and Engineering Advances in Life Sciences and Oncology*, Science Policy Reports,
DOI 10.1007/978-3-319-17930-8_3

vide an excellent review and additional information regarding this topic (Lu et al. 2012).

Low oxygen tension (i.e., hypoxia) is a common feature of the tumor microenvironment. Hypoxic regions in the tumor occur as a result of inconsistencies in blood flow patterns (i.e., intermittent or acute hypoxia) (Dewhirst 2009; Kimura et al. 1996), which leads to masses of rapidly dividing cells that move beyond the limits of oxygen diffusion (i.e., chronic hypoxia) (Brown and Giaccia 1998). Intermittent and chronic hypoxia both support unique aspects of tumor progression. However, intermittent hypoxia is known to enhance angiogenic responses in tumors (Martinive et al. 2006; Rofstad et al. 2010) and stabilize the expression of hypoxia response factors such as hypoxia inducible factor-1alpha (HIF-1α) (Dewhirst et al. 2008).

Through these external cues, cells gather information about the chemical and physical nature of their microenvironment. They integrate and interpret this data and then generate an appropriate physiological response. Understanding how cells sense different cues and make such "intelligent decisions" is necessary to understanding the basics of cancer progression. This information will lead to the development of new and effective therapeutic agents.

The use of advanced microengineering approaches in cancer research offers investigators the ability to control key spatiotemporal features reminiscent of the tumor environment. These engineered technologies are allowing researchers to recapitulate conditions present within the tumor and to study the complex cell-cell and cell-ECM interactions that take place in the tumor environment. These systems are pivotal in accelerating research into the complex mechanisms regulating tumorigenesis in a relevant *in vitro*-mimetic environment. This chapter reviews several engineering approaches that are being used to investigate tumor growth in the United States, Europe, and Asia. Researchers at several of the sites we visited in Europe and Asia described the evolution of complex technologies that are able to detect single to multiple signals.

Research

Recapitulating Single Cues in the Microenvironment

Cell adhesion to the ECM and to neighboring cells is a complex, tightly regulated process that plays a crucial role in fundamental cellular functions, including cell migration, proliferation, differentiation, and apoptosis. Various approaches to control cell adhesion *in vitro* have been developed throughout the years in the United States, Europe, and Asia.

Joachim Spatz's laboratory at the University of Heidelberg, Germany (site report, Appendix B), pioneered a method to precisely define spatial distribution of ligands on an otherwise inert substrate to enable the understanding of how cell adhesion and signaling depend on the composition, size, and distribution of specific adhesion sites (Arnold et al. 2004). Utilizing micelle diblock copolymer lithography technology enabled the modification of substrates for the presentation of adhesive ligands

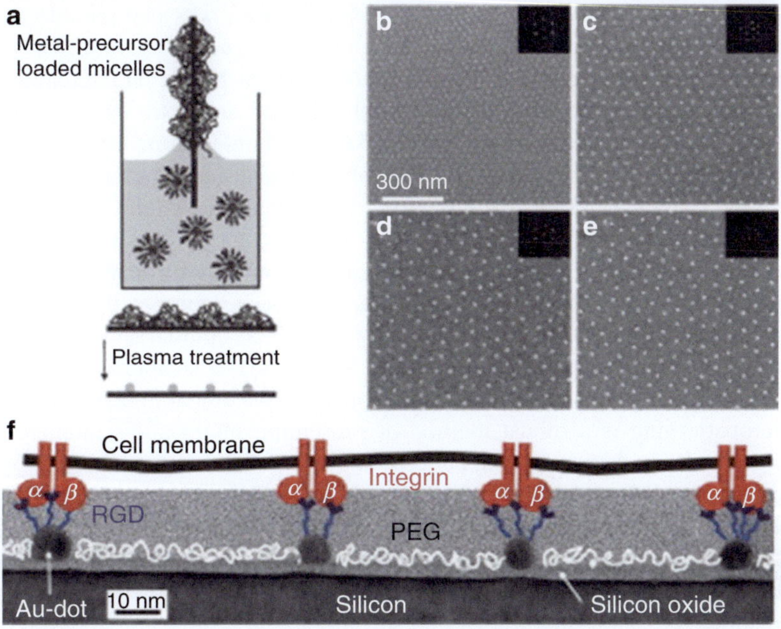

Fig. 3.1 Micellar block copolymer lithography and biofunctionalization (From Spatz and Geiger 2007)
(**a**) Scheme of diblock copolymer micelle lithography. (**b–e**) Extended Au nanodot patterns are displayed by scanning electron microscope. Uniform nanodots (*bright spots*) spacing of (**b**) 3 nm by PS(190)-b-P[2VP(HAuCl4)0.5](190); (**c**) 5 nm by PS(500)-b-P[2VP(HAuCl4)0.5](270); (**d**) 6 nm by PS(990)-b-P[2VP(HAuCl4)0.5](385); and (**e**) 8 nm by PS(1350)-b-P[2VP(HAuCl4)0.5](400) deposited onto Si-wafers. The number in brackets refers to the number of monomer units in each block that control the separation between Au dots. These varied between (**b**) 28, (**c**) 58, (**d**) 73, and (**e**) 85 nm. The nanodots form extended, nearly perfect, hexagonally close-packed patterns as indicated by the Fourier transform images (*inset*), which show second-order intensity spots. (**f**) Biofunctionalization of the nanodots pattern (Arnold et al. 2004). Since the nanodot is sufficiently small, it is likely that only one integrin transmembrane receptor directly interacts with one dot. The nanodots are presented as side-view micrographs taken with a high-resolution transmission electron microscope

in defined spacing at a length scale of 10–200 nm (Fig. 3.1). This approach is based on the self-assembly of diblock copolymers of polystyrene-block-poly (2-vinylpyridine) (PS-b-P2VP) into reverse micelles in toluene. Using this approach, the size of the biofunctionalized nanoparticles may vary between 1 and 20 nm. The spacing between the nanoparticles may also be adjusted from 15 to 250 nm (Arnold et al. 2004; Spatz and Geiger 2007).

By using this platform Spatz and colleagues found that cell adhesion is receptor-space specific. Surfaces of nanoparticles functionalized with RGD—an ECM adhesive peptide that is recognized by αvβ3-integrin with high affinity—were generated with spacing of 28, 58, 73, and 85 nm. Different cell types, including MC3T3 osteoblasts, REF52 fibroblasts, 3 T3 fibroblasts, and B16 melanocytes, seeded on these surfaces spread very well on the 58-nm pattern. This was comparable to their spreading on uniform RGD- or fibronectin-coated surfaces. A separation of $\geq$73 nm between the adhesive dots resulted in limited cell attachment and spreading, and dramatically reduced the formation of focal adhesion and actin stress fibers. The researchers suggest that the range between 58 and 73 nm is a universal length scale for integrin clustering and activation. The lab created a spacing gradient surface to explore the role of adhesive spacing during migration (Hirschfeld-Warneken et al. 2008). It was found that MC3T3 osteoblasts' morphology, adhesion area, actin, and vinculin distribution, as well as cell body polarization were influenced by the peptide patch spacing gradient. As a consequence, these adhesive ligand gradients induce MC3T3 osteoblasts orientation towards their optimal adhesion spacing (Hirschfeld-Warneken et al. 2008). Chapter 5 discusses these approaches to understanding cancer development and growth from the time domain perspective.

Micropatterning is a robust method of presenting micro and nanometer scale ECM molecules in distinct spatial patterns and allows for the control of various cell responses. Throughout the years, numerous approaches to generate micropatterns have been developed in the United States, Europe, and Asia, including direct printing techniques such as bioprinting or dip pen lithography. These approaches use inkjet/laser printers or atomic force microscope tips as well as micropattern fabrication methods like photolithography and soft lithography. An example of the role of cell adhesion in the formation of normal vs. abnormal tissue architecture was given by Dr. Wilhelm Huck at the University Nijmegen Medical Centre, Netherlands (site report, Appendix B) in collaboration with Dr. Fiona Watt (Cambridge, U.K.). Surfaces printed with collagen I in 10-μm rings containing non-adhesive disks (ranging from 0 to 60 mm) were generated with a polymer brush to examine the formation of the micro-epidermis. Human epidermal keratinocytes seeded on these surfaces were examined for cell differentiation vs. proliferation. For rings with a non-adhesive center of up to 40 mm diameter, cell-cell and cell-matrix adhesive interactions result in correct micro-epidermis assembly. Using this platform, cells isolated from oral cavity tumors (squamous cell carcinomas that have increased proliferation and reduced differentiation with disturbed tissue architecture) were cultured on these micropatterned islands and found to exhibit disturbed architecture that correlates with the original tumor characteristics (Fig. 3.2).

Tumors are "stiffer" than their normal tissue counterparts. The mechanical stiffness of breast carcinomas has been reported to be as high as 42.5 kilopascals (kPa) for high-grade invasive ductal carcinomas, as opposed to 3.25 kPa for non-malignant mammary tissues (Samani et al. 2007). Similarly, lymph nodes harboring metastatic tumor foci had a mechanical stiffness of 3.35 ± 1.57 g/cm (i.e., 329 Pa) versus non-tumor bearing lymph nodes, which have a mechanical stiffness of 1.23 g/cm (i.e., 121 Pa) (Miyaji et al. 1997).

Tumors are "stiffer" than their normal tissue counterparts. The mechanical stiffness of breast carcinomas has been reported to be as high as 42.5 kilopascals (kPa) for high-grade invasive ductal carcinomas, as opposed to 3.25 kPa for non-malignant mammary tissues (Samani et al. 2007). Similarly, lymph nodes harboring metastatic tumor foci had a mechanical stiffness of 3.35 ± 1.57 g/cm (i.e., 329 Pa) versus non-

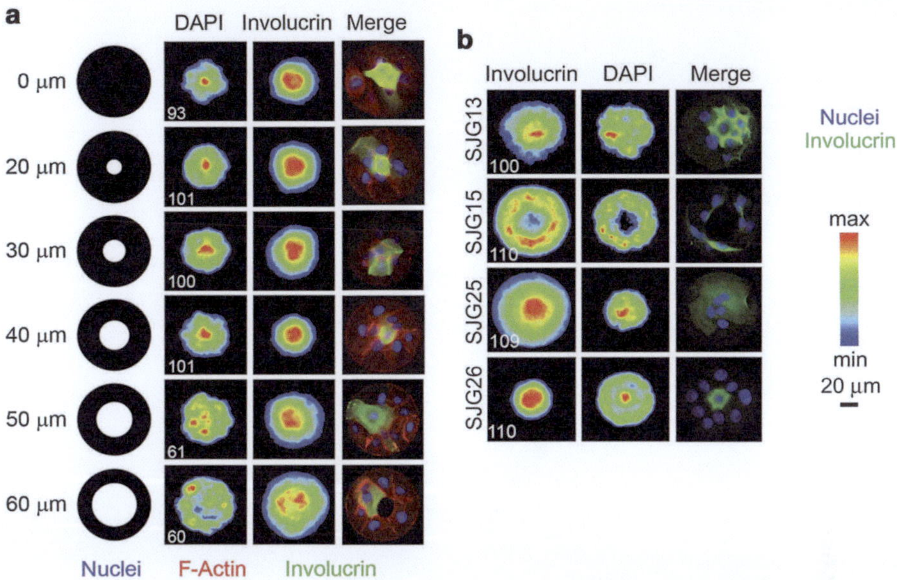

Fig. 3.2 Formation of a micro-epidermis on a collagen island (Gautrot et al. 2012)
(**a**) Undifferentiated keratinocytes adhered to collagen I disks and rings for 24 h before fixation and immunostaining. Adhesive island geometries (*in black*, *far left*) for collagen rings with protein resistant patches ranging from 0 to 60 mm in diameter. Heat maps (*left and middle columns*) and single cluster images (*right column*); involucrin and for the two *bottom* rows, pan keratins in *green* (*middle ring*), F-actin in *red* (*center*), and DAPI in *blue* (*outer ring*). White numbers (*bottom left* of heat maps) are the number of images overlaid. Images demonstrate that cells could assemble a normal microepidermis structure on 100-mm-diameter disks with ≤40 mm rings. (**b**) Cancer cells (SJG13, 15, 25, and 26, corresponding to the tumor from which the cells were isolated) were allowed to adhere to collagen rings (40 mm non-adhesive patch) for 24 h before fixation and immunostaining as detailed in (**a**). Images show disturbed tissue architecture of SJG13, 15, and 26 with relatively normal behavior of SJG26 cells. This correlated well with the lesion histology from which the cells were derived

tumor bearing lymph nodes, which have a mechanical stiffness of 1.23 g/cm (i.e., 121 Pa; Miyaji et al. 1997).

Another important aspect of the microenvironment is the indirect interactions among the various cell types in the cancerous niche. Specifically, paracrine signaling in which growth factors secreted by resident cells are sequestered in the ECM and affect cancer cell fate, play an important role in tumor development. Dr. Chau-Hwang Lee's laboratory at the Institute of Biophotonics, National Yang-Ming University Research Center for Applied Sciences Academia Sinica, Taiwan (site report, Appendix C), employed a microfluidic cell culture chip to investigate the timing sequences of the paracrine loop between cancer cells and fibroblasts (Fig. 3.3). In this well-controlled interaction sequence, the group measured the migration speeds of cancer cells and the aspect ratios of fibroblasts, which reflect the phenotype of myofibroblasts. They found that cytokines from cancer cells effectively stimulate the fibroblasts into myofibroblasts and that the cytokines from myofibro-

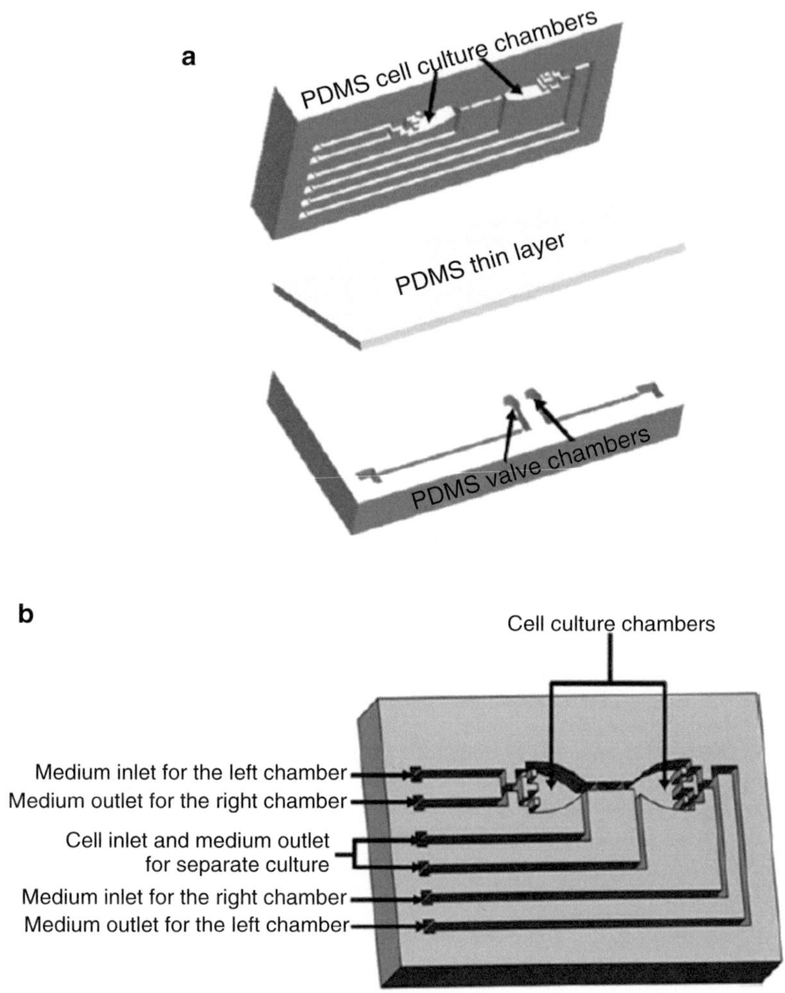

Fig. 3.3 The microfluidic cell culture chip used in this study. (**a**) Structure and (**b**) fluidic channels of the cell culture chip with pneumatic microvalves (From Hsu et al. 2011)

blasts increase the migration speeds of cancer cells. They further demonstrated that transforming growth factor-b1 (TGF-b1) is involved in these interactions (Hsu et al. 2011). In recent work, the group added complexity to the system by introducing macrophages. They have found that both myofibroblasts and macrophages have synergistic effects in accelerating the migration of cancer cells, while pre-treating the myofibroblasts with the macrophages' secreted cytokines reduced the myofibroblasts' ability to enhance the migration of cancer cells. Indeed, TGF-b1 was found to be key regulator in this process (Hsu et al. 2012).

Engineering the mechanics of matrices for the 2D and 3D cell culture offers the opportunity to study cellular responses and the regulatory mechanisms involved. In recent years, there has been an increase in research of human stem cells towards their translational use. This has tremendously boosted studies focusing on the development and utilization of biomaterials to control cellular behaviors and advance cancer research in the United States, Europe, and Asia. In fact, the approaches to using biomaterials that we were shown during our visits in Europe are mostly derived from studies aiming at guidance of stem cell fate. Overall, research focus and advancement in technology seems to be comparable between the United States and Europe. Dr. Matthias Lutolf's lab at the École Polytechnique Fédérale de Lausanne, Switzerland (site report, Appendix B), has been studying the effect of stiffness on stem cell behavior. In a landmark study, Lutolf and colleagues at Stanford University utilized tunable polyethylene glycol (PEG) hydrogels as substrates and showed that when cultured on soft hydrogel substrates that mimic the elasticity of muscle (12 kPa), muscle stem cells self-renew *in vitro* and contribute extensively to muscle regeneration when subsequently transplanted into mice (Gilbert et al. 2010). Continuing this line of research, the Lutolf laboratory developed microarrays for the high-throughput analysis of cellular responses to various biochemical and physical cues of the microenvironment (Gobaa et al. 2011). This microengineered platform consists of PEG hydrogel microwell arrays with individual microwells that are functionalized with combinations of proteins spotted by robotic technology. By varying the PEG precursor concentrations, they were able to generate microwells with a range of stiffness (shear moduli of 1–50 kPa) and the same adhesive functionality. This allowed the researchers to examine cellular responses to stiffness independent of adhesion. As proof-of-principle, they have tested the effect of substrate stiffness on osteogenic (bone) differentiation of human mesenchymal stem cells (MSCs). As expected, increasing the elastic modulus of the substrate resulted in increased osteogenic differentiation independent of the specific adhesion motif (Gobaa et al. 2011). This system can be used to deconstruct various parameters in the stem and cancer cell niche to examine their effect on differentiation vs. self-renewal.

Similarly, the Nicola Elvassore group at the University of Padua, Italy (site report, Appendix B), used photo cross-linkable elastic polyacrylamide (PAA) hydrogel to generate substrates with stiffness ranging from 12 to 21 kPa. They showed that sarcomere formation during myoblast differentiation occurs on a softer substrate (15 kPa) that enables the maturation and functionality of myotubes (Serena et al. 2010). This model is being used to study muscular dystrophy, where the soft surrounding of the diseased muscle results in dysfunctional myotubes with impaired calcium release and upregulation in dystrophin expression.

Variations in oxygen concentrations lead to a range of cellular responses, depending on the cell type and microenvironment at every stage of embryogenesis, in different ischemic regions of diseased tissue (e.g., cardiac), and in developing tumors. Many cell types respond differently, but collectively to the changes in oxygen equilibrium through specialized sensing mechanisms and effectors to maintain homeostasis. Yet, to recreate *in vitro* ischemia pathological models, accurate gas concentration dynamics is one of the most difficult parameters to control. In fact, most of the *in vitro* studies that investigate the effect of hypoxia on cells have adopted one of the following two approaches: (1) cells are cultured in macroscopic environments maintained at low oxygen levels in order to activate signaling pathways involving hypoxia-inducible factors (HIFs) (Ezashi et al. 2005; Niebruegge et al. 2008; Purpura et al. 2008) or (2) the expression of HIFs is genetically enforced in order to investigate the resulting molecular pathways (Jiang et al. 2008a, b). However, these two methods do not take into account the complex and dynamic feedback between the cells and their microenvironment.

Advances in microfluidics and sensor technology offer the unprecedented opportunity to perform on-line measurement and analysis of dissolved oxygen dynamics in the microenvironment of adherent cell cultures, and to correlate them with various cellular responses. Moreover, microfluidic systems with low gas permeability are ideally suited to perform hypoxic experiments using routine tissue culture cabinets and incubators. A number of microfluidic devices have been developed to monitor and control oxygen tension in cell cultures. Mehta and colleagues developed a system that prevents the development of oxygen gradients along the culture channel (Mehta et al. 2007). However, the control over oxygen tension in their device is directly coupled to the shear stress acting on the cells. Therefore, the interplay between these microenvironmental cues cannot be eliminated.

In order to address this problem, laboratories throughout the United States, Europe, and Asia apply microfluidic devices to control oxygen concentration. New cell culture microdevices use a two-channel approach where oxygen is supplied to the cell culture from an independent channel, separated from the culture channel by a gas permeable membrane (Kane et al. 2006; Lam et al. 2009; Leclerc et al. 2004; Lo et al. 2010; Polinkovsky et al. 2009). A simple and versatile system based on the two-channel approach was recently developed to enable long-term cell culture studies while accurately controlling and continuously on-line monitoring the dissolved oxygen level in the cell microenvironment (Abaci et al. 2011). The Elvassore lab is using microfluidic technology to control oxygen tension in media of cultured cells. The newly developed microfluidic device includes a gas exchanger for control over dissolved oxygen levels in the study of the hypoxic effect on calcium transients in response to electrical stimulation of muscle cell derivatives (Fig. 3.4). The device was designed to enable on-line confocal microscopy analysis while finely tuning the oxygen concentration in the culture medium and maintaining the proximal cell environment isolated from atmospheric conditions. In addition, the device allows cells to be exposed to fast transient changes in the environment without perturbing the data acquisition process. Results demonstrated that exposure of neonatal rat cardiomyocytes to hypoxic conditions induced changes in intracellular Ca^{2+} transients. This event was reversible for hypoxic levels below 5 % oxygen partial pressure.

Dr. Tung's group at the Research Center for Applied Sciences at Academia Sinica (site report, Appendix C) recently develop a microfluidic cell culture array platform capable of performing cell culture under various oxygen tensions simultaneously. Using polydimethylsiloxane (PDMS) and the well-developed multi-layer soft lithography technique the device has 4×4 wells in which the oxygen tensions are

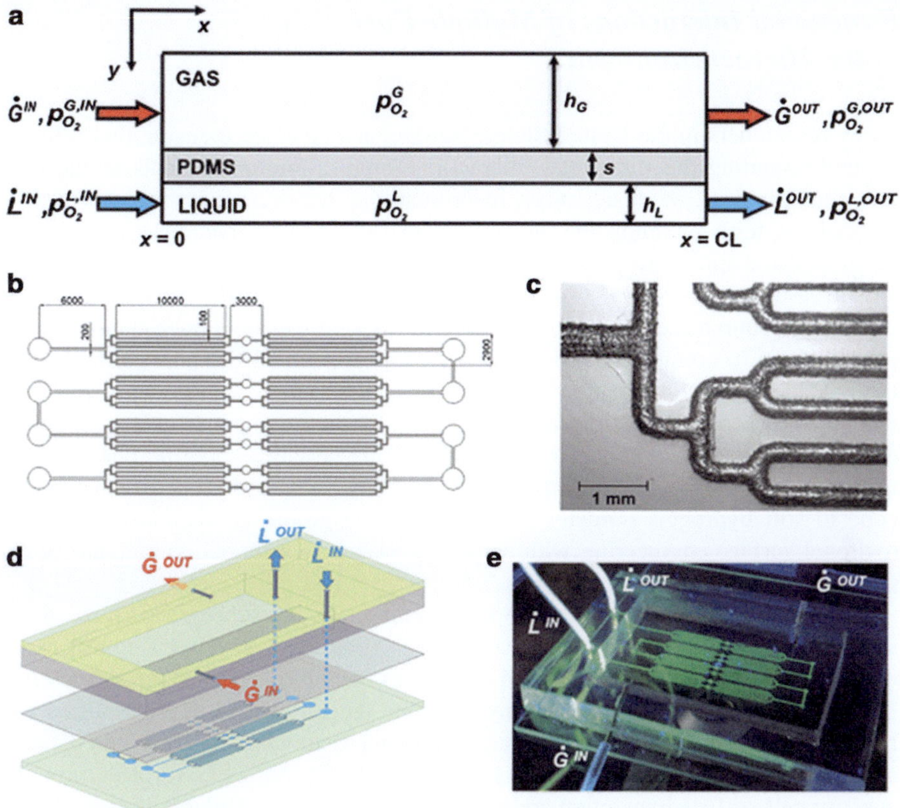

Fig. 3.4 Microfluidic gas exchanger (From Martewicz et al. 2011)
(**a**) Schematic representation of the three-layered microfluidic system; inlet/outlet flow rates and oxygen partial pressure are shown for both the gas, G, and liquid, L, phase. (**b**) Top view of the fluidic layer channel network (all dimensional values are in μm). (**c**) Image of a glass-etched microfluidic channel network obtained with a wet-etching technique and observed under an inverted optical microscope. (**d**) Schematic view of the three different layers of the gas exchanger. *Red* and *blue arrows* show gas and liquid phase inlet and outlet inside the platform. (**e**) Image of the gas-exchanger with inlet/outlet connections for liquid and gas phase perfusion. The microfluidic channels are perfused with 1 mM fluorescein solution

controlled by spatially confined oxygen scavenging chemical reactions underneath the wells using microfluidics. Utilizing the developed platform, drug testing using an anti-cancer drug, triapazamine, was performed on adenocarcinomic human alveolar basal epithelial cell line (A549) under three oxygen tensions ranging from 1.4 % to normoxia. They have found that oxygen tension plays an essential role to not only regulate cell behaviors but also affect the effectiveness of drugs according to their modes of action. The developed platform is thus promising to provide a more meaningful *in vitro* cell model for various biomedical applications while maintaining desired high throughput capabilities (Peng et al. 2013).

Functional Interactions of Multiple Cues in the Microenvironment

Many researchers in the United States, Europe, and Asia are focusing their efforts on understanding the functional interactions among multiple signals in the cell microenvironment. With advances in biomaterial sciences and microfabrication technology, tools enabling the decoupling of two or more parameters are being developed and applied to study cellular responses to surrounding stresses in healthy and diseased tissue.

Spatz's group has extended the function of their block copolymer nanolithography platform to arrange a defined number of biofunctionalized nanoparticles in a designated pattern. Such patterns can be transferred to almost any type of soft surface, thereby replacing a stiff glass or silicon oxide surface with an elastic or viscoelastic polymer surface (Graeter et al. 2007). This provides opportunities to study adhesions on surfaces ranging from rigid hydrophobic polystyrene to elastic silicone to soft hydrogels (Graeter et al. 2007). The group has demonstrated PEG hydrogel surface engineering with respect to elasticity, nanopatterning, and functionalization with biomolecules (Fig. 3.5).

Biomolecule arrangement on the nanometer scale and substrate stiffness were shown to vary independently from each other (Aydin et al. 2010). Using this technology, Young's moduli can be tuned over four orders of magnitude, and structured hydrogels can be used to pattern any histidine-tagged protein as exemplified for his-protein A as an acceptor for immunoglobulin (Aydin et al. 2010). When the cell adhesion-promoting peptide RGD is used selectively, the PEG surfaces provide cues for cell-surface interaction and allow for the study of cellular adhesion modulation by the environmental mechanical properties. Therefore, these substrates represent a unique multipurpose platform for studying receptor/ligand interactions with adhering cells, mechanotransduction, and cell adhesion-dependent signaling.

Dr. Main Long and colleagues at the Institute of Mechanics, Chinese Academy of Sciences, China (site report, Appendix C), have recently utilized a microfabricated polyacrylamide hydrogel substrate with two elasticities, two topographies, and three dimensions to systematically test MSC responses (Fig. 3.6; Li et al. 2013). They have found that substrate stiffness or dimension is predominant in regulating MSC proliferation while topography is a key factor for manipulating MSC morphology and spreading. Furthermore, MSC differentiation was mainly affected by substrate stiffness and to lesser extent by substrate topography or dimension.

To enable control over rigidity and adhesion in a 3D setting, Dr. Keng-hui Lin's group at the Institute of Physics Academia Sinica (site report, Appendix C) have recently developed an approach in which uniform pores in compliant gels are generated followed by their coating with fibronectin to pattern ECM proteins as spherical shells. The rigidity of the 3D microenvironment is controlled by the choice of base gels used to assemble the scaffolds. Using 3D polyacrylamide scaffolds, Dr. Lin's group found that fibroblasts sense the local rigidity of the scaffold and exhibit a 3D

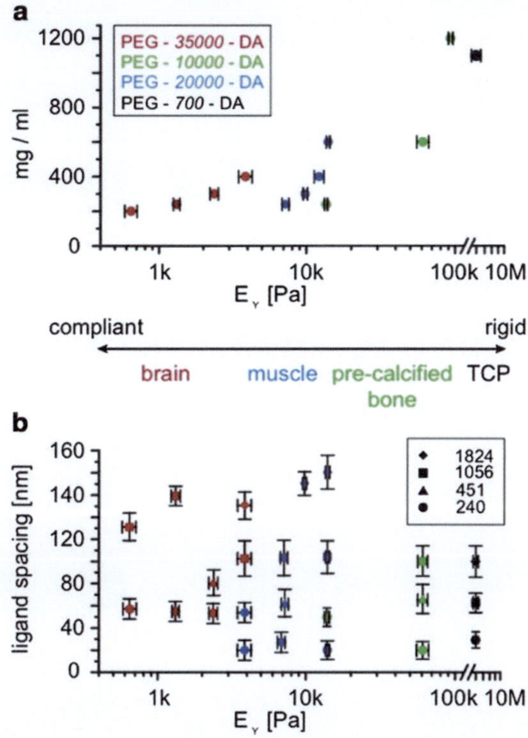

Fig. 3.5 Compliance and nanoparticle decoration properties of PEG- diacrylate (DA) hydrogels (From Aydin et al. 2010)
(**a**) The Young's modulus (Eγ) of PEG-10000-DA, PEG-20000-DA, PEG-35000-DA, and PEG-700-DA hydrogels polymerized with different initial water content is compared to the Young's modulus of different tissues. The complete range of biologically relevant material compliance can be covered. (**b**) Overview of the adjustable material properties of nanopatterned PEG-DA hydrogels. The variations of bioactive ligand spacing on the hydrogel and the compliance of the material are fully independent of each other. The number associated to the various symbols denotes the polymer used to prepare the micellar solutions

distribution of actin cytoskeleton and adhesions that became more pronounced as the pore size was reduced. In small pores, they observed that elongated adhesions can exist without attachment to any solid support (Li et al. 2013).

Very recently, Lutolf's group developed an approach to pattern biomolecules in 3D hydrogels. The group utilized an approach that masked an enzymatic peptide substrate with a photolabile cage. The enzyme-mediated bioconjugation could be spatiotemporally controlled by light exposure (Fig. 3.7). By covalently incorporating these caged substrates into a hydrogel network, subsequent local photoactivation was expected to enable highly localized enzymatic biomolecule tethering. This approach allows control over the site-specific immobilization of any desired protein within a synthetic hydrogel network. Using this technology, the group showed that

Fig. 3.6 Schematic of experimental set-up. Systematic testing of the effect of PA hydrogel substrate stiffness, topography, and dimension on MSCs behaviors (From Li et al. 2013)

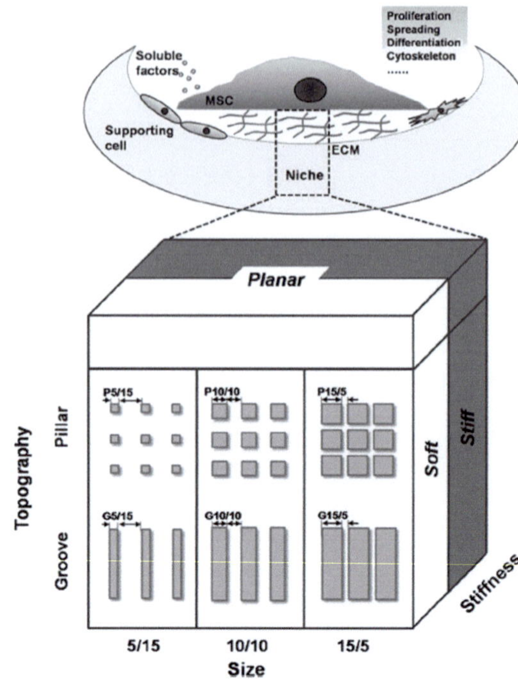

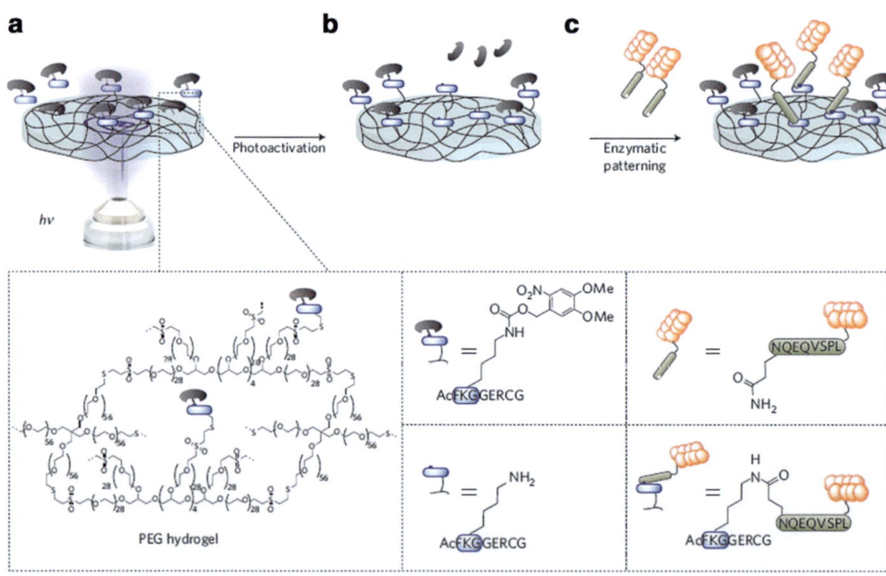

Fig. 3.7 Concept of light-controlled enzymatic biomolecule patterning of hydrogels (From Mosiewicz et al. 2013)
(**a**) A photolabile, caged, and therefore inactive enzymatic peptide substrate is covalently incorporated into PEG hydrogels and can be activated by light. (**b**) Localized cleavage of the cage by controlled light exposure from a confocal laser allows reactivation of the enzyme substrate. (**c**) Enzyme-catalyzed (here, the transglutaminase factor XIII) reaction of the uncaged substrate with a counter-reactive substrate on a biomolecule of interest allows covalent biomolecule tethering in a highly localized, user-defined pattern

the 3D invasion of human MSCs can be spatiotemporally controlled by micropatterning the hydrogel with desired ECM proteins and growth factors (Mosiewicz et al. 2013).

Spatz and his team used their block copolymer nanolithography platform to arrange biofunctionalized nanoparticles with either biotinylated RGD, epidermal growth factor (EGF), or both on an inert surface (Shahal et al. 2012). With this surface, they investigated the coregulation of RGD- and immobilized/soluble EGF-mediated signaling in the adhesion of A431 epidermoid carcinoma cells. A synergism was found between integrin and the EGF receptor in an RGD and EGF density- or concentration-dependent manner. The effect of immobilized EGF differed from the effect of soluble EGF above a certain EGF density. The researchers speculate that the critical EGF density is most likely required for the induction of EGF receptor dimerization, and lies within the range of the RGD density. This suggests a role for EGFR-integrin cooperativity in EGF-mediated adhesion response (Shahal et al. 2012).

It is thus becoming apparent that in addition to growth factor sequestration, ECM-growth factor interactions also directly modulate growth factor signaling through a co-association of integrins with growth factor receptors. In this context, complexes between ECM proteins and growth factors can mediate enhanced growth factor receptor-integrin signaling by the formation of clusters between growth factor receptors and integrins (Comoglio et al. 2003; Giancotti and Tarone 2003; Guo and Giancotti 2004).

The Jeffrey Hubbell group at École Polytechnique Fédérale de Lausanne, Switzerland (site report, Appendix B), is moving towards a 3D scaffold platform. They recently explored whether growth factor-induced tissue response could be strongly enhanced when growth factors are delivered within a hydrogel with a well-defined microenvironment designed to trigger synergistic signaling between growth factor receptors and integrins (Martino et al. 2011). To achieve this, a multifunctional recombinant fibronectin fragment was engineered to display integrin-binding domains linked to growth factor binding domains. The sequence consisted of a coagulation transglutaminase enabling polymerization of fibrin (Fig. 3.8). Such engineered microenvironments allow the sequestration of multiple growth factors while promoting joint integrin-growth factor receptor signaling. The group showed that proliferation and migration of endothelial cells, smooth muscle cells, and MSCs is enhanced by co-delivery of FnIII9-10/12-14 and VEGF, PDGF-BB and BMP-2, respectively. Moreover, the FN domains enhanced growth factor-induced morphogenesis in the fibrin matrices. Finally, the engineered fibrin scaffolds were shown to improve growth factor efficiency in tissue repair *in vivo* (i.e., diabetic wound healing and critical-size bone defect) (Martino et al. 2011). Chapter 5 discusses the use of such fibrin gels for the analysis of lymphatic drainage during tumor growth.

A recent collaboration between Stefano Piccolo's lab at the University of Padua, Italy (site report, Appendix B), and the Elvassore lab identified the YAP/TAZ [Yorkie-homologues YAP (Yes-associated protein) and TAZ (transcriptional coactivator with PDZ-binding motif)] as nuclear relays of mechanical signals exerted by extracellular rigidity and cell shape, independently from Hippo pathways

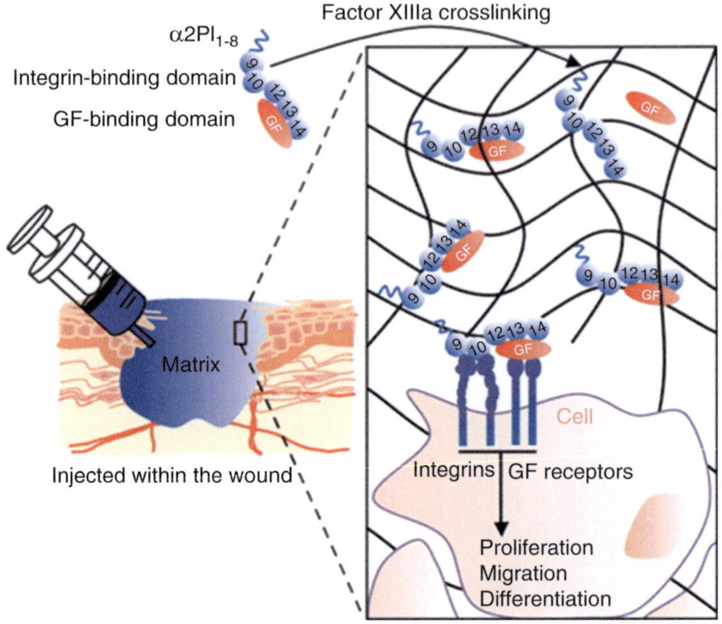

Fig. 3.8 Engineered 3D microenvironment (From Martino et al. 2011)
A multifunctional recombinant fibronectin (FN) fragment is engineered to display the integrin-binding domain (FN III9-10) linked to the GF-binding domain (FN III12-14), and to comprise the substrate sequence $\alpha 2PI_{1-8}$ for factor XIIIa. The fragment is covalently cross-linked into a fibrin matrix during the natural polymerization process of fibrin via the transglutaminase activity of factor XIIIa. The engineered matrix allows sequestration of GFs and joint integrin-GF receptor signaling, thus leading to cell recruitment, proliferation, and differentiation

(Dupont et al. 2011). The groups used micropillar substrates with varying rigidity to decouple matrix stiffness and adhesion. First, following bioinformatic analysis that suggested YAP/TAZ involvement, they examined the expression levels and localization in mammary epithelial cells cultured on fibronectin-coated PAA hydrogel of varying stiffness. YAP/TAZ activity increased and was localized to the nucleus on stiff hydrogels. Next, they used micropatterning of fibronectin islands of different sizes to show that in large adhesion/cell spreading areas, the cells experience broader cell-ECM contact area with nuclear YAP/TAZ. Micropillar surfaces with fibronectin deposited on the tip of the micropillars were used to determine if the cell spreading affected YAP/TAZ expression independently of the total amount of ECM. The researchers found that nuclear localization of YAP/TAZ is regulated by cell spreading. Using the same technology with modified pillar elasticity further demonstrated that the localization correlated with higher cytoskeleton tension. They demonstrated how these effects are independent of Hippo pathways (Dupont et al. 2011). This technology has enabled the groups to address the co-regulatory mechanism by which several cues in the microenvironment modulate cellular responses.

Discussion

Future Challenges

In order for us to better understand the regulatory mechanism by which the microenvironment modulates cancer growth and development, we must better recapitulate the dynamically changing cellular surroundings. Advances in miniaturization technologies will allow the high-resolution analysis of the engineered microenvironments and the resulting cellular behaviors. One example of analyzing responses to the "bulk" vs. "microscale" is a recent study conducted in four European laboratories led by Huck and Watt. They challenge that MSC differentiation is affected by stiffness in the bulk scale (Trappmann et al. 2012). They repeated previous data where substrate stiffness affects MSC spreading and adipocyte (fat) and osteoblast differentiation by using PAA. However, using PDMS substrate, the elastic modulus did not seem to affect cell spreading and differentiation. Therefore, they looked into the deposition of the coating protein, collagen I, on each of these surfaces. They found that it is differently deposited on PAA substrates with smaller pores in stiffer hydrogels. This suggests that adhesion to the collagen coating regulates cell response, not the bulk stiffness. They also controlled the density of adhesions using the same stiffness and demonstrated the different cellular responses by applying covalent bonding (Trappmann et al. 2012). This approach utilizing technological capabilities in laboratories in different institutions and countries throughout Europe is a noticeable strength.

This work raises new questions and indicates the need for a more complex approach, which will take into account the presentation of adhesions at the microscale and how this can relate to mechanics. Advances in technology should also be made to enable better imitation of physiological conditions. For example, most technologies currently capture a specific time point in the microenvironment and allow the study of cellular behaviors in response to a given cue. The ability to generate dynamic environments that respond to feedback from the tissue/cells during long culture periods is envisioned to allow the analysis of cellular responses to the dynamically changing cancerous surroundings.

In a step toward this goal, a recent publication by Guvendiren and Burdick in the United States describes how new hydrogels were developed to enable dynamic stiffening along the culture period (Guvendiren and Burdick 2012). Sequential crosslinking—first gelation by an addition reaction (of DTT) to methacrylated HA to generate soft environments in the presence of cells followed by radical polymerization (crosslinking of the remaining group by exposure to UV) created a stiffening niche for the encapsulated cells. The Burdick lab has shown the response of MSCs to the stiffening environment from non-spreading to spreading morphology. They then demonstrated that differentiation of MSCs during a 14-day span can be controlled by temporal stiffening. While stiff hydrogels supported bone differentiation and soft hydrogels supported fat differentiation, the stiffening hydrogel induced mixed adipogenic/osteogenic population, suggesting that not all cells were responsive to changes in mechanical properties along the culture period (Guvendiren and Burdick 2012).

Again, new questions arise with the use of more "biologically" accurate biomaterials. Another venue is the generation of different cue gradients *in vitro*, similar to their presence within the growing tumor and its surroundings, such as cytokines and oxygen tension. Finally, decoupling parameters in 3D still presents a major challenge and worldwide advances in biomaterial synthesis are needed to enable progress.

References

Abaci, H.E., Devendra, R., Smith, Q., Gerecht, S., Drazer, G.: Design and development of microbioreactors for long-term cell culture in controlled oxygen microenvironments. Biomed. Microdevices **14**, 145–152 (2011)

Alowami, S., Troup, S., Al-Haddad, S., Kirkpatrick, I., Watson, P.H.: Mammographic density is related to stroma and stromal proteoglycan expression. Breast Cancer Res. **5**, R129–R135 (2003)

Arnold, M., Cavalcanti-Adam, E.A., Glass, R., Blummel, J., Eck, W., Kantlehner, M., Kessler, H., Spatz, J.P.: Activation of integrin function by nanopatterned adhesive interfaces. Chemphyschem **5**, 383–388 (2004)

Aydin, D., Louban, I., Perschmann, N., Blummel, J., Lohmuller, T., Cavalcanti-Adam, E.A., Haas, T.L., Walczak, H., Kessler, H., Fiammengo, R., Spatz, J.P.: Polymeric substrates with tunable elasticity and nanoscopically controlled biomolecule presentation. Langmuir **26**, 15472–15480 (2010)

Boyd, N.F., Lockwood, G.A., Byng, J.W., Tritchler, D.L., Yaffe, M.J.: Mammographic densities and breast cancer risk. Cancer Epidemiol. Biomarkers Prev. **7**, 1133–1144 (1998)

Boyd, N.F., Martin, L.J., Stone, J., Greenberg, C., Minkin, S., Yaffe, M.J.: Mammographic densities as a marker of human breast cancer risk and their use in chemoprevention. Curr. Oncol. Rep. **3**, 314–321 (2001)

Brown, J.M., Giaccia, A.J.: The unique physiology of solid tumors: opportunities (and problems) for cancer therapy. Cancer Res. **58**, 1408–1416 (1998)

Comoglio, P.M., Boccaccio, C., Trusolino, L.: Interactions between growth factor receptors and adhesion molecules: breaking the rules. Curr. Opin. Cell Biol. **15**, 565–571 (2003)

Dewhirst, M.W.: Relationships between cycling hypoxia, HIF-1, angiogenesis and oxidative stress. Radiat. Res. **172**, 653–665 (2009)

Dewhirst, M.W., Cao, Y., Moeller, B.: Cycling hypoxia and free radicals regulate angiogenesis and radiotherapy response. Nat. Rev. Cancer **8**, 425–437 (2008)

Dupont, S., Morsut, L., Aragona, M., Enzo, E., Giulitti, S., Cordenonsi, M., Zanconato, F., Le Digabel, J., Forcato, M., Bicciato, S., Elvassore, N., Piccolo, S.: Role of YAP/TAZ in mechanotransduction. Nature **474**, 179–183 (2011)

Eccles, S.A.: Targeting key steps in metastatic tumour progression. Curr. Opin. Genet. Dev. **15**, 77–86 (2004)

Entschladen, F., Drell, T.L., Lang, K., Joseph, J., Zaenker, K.S.: Tumour-cell migration, invasion, and metastasis: navigation by neurotransmitters. Lancet Oncol. **5**, 254–258 (2004)

Ezashi, T., Das, P., Roberts, R.M.: Low O_2 tensions and the prevention of differentiation of hES cells. Proc. Natl. Acad. Sci. U. S. A. **102**, 4783–4788 (2005)

Gautrot, J.E., Wang, C., Liu, X., Goldie, S.J., Trappmann, B., Huck, W.T., Watt, F.M.: Mimicking normal tissue architecture and perturbation in cancer with engineered micro-epidermis. Biomaterials **33**, 5221–5229 (2012)

Giancotti, F.G., Tarone, G.: Positional control of cell fate through joint integrin/receptor protein kinase signaling. Annu. Rev. Cell Dev. Biol. **19**, 173–206 (2003)

Gilbert, P.M., Havenstrite, K.L., Magnusson, K.E., Sacco, A., Leonardi, N.A., Kraft, P., Nguyen, N.K., Thrun, S., Lutolf, M.P., Blau, H.M.: Substrate elasticity regulates skeletal muscle stem cell self-renewal in culture. Science **329**, 1078–1081 (2010)

Gobaa, S., Hoehnel, S., Roccio, M., Negro, A., Kobel, S., Lutolf, M.P.: Artificial niche microarrays for probing single stem cell fate in high throughput. Nat. Methods **8**, 949–955 (2011)

Graeter, S.V., Huang, J., Perschmann, N., Lopez-Garcia, M., Kessler, H., Ding, J., Spatz, J.P.: Mimicking cellular environments by nanostructured soft interfaces. Nano Lett. **7**, 1413–1418 (2007)

Guo, W., Giancotti, F.G.: Integrin signalling during tumour progression. Nat. Rev. Mol. Cell Biol. **5**, 816–826 (2004)

Guo, Y.P., Martin, L.J., Hanna, W., Banerjee, D., Miller, N., Fishell, E., Khokha, R., Boyd, N.F.: Growth factors and stromal matrix proteins associated with mammographic densities. Cancer Epidemiol. Biomarkers Prev. **10**, 243–248 (2001)

Guvendiren, M., Burdick, J.A.: Stiffening hydrogels to probe short- and long-term cellular responses to dynamic mechanics. Nat. Commun. **3**, 792 (2012)

Hirschfeld-Warneken, V.C., Arnold, M., Cavalcanti-Adam, A., Lopez-Garcia, M., Kessler, H., Spatz, J.P.: Cell adhesion and polarisation on molecularly defined spacing gradient surfaces of cyclic RGDfK peptide patches. Eur. J. Cell Biol. **87**, 743–750 (2008)

Hsu, T.H., Xiao, J.L., Tsao, Y.W., Kao, Y.L., Huang, S.H., Liao, W.Y., Lee, C.H.: Analysis of the paracrine loop between cancer cells and fibroblasts using a microfluidic chip. Lab Chip **11**, 1808–1814 (2011)

Hsu, T.H., Kao, Y.L., Lin, W.L., Xiao, J.L., Kuo, P.L., Wu, C.W., Liao, W.Y., Lee, C.H.: The migration speed of cancer cells influenced by macrophages and myofibroblasts co-cultured in a microfluidic chip. Integr. Biol. (Camb.) **4**, 177–182 (2012)

Jarvelainen, H., Sainio, A., Koulu, M., Wight, T.N., Penttinen, R.: Extracellular matrix molecules: potential targets in pharmacotherapy. Pharmacol. Rev. **61**, 198–223 (2009)

Jiang, M., Wang, B., Wang, C., He, B., Fan, H., Guo, T.B., Shao, Q., Gao, L., Liu, Y.: Angiogenesis by transplantation of HIF-1 alpha modified EPCs into ischemic limbs. J. Cell. Biochem. **103**, 321–334 (2008a)

Jiang, M., Wang, B., Wang, C., He, B., Fan, H., Shao, Q., Gao, L., Liu, Y., Yan, G., Pu, J.: *In vivo* enhancement of angiogenesis by adenoviral transfer of HIF-1alpha-modified endothelial progenitor cells (Ad-HIF-1alpha-modified EPC for angiogenesis). Int. J. Biochem. Cell Biol. **40**, 2284–2295 (2008b)

Kane, B.J., Zinner, M.J., Yarmush, M.L., Toner, M.: Liver-specific functional studies in a microfluidic array of primary mammalian hepatocytes. Anal. Chem. **78**, 4291–4298 (2006)

Kimura, H., Braun, R.D., Ong, E.T., Hsu, R., Secomb, T.W., Papahadjopoulos, D., Hong, K., Dewhirst, M.W.: Fluctuations in red cell flux in tumor microvessels can lead to transient hypoxia and reoxygenation in tumor parenchyma. Cancer Res. **56**, 5522–5528 (1996)

Lam, R.H., Kim, M.C., Thorsen, T.: Culturing aerobic and anaerobic bacteria and mammalian cells with a microfluidic differential oxygenator. Anal. Chem. **81**, 5918–5924 (2009)

Leclerc, E., Sakai, Y., Fujii, T.: Microfluidic PDMSMS (polydimethylsiloxane) bioreactor for large-scale culture of hepatocytes. Biotechnol. Prog. **20**, 750–755 (2004)

Li, T., Sun, L., Miller, N., Nicklee, T., Woo, J., Hulse-Smith, L., Tsao, M.S., Khokha, R., Martin, L., Boyd, N.: The association of measured breast tissue characteristics with mammographic density and other risk factors for breast cancer. Cancer Epidemiol. Biomarkers Prev. **14**, 343–349 (2005)

Li, Z., Gong, Y., Sun, S., Du, Y., Lü, D., Liu, X., Long, M.: Differential regulation of stiffness, topography, and dimension of substrates in rat mesenchymal stem cells. Biomaterials **34**, 7616–7625 (2013)

Lo, J.F., Sinkala, E., Eddington, D.T.: Oxygen gradients for open well cellular cultures via microfluidic substrates. Lab Chip **10**, 2394–2401 (2010)

Lu, P., Weaver, V.M., Werb, Z.: The extracellular matrix: a dynamic niche in cancer progression. J. Cell Biol. **196**, 395–406 (2012)

Martewicz, S., Michielin, F., Serena, E., Zambon, A., Mongillo, M., Elvassore, N.: Reversible alteration of calcium dynamics in cardiomyocytes during acute hypoxia transient in a microfluidic platform. Integr. Biol. (Camb.) **4**, 153–164 (2011)

Martinive, P., Defresne, F., Bouzin, C., Saliez, J., Lair, F., Gregoire, V., Michiels, C., Dessy, C., Feron, O.: Preconditioning of the tumor vasculature and tumor cells by intermittent hypoxia: implications for anticancer therapies. Cancer Res. **66**, 11736–11744 (2006)

Martino, M.M., Tortelli, F., Mochizuki, M., Traub, S., Ben-David, D., Kuhn, G.A., Muller, R., Livne, E., Eming, S.A., Hubbell, J.A.: Engineering the growth factor microenvironment with fibronectin domains to promote wound and bone tissue healing. Sci. Transl. Med. **3**, 100ra189 (2011)

McCormack, V.A., dos Santos Silva, I.: Breast density and parenchymal patterns as markers of breast cancer risk: a meta-analysis. Cancer Epidemiol. Biomarkers Prev. **15**, 1159–1169 (2006)

Mehta, G., Mehta, K., Sud, D., Song, J.W., Bersano-Begey, T., Futai, N., Heo, Y.S., Mycek, M.A., Linderman, J.J., Takayama, S.: Quantitative measurement and control of oxygen levels in microfluidic poly(dimethylsiloxane) bioreactors during cell culture. Biomed. Microdevices **9**, 123–134 (2007)

Miyaji, K., Furuse, A., Nakajima, J., Kohno, T., Ohtsuka, T., Yagyu, K., Oka, T., Omata, S.: The stiffness of lymph nodes containing lung carcinoma metastases: a new diagnostic parameter measured by a tactile sensor. Cancer **80**, 1920–1925 (1997)

Mosiewicz, K.A., Kolb, L., van der Vlies, A.J., Martino, M.M., Lienemann, P.S., Hubbell, J.A., Ehrbar, M., Lutolf, M.P.: *In situ* cell manipulation through enzymatic hydrogel photopatterning. Nat. Mater. **12**, 1072–1078 (2013)

Niebruegge, S., Bauwens, C.L., Peerani, R., Thavandiran, N., Masse, S., Sevaptisidis, E., Nanthakumar, K., Woodhouse, K., Husain, M., Kumacheva, E., Zandstra, P.W.: Generation of human embryonic stem cell-derived mesoderm and cardiac cells using size-specified aggregates in an oxygen-controlled bioreactor. Biotechnol. Bioeng. **102**, 493–507 (2008)

Peng, C.C., Liao, W.H., Chen, Y.H., Wu, C.Y., Tung, Y.C.: A microfluidic cell culture array with various oxygen tensions. Lab Chip **13**, 3239–3245 (2013)

Polinkovsky, M., Gutierrez, E., Levchenko, A., Groisman, A.: Fine temporal control of the medium gas content and acidity and on-chip generation of series of oxygen concentrations for cell cultures. Lab Chip **9**, 1073–1084 (2009)

Purpura, K.A., George, S.H., Dang, S.M., Choi, K., Nagy, A., Zandstra, P.W.: Soluble Flt-1 regulates Flk-1 activation to control hematopoietic and endothelial development in an oxygen-responsive manner. Stem Cells **26**, 2832–2842 (2008)

Rofstad, E.K., Gaustad, J.V., Egeland, T.A., Mathiesen, B., Galappathi, K.: Tumors exposed to acute cyclic hypoxic stress show enhanced angiogenesis, perfusion and metastatic dissemination. Int. J. Cancer **127**, 1535–1546 (2010)

Samani, A., Zubovits, J., Plewes, D.: Elastic moduli of normal and pathological human breast tissues: an inversion-technique-based investigation of 169 samples. Phys. Med. Biol. **52**, 1565–1576 (2007)

Serena, E., Zatti, S., Reghelin, E., Pasut, A., Cimetta, E., Elvassore, N.: Soft substrates drive optimal differentiation of human healthy and dystrophic myotubes. Integr. Biol. (Camb.) **2**, 193–201 (2010)

Shahal, T., Geiger, B., Dunlop, I.E., Spatz, J.P.: Regulation of integrin adhesions by varying the density of substrate-bound epidermal growth factor. Biointerphases **7**, 23 (2012)

Spatz, J.P., Geiger, B.: Molecular engineering of cellular environments: cell adhesion to nano-digital surfaces. Methods Cell Biol. **83**, 89–111 (2007)

Trappmann, B., Gautrot, J.E., Connelly, J.T., Strange, D.G., Li, Y., Oyen, M.L., Cohen Stuart, M.A., Boehm, H., Li, B., Vogel, V., Spatz, J.P., Watt, F.M., Huck, W.T.S.: Extracellular-matrix tethering regulates stem-cell fate. Nat. Mater. **11**, 642–649 (2012)

Chapter 4
Cancer Cell Mechanics

Cynthia A. Reinhart-King

Introduction

Cell mechanics research is rapidly expanding within the cancer community. Its increasing recognition within cancer biology likely stems from the fact that metastasis is an inherently physical process, involving the pushing/pulling of cells away from the primary tumor and through the surrounding stromal environment (Friedl and Alexander 2011; Friedl and Wolf 2003; Wirtz et al. 2011). While cell mechanics has a long history, it has had a significant recent resurgence, largely enabled by new and better technologies for interrogating cells and molecules.

"Cell mechanics" is in some ways a misnomer for a field that encompasses much more than simply cell-scale behaviors and properties. It includes not only the mechanics of individual cells, but also the mechanical forces and mechanical properties at the molecular and tissue scales (Fig. 4.1).

In this chapter, the focus is primarily on mechanics research being performed in Europe, Asia and the United States at the cellular scale; however, it is important to note that the multi-scale contributions of subcellular structures and supracellular tissue properties cannot be overlooked. At the subcellular scale, the cytoskeleton organizes to exert intracellular forces that translate into cell behaviors such as mitosis, intracellular transport, lamellipodial extension, and cell migration. Changes in various molecules within cells are closely tied to mechanical changes in the cell. For example, the mechanical properties and kinetics of cytoskeletal assembly at the molecular level contribute to changes in the mechanical properties of the cell (Kraning-Rush et al. 2011). Similarly, at the supracellular scale, the mechanical properties and architecture of a tissue contribute to cell function and dysfunction. In

C.A. Reinhart-King (✉)
Department of Biomedical Engineering, Cornell University,
302 Weill Hall, Ithaca, NY 14853, USA
e-mail: CAK57@cornell.edu

P. Janmey et al. (eds.), *Physical Sciences and Engineering Advances in Life Sciences and Oncology*, Science Policy Reports,
DOI 10.1007/978-3-319-17930-8_4

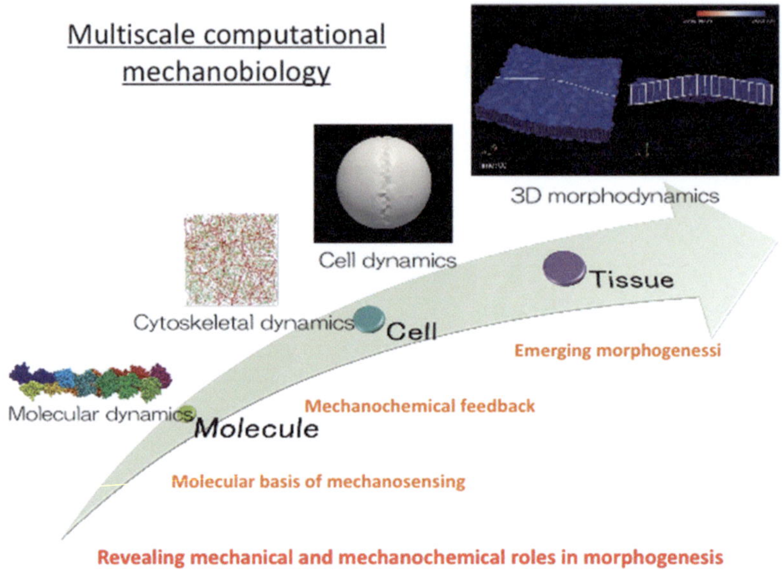

Fig. 4.1 Multiscale computational mechanobiology of epithelial tissue morphogenesis (Courtesy of Y. Inoue, Kyoto University)

the cancer field, this is particularly relevant as it is well-established that most solid tumors are stiffer than normal tissue. This stiffening at the tissue level is not only the basis for many diagnostic methods, but can also contribute to malignancy at the cellular level (Paszek et al. 2005). As such, "cell mechanics" is not limited to the cellular-scale because it spans the molecular, cellular, and tissue level scales. To fully understand and manipulate mechanics at the cellular level, one must consider the mechanics at both the molecular and tissue scales. Biologically, it is this integration that leads to changes in cell function and dysfunction.

Cell mechanics can be divided into three separate subfields: (1) cellular mechanical properties; (2) mechanotransduction (cellular response to forces imposed on cells by the external environment); and (3) cell-generated forces (Fig. 4.2). Studies of cellular mechanical properties have largely focused on either the elastic modulus or deformability of the whole cell or the rheology of the cytoplasm. These measurements are important within the cancer field because they implicate if and how cells will migrate and squeeze through the matrix during metastatic invasion through the stromal matrix and into the vasculature. Mechanotransduction research in cancer has been centered on the response of cells to imposed pressures and fluid shear stresses within the tumor microenvironment which are known to influence tumor growth and metastasis. Lastly, studies of cell-generated forces are critical to our understanding of how cells adhere, traverse, and sense their microenvironment. All of these facets of cell mechanics (mechanical properties, mechanotransduction, and cell-generated forces) influence each other, demonstrating the integration of inside and outside signals. This chapter will address all three of these subfields in oncology

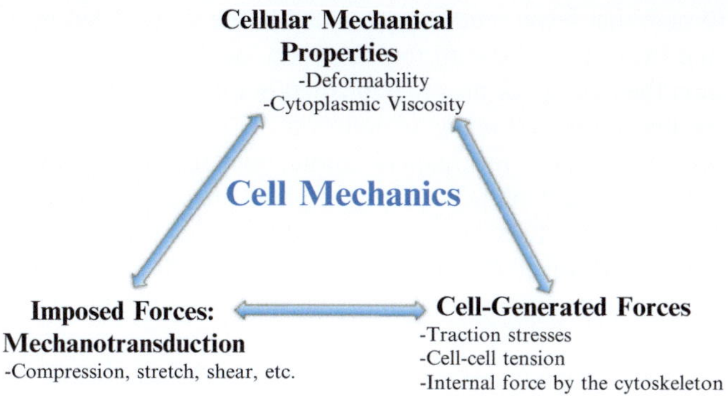

Fig. 4.2 The field of cell mechanics can be divided into three main subsections: cellular mechanical properties, imposed forces, and cell-generated forces (Courtesy of C. Reinhart-King)

and the efforts underway in the United States and those observed during the APHELION study tours in Europe and Asia.

Research

A Brief History of Cell Mechanics

The field of cell mechanics is often thought of as an emerging field, likely because most of the literature on this topic has been published since the 1990s. However, cell mechanics has a long history that dates back almost as far as the invention of the first microscopes. Early reports of cilia by Anton van Leeuwenhoek in the early 1700s noted the movement of particles in the liquid surrounding cells, a phenomenon that became the basis for microrheology studies and the characterization of cytoplasmic viscosity (Pelling and Horton 2008). At the same time, Isaac Newton was conducting separate experiments to define viscosity and descriptors of fluid properties. Even now we consider magnetic tweezers a state-of-the-art technique, when, in fact, the first microscopes equipped with magnetic micromanipulators were reported in the 1920s (Seifriz 1924). Early use of these magnetic systems included measurement of the cytoplasmic viscosity of cells, an active area of oncology research today (Baker et al. 2009; Wu et al. 2012). In 1950, Francis Crick—one of the fathers of molecular biology—published his first two papers on the mechanical properties of chick fibroblasts using a magnetic system with Arthur Hughes (Logothetis 2004). As these papers were published, Crick moved to the Cavendish labs where he switched scientific interests and began his work on the structure of proteins.

Crick was certainly not the only scientist working on questions related to cell mechanics in the pre-genome era. Interest in protoplasm dynamics and viscosity and questions in cell motility appear frequently in the pre-1950 literature. It is interesting to note that fewer studies performed in the area of cell mechanics were reported in the literature following the discovery of the structure of DNA (determined based on the number of papers published in this time). It is likely that it was this discovery that prompted many scientists to indirectly follow in Crick's footsteps and move away from mechanical studies and into molecular biology. The discovery of the structure of DNA brought about the era of molecular biology, which in many ways may have suppressed what had been the growing field of cancer cell mechanics. However, the importance of cancer cell mechanics has resurged in a very significant way, and it is making great strides in the United States, Europe, and Asia.

Cancer Progression and Metastasis: An Inherently Physical Process

Tumor growth and spread, in addition to being stimulated by genetic, epigenetic, and microenvironmental changes, is a very physical process (Fig. 4.3). From a biophysical perspective, metastasis occurs as cells dissociate from the primary tumor,

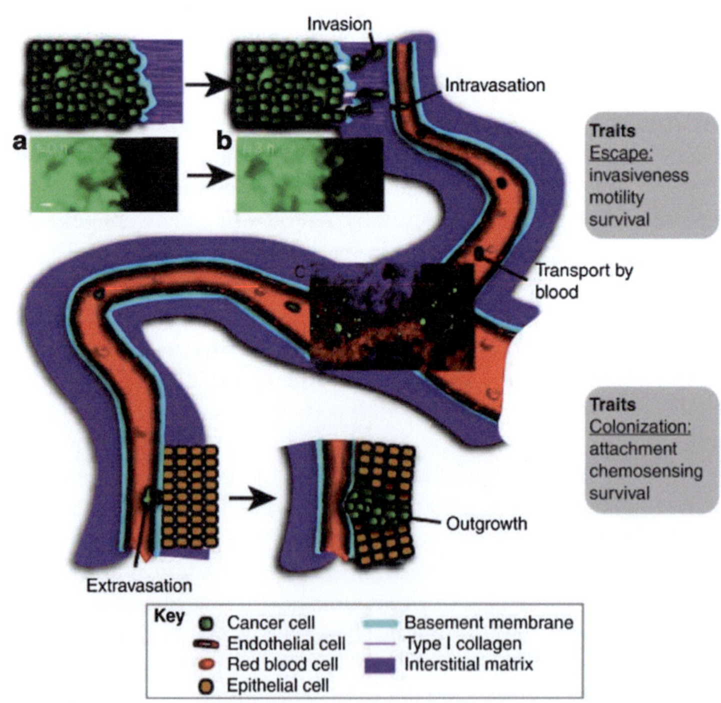

Fig. 4.3 The metastatic cascade (**a**, **b**) Fluorescence micrograph of invasion of a polyoma middle T (PyMT) mammary tumor. (**c**) Fluorescence images of tumor cells within a blood vessel (From Beerling et al. 2011)

breaking cell-cell adhesion bonds. As cells migrate through the matrix-dense stroma, they must push, pull, and degrade matrix fibers to navigate through the fibrous protein mesh. Simultaneously, the cells squeeze and deform to move through the pores within the matrix. As cells intravasate from the matrix into the vasculature, they must squeeze through the vessel wall. Once they are in the circulation, the cells must survive the forces imposed by blood flow. Finally, to colonize a secondary site, metastatic cells must adhere to the lumen of the vessel wall and squeeze through the vasculature during transmigration to seed within a secondary tissue. At each of these major steps in the metastatic cascade, cells both exert force and are exposed to externally imposed forces. The physical nature of these steps naturally leads to numerous questions regarding the mechanical forces involved in cancer metastasis.

The Role of Cell Deformability in Cancer Progression

As cells metastasize, they deform to squeeze through the fibers of the matrix-dense stroma. Their ability to deform is related to their mechanical properties (viscoelasticity), and as such there has been increasing interest in how the mechanical properties of metastatic cells differ from non-metastatic or normal cells. One prevailing hypothesis is that metastatic cells are more deformable, which aids in their invasion and motility.

There are multiple methods to measure the mechanical properties of cells: atomic force microscopy, micropipette aspiration (Bao and Suresh 2003), glass cantilevers (Mitrossilis et al. 2010), particle-tracking microrheology (Wirtz 2009), and more recently optical stretching (Fig. 4.4). Measurements of cell deformability have improved significantly over the past several years; however, they are not necessarily new to biology. Early studies of protoplasm viscosity were reported in the 1920s by Heilbrunn (Heilbrunn 1921). He also studied the effects of chemotherapeutics on cellular viscosity in 1957 (Wilson 1957) and showed that ethyl urethane increases the viscosity of cells. This finding is relevant and interesting today in light of newer data regarding the relationship between deformability and metastatic potential. Using an optical stretching device (see Chap. 7 for a description), Guck and colleagues from the University of Leipzig, Germany, showed that deformability increases with metastatic potential in breast cancer cell lines (Fig. 4.5; site report, Appendix B). This effect was later confirmed with cells from human primary tumors. Using atomic force microscopy, Cross and colleagues showed that tumor cells from patients are more compliant than their normal counterparts (Cross et al. 2007). Dr. Sylvie Hénon at the University of Paris, Diderot, France (site report, Appendix B), investigated how cellular mechanical properties change in response to imposed mechanical forces, such as tension, showing that cells can actively stiffen due to imposed forces by recruiting and polymerizing actin (Icard-Arcizet et al. 2008). Together, these data indicate that the mechanical properties of the cell may

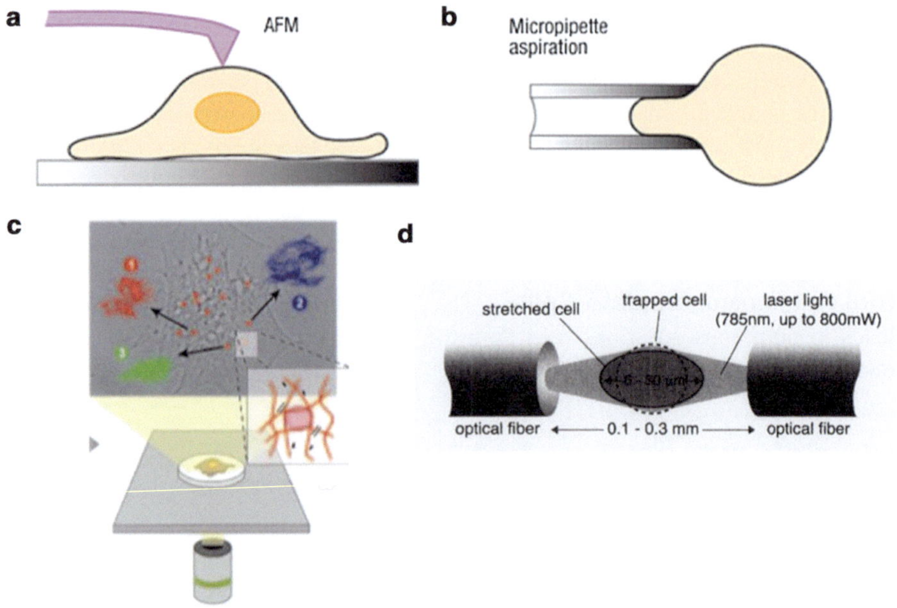

Fig. 4.4 Methods to measure the mechanical properties of cells. (**a**) Atomic force microscopy (Bao and Suresh 2003); (**b**) micropipette aspiration (Bao and Suresh 2003); (**c**) particle-tracking microrheology (Wirtz 2009); and (**d**) optical stretching (Guck et al. 2001)

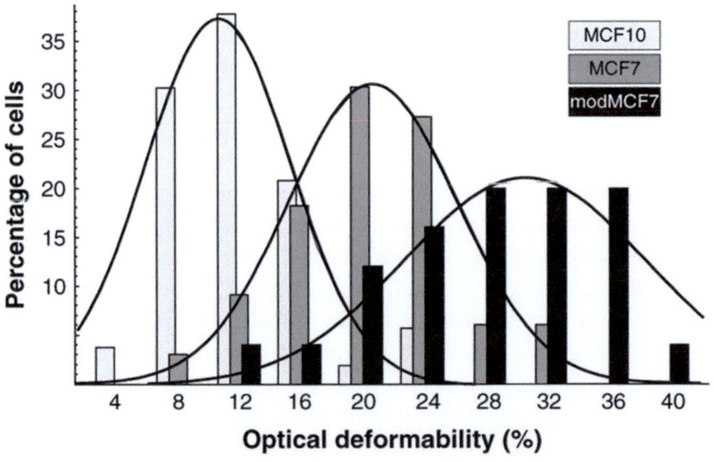

Fig. 4.5 Deformability of cells increases with metastatic potential, as measured using an optical stretcher (From Guck et al. 2005)

be predictors of metastatic potential and have the ability to actively remodel due to externally imposed forces.

The implications of cell stiffening may be both positive and negative with respect to cancer treatments. The data seem to suggest that stiffer cells are less metastatic and that chemotherapeutics stiffen cells, which is a positive effect. However, this stiffening may also have deleterious consequences. Recently, the Fletcher lab (author, Chap. 7) has shown that chemotherapeutic treatments can stiffen leukemic cells that then plug microfluidic channels (Lam et al. 2007). These data suggest that chemotherapy could result in vascular occlusion in capillaries by cancer cells. Therefore, it cannot yet be said that stiffening cells is necessarily a viable, universal method to prevent metastasis due to potentially significant side effects.

For this work to have impact, it must be translated into the clinic. Work being done by Josef Käs' lab at the University of Leipzig, Germany (site report, Appendix B), is continuing research in the area of cell deformability by testing clinical samples and working to move the optical stretching methodology to the clinic. For it to be used in the clinic, the method must be both user-friendly and high throughput. While the method of optical stretching is technically complicated, significant strides are being made to make it tractable to clinical laboratory technicians as a potential mechanism to diagnose the likelihood of metastasis.

In addition to studies of whole cell deformability, data are also emerging on the role of nuclear deformability in invasion. Since the nucleus is the largest organelle within the cell, nuclear mechanics could be a limiting factor in allowing cells to permeate through the pores of a matrix (Friedl et al. 2011). If the nucleus is unable to squeeze through a pore, then the cell cannot invade unless the matrix is degraded. The Lammerding lab at Cornell University has been probing the role of nuclear deformability in 3D cell migration directly using microfabricated platforms (Fig. 4.6). Using a series of microfabricated, constricted channels through which cells can migrate, they have shown that manipulation of the nuclear envelope and lamin A expression alters whether cells can easily pass through the constrictions (Rowat et al. 2013). Given that the nucleus is a limiting factor in the translocation of the cells through tight spaces, a natural extension of this work is to then ask whether the cytoskeleton pushes or pulls the nucleus through constrictions (Isermann and Lammerding 2013).

Recent data from the Wirtz lab at Johns Hopkins University suggests that the nucleus is not only important in migration due to its deformability, but also because its connection to the cytoskeleton. The nucleus connects to the cytoskeleton through the LINC complex and their recent data suggest that this connection plays a critical role in pseudopodial extension during 3D migration (Khatau et al. 2012). Notably, this effect is only observed in 3D migration and not in cells on planar substrates. While there is evidence mounting regarding the importance of the nucleus and nuclear-cytoskeletal coupling in invasion and migration, research on the role of nuclear mechanics in cancer progression is still in its infancy and requires further investigation.

The role of nuclear deformation in cell physiology extends beyond its role in being a physical impairment in the ability of cells to invade and migrate through small pores. Recent work from G.V. Shivashankar at the Mechanobiology Institute in Singapore has shown that nuclear deformations due to actomyosin contractility

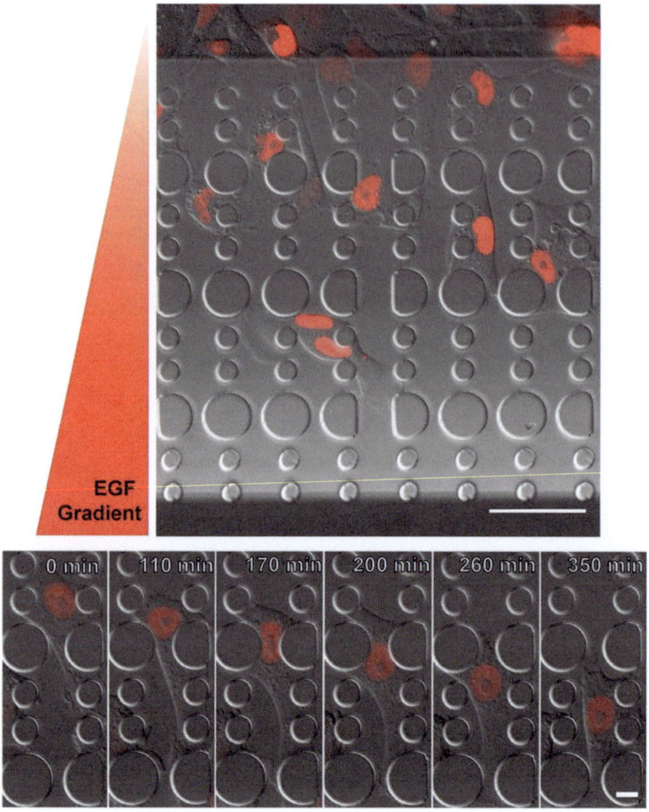

Fig. 4.6 Breast cancer cells (MDA-MB-231) migrating along chemotactic gradient (EGF) through microfluidic constriction channels. *Top panel*: overview of the device with constrictions of decreasing width. Scale bar: 50 um. *Bottom*: time-lapse sequence of cells with fluorescently labeled nucleus (mCherry-Histone-4) passing through 5 um wide constriction, displaying substantial nuclear deformations in the process. Scale bar: 10 um (Figure courtesy of Celine Denais, Lammerding Lab)

and geometric constraints can alter gene expression (Jain et al. 2013; Gupta et al. 2012; site report, Appendix C). Changing cell shape and nuclear size correlates with changes in global acetylation levels and transcriptional profiles (Fig. 4.7). These data suggest that the shape changes cells undergo during metastatic invasion may alter transcription profiles.

Cell-Generated Forces in Adhesion and Migration

Cells generate traction stresses against their matrix to adhere and migrate. These forces aid in remodeling the matrix and propelling cells forward during migration. Cell migration is necessary for metastatic invasion, and as such there has been recent interest in characterizing and understanding cell-generated traction stresses.

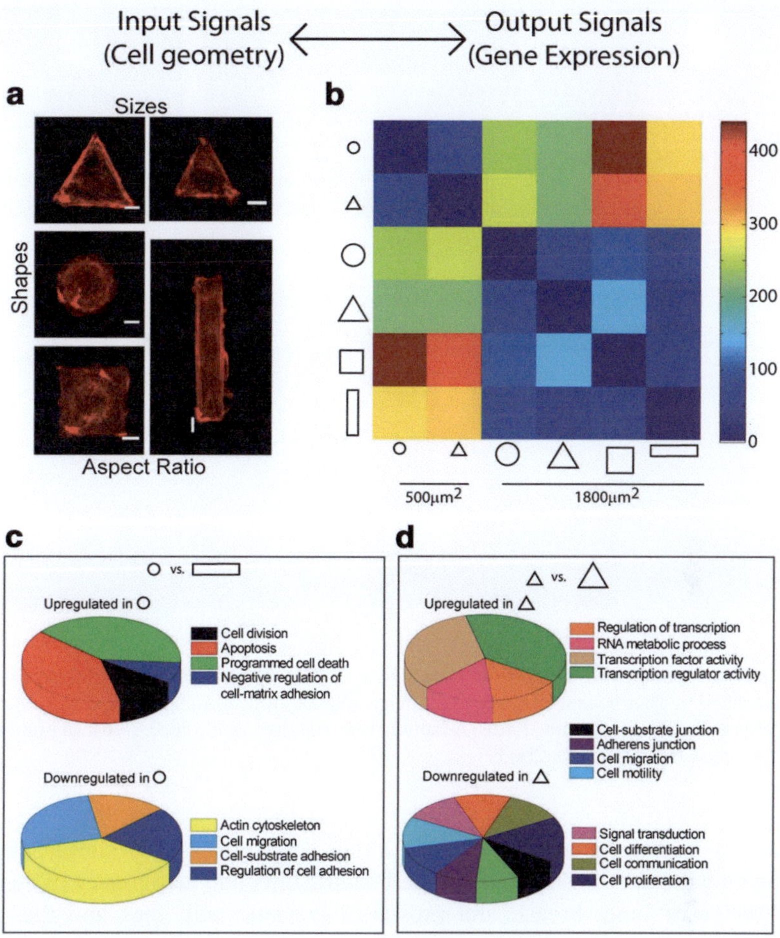

Fig. 4.7 Gene expression is regulated in part by cell geometry (From Jain et al. 2013) (**a**) Cells can be pattered in various, sizes, shapes, and aspect ratios. (**b**) Color-coded matrix showing differential gene regulation as a function of cell shape and size. (**c, d**) Gene ontology analysis from (**c**) cells in small circle compared with larger cells and (**d**) small triangle compared with large triangle

There are a number of methods that have been developed to measure cellular traction stresses, including wrinkling substrates (Harris et al. 1980), traction force microscopy (Dembo and Wang 1999), micropatterned elastomeric substrates (Balaban et al. 2001), and micropillar arrays (mPADS) (Fig. 4.8, Tan et al. 2003). Original versions of traction force microscopy allow for only the measurement of individual cells or cell pairs. More recent modification by Dr. Xavier Trepat, University of Barcelona, Spain (site report, Appendix B), allow for the measurement of forces in cell sheets (Tambe et al. 2011). These methodologies have enabled several interesting insights in the area of cell migration and adhesion.

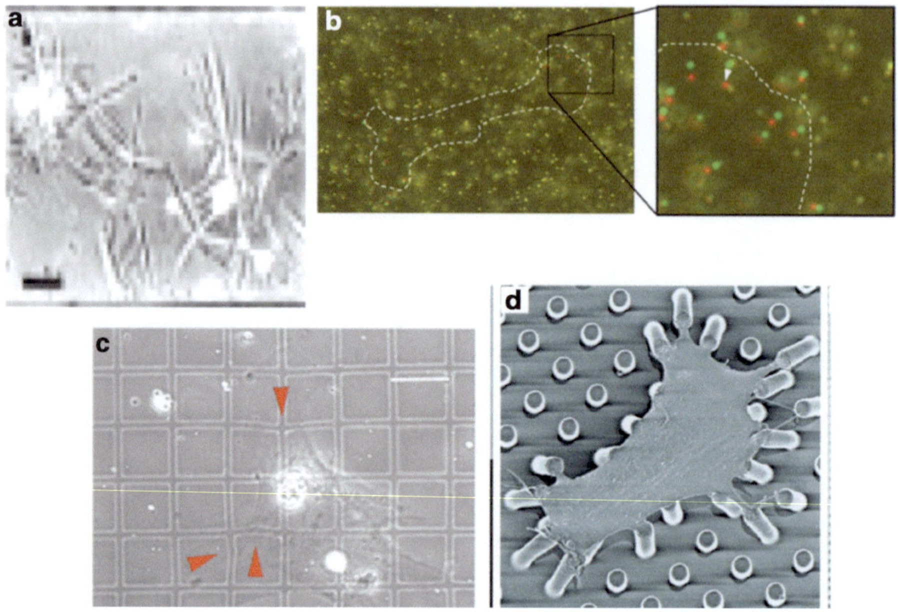

Fig. 4.8 Methods to measure cell-generated traction stresses
(**a**) Wrinkling substrates (Adapted from Harris et al. 1980); (**b**) Traction force microscopy, originally described by Dembo and Wang (1999); image adapted from Kraning-Rush et al. (2012b); (**c**) micropatterned elastomeric substrates (Adapted from Balaban et al. 2001); and (**d**) micropillar arrays (Adapted from Tan et al. 2003)

Traction stresses play a key role in migration, and there has been significant work investigating the relationship between cell contractility and malignancy. When traction stresses of lung, breast, and prostate metastatic cell lines and their non-metastatic counterparts were measured, it was reported that metastatic cells exert increased forces as compared to the non-metastatic cells (Fig. 4.9). These forces increase with matrix stiffness (Kraning-Rush et al. 2012a). Similar to the results discussed earlier on cell deformability, these data suggest that traction stresses may also be a mechanical biomarker of metastasis. A study published recently by Ben Fabry's lab at the University of Erlangen-Nuremberg, Germany (site report, Appendix B), indicates that in addition to the force magnitude, the anisotropy and polarization of the force may also be an important indicator of metastatic potential using a similar panel of metastatic and non-metastatic cell lines (Koch et al. 2012).

Cell polarity during cell migration dictates the direction and persistence of movement. Samuel Safran's group at the Weizmann Institute of Science, Israel (site report, Appendix B), has explored the effects of the microenvironment on polarity, showing that it increases in single cells on stiff substrates (De et al. 2008, 2010; Safran and De 2009). Asymmetries that develop during cell spreading can affect actin-myosin polarity and the extent of alignment of forces in response to matrix rigidity (Zemel et al. 2010). This polarity is often accompanied by polarized

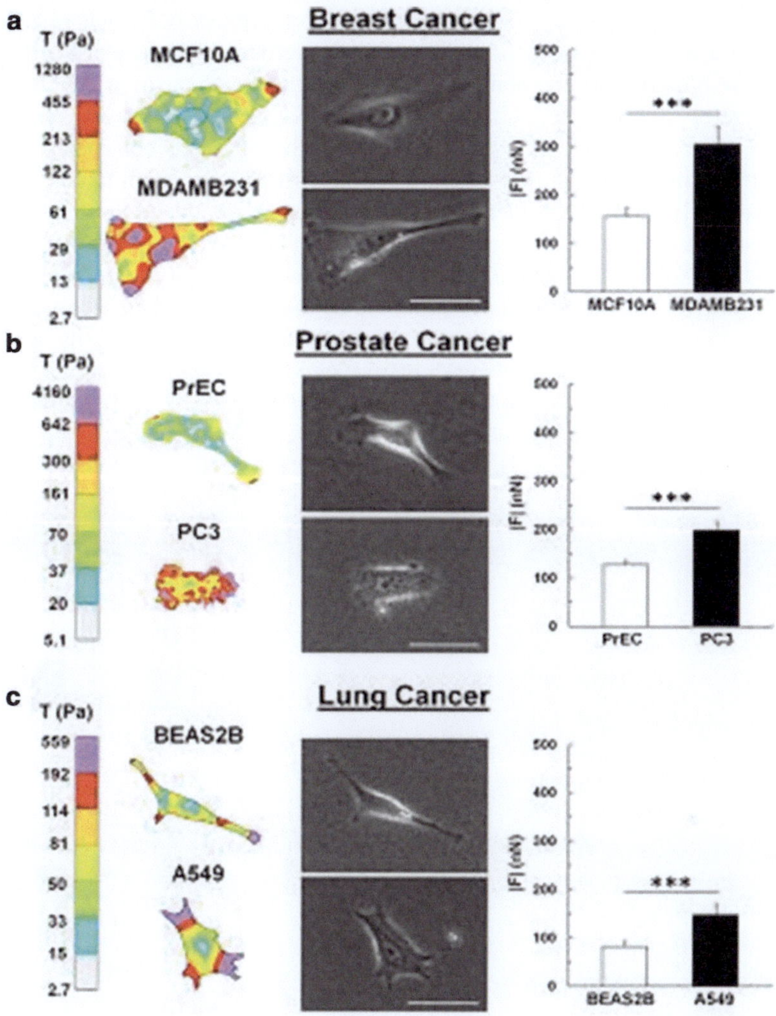

Fig. 4.9 Traction stresses increase with metastatic potential (Kraning-Rush et al. 2012a) Traction maps and corresponding phase images of metastatic and non-metastatic cells from (**a**) breast, (**b**) prostate, and (**c**) lung

remodeling of the matrix and may be important to our understanding of the role of elongation and polarity in mesenchymal modes of metastatic cell migration.

Studies of early polarization events in single cells have also been extended into studies of collective cell movements. For example, du Roure and colleagues have shown that cells along the edge of an adherent population exert the highest forces during collective migration (du Roure et al. 2005). These forces are significantly higher than forces at the edge of single cells. In light of the leader-follower migration dynamics described by Peter Friedl and colleagues at the Radboud University

Nijmegen Medical Centre, Netherlands (site report, Appendix B), where cell invasion from a tumor can occur in a sheet-like movement with "leader" cells at the front, these data may also explain a more cooperative mechanism of cell migration, where force is concentrated within leader cells (Khalil and Friedl 2010; Lee et al. 2012). In fact, Carey recently used a spheroid co-culture of non-metastatic, low force-producing epithelial cells mixed with highly metastatic, high force-producing cells to directly probe the leader-follower behavior. When the co-culture spheroids are embedded into collagen, the highly metastatic cells move to the outside of the spheroid, reorganize the surrounding collagen into tracks that run perpendicular to the spheroid, and invade into the surrounding collagen followed closely by the typically non-invasive, non-metastatic cells. The high-force producing cells emerge as early leaders in the leader-follower dynamic, remodeling the matrix to enable collective movements (Fig. 4.10). Notably, inhibition of force or MMP activity using Y27632 or GM6001 respectively ablates these effects. With pharmacological inhi-

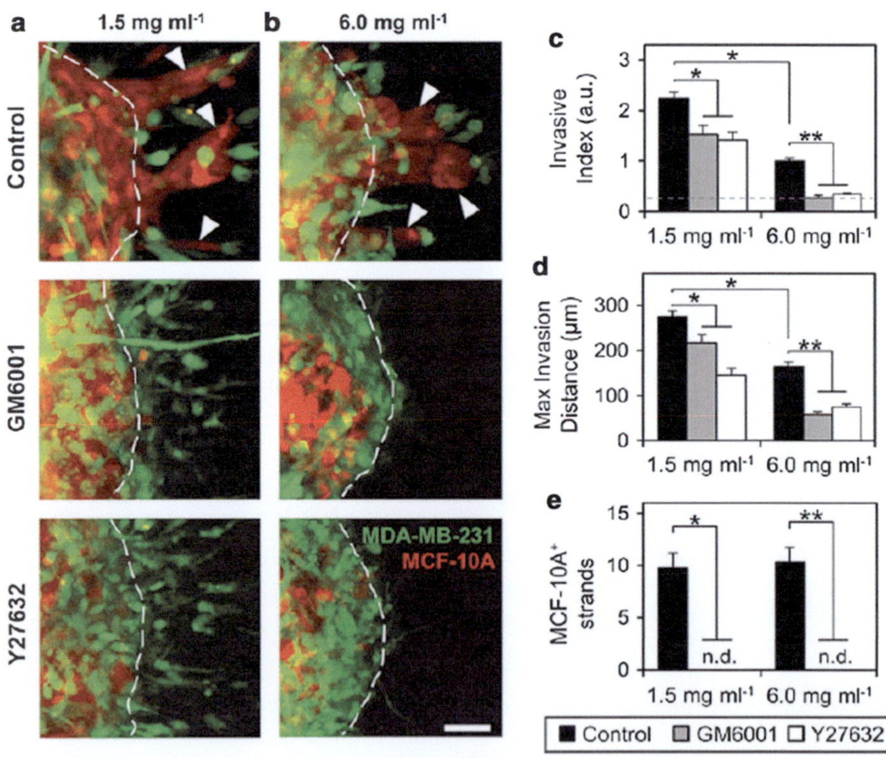

Fig. 4.10 Co-culture spheroid model of invasion (From Carey et al. 2013)
Highly invasive epithelial cells (MDA-MB-231, *green*) were mixed with non-invasive epithelial cells (MCF-10A, *red*) and embedded in (**a**) 1.5 or (**b**) 6.0 mg/ml collagen and treated with GM6001 or Y27632. *Dashed lines* indicate original spheroid boundary and arrowheads indicate invasive strands containing non-metastatic cells. Scale bar equals 50 μm. Invasion was quantified in terms of (**c**) invasive index, (**d**) maximum invasion distance, and (**e**) number of invasive strands containing non-metastatic MCF10A cells

bition, invasive cells move into the surrounding matrix but there is no matrix remodeling, and non-invasive cells do not move into the surrounding matrix. Together, these studies implicate force and polarization of force as key events to collective cell movements both in planar cultures and 3D spheroid cultures.

In addition to a clear focus on the role of forces in migration, there has also been significant interest in the basic biology of force generation and the genesis and transmission of these forces at focal adhesions. Work to quantify the force-transducing ability of individual focal contacts and focal adhesions has been largely pioneered by Drs. Benjamin Geiger and Alexander Bershadsky at the Weizmann Institute, Israel (site report, Appendix B). Their early work in this area made the first attempts at correlating focal adhesion and size with force magnitude (Balaban et al. 2001). A number of studies have followed, including work by Dr. Ulrich Schwarz at the University of Heidelberg, Germany (site report, Appendix B), investigating adhesion size and geometry with respect to force generation, and it continues to be an area of intense interest (Schwarz et al. 2002, 2006; Stricker et al. 2010). Additional work by Dr. Michael Sheetz has provided unique insights into the role of individual focal adhesion proteins in adhesion and contraction of adherent cells (site report, Appendix C). Using nanopatterned substrates, they investigated the time course of integrin clustering, intracellular focal adhesion protein clustering, and cells contraction (Yu et al. 2011; Roca-Cusachs et al. 2013). This approach, focused on identifying the critical molecular players for adhesion and force generation, may lead to the identification of therapeutic targets to prevent cell migration during metastasis.

In addition to examining the forces generated by cells against their substrate, several groups have researched forces at cell-cell contacts and their roles in collective cell movements and the initiation of signaling at junctions. Significant work in this area has been done by Chen and colleagues at the University of Pennsylvania using mPADs (Liu et al. 2010), and Trepat using a modified version of traction force microscopy (Trepat and Fredberg 2011). Trepat has shown that cell-cell contacts contribute to epithelial collective migration and that cells in a cluster can each contribute to collective movements (site report, Appendix B). More recently, an international collaboration between Ladoux (France), Trepat (Spain), and Lim (Singapore) have shown that cell-cell contacts can dominate in areas of low cell-matrix adhesion and can function as bridges to maintain cell sheet integrity (Vedula et al. 2014).

These experimental approaches to investigate cellular force generation and its effects on cell-cell dynamics have been complemented by computational approaches to understand how individual mechanical interactions at the cellular level translate into 3D morphogenesis events. Mechanobiologists at Kyoto University, Japan, have developed models that span from cell to tissue to understand morphogenesis (Okuda et al. 2013; site report, Appendix C). Treating cells as interacting polyhedrons with formal boundaries and taking into consideration the mechanical interactions of the cell with their surroundings and each other, Okuda and colleagues have recapitulated the large-scale tissue deformations that occur during morphogenesis *in silico* (Fig. 4.11). In an analogous *in silico* approach, Dr. Yoshikiro Morishita's group at the RIKEN Center for Developmental Biology, Japan, has taken information regarding large-scale tissue-level deformations and used Bayesian statistical models to predict the individual cell movements and deformations and morphogen gradients

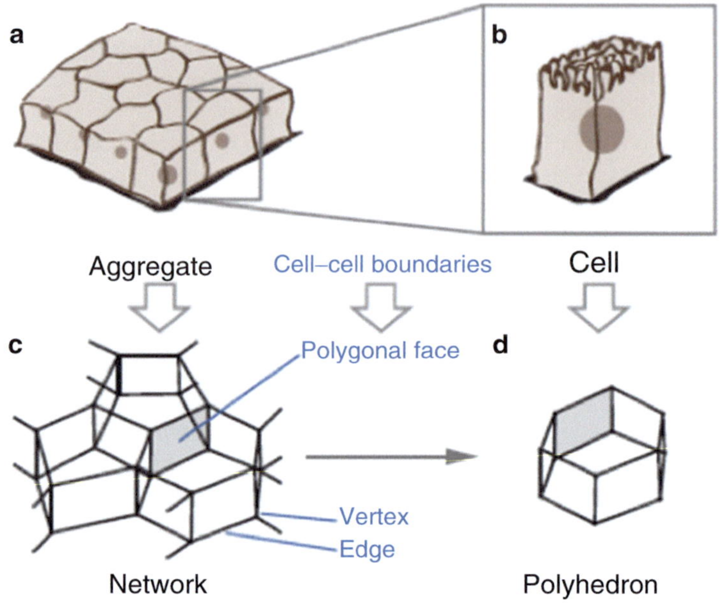

Fig. 4.11 Model shapes of a multicellular aggregate (From Okuda et al. 2013)
(**a**) Aggregate composed of cells. (**b**) Single cell. (**c**) Network that represents a cell aggregate composed of vertices and edges (*solid lines*). Polygonal faces (*gray area*) compartmentalize the network. (**d**) A polyhedron represents a single cell

that cause these tissue-level changes (Hironaka and Morishita 2012; Morishita and Iwasa 2011). These computational models have a specific benefit in their ability to capture the multi-scale (molecular, cellular, tissue) contributions of mechanobiology to tissue structure. Adapting these models to tumor formation and spread may prove to be a powerful approach to integrating the multi-scale contributions of the microenvironment to tumor biology.

More recently, there has been increasing attention on experimentally measuring cellular forces of cells within fully 3D matrices. Chen and colleagues at the University of Pennsylvania have published the first work in this area, developing both PEG-based hydrogel materials and a computational approach for the measurement of forces exerted by cells in 3D matrices (Fig. 4.12). Numerous groups are developing alternate methods for 3D traction measurements. Keng-hui Lin's lab at Academia Sinica, Taiwan, for instance, described 3D foam-like materials containing spherical cavities to study 3D cell contractility (Lee et al. 2013; site report, Appendix C). The cavities are on the cell-size scale, and as such, cells can adhere to and pull on the walls of the cavity (Fig. 4.13). The use of well-defined, flexible materials lays the foundation for the calculation of 3D forces based on the deformations of the cavity walls due to cell contraction.

Measurements of cell-generated forces have resulted in both the identification of cellular force as a mechanical biomarker of metastasis and key insights into how cells move. Further investigation into the mechanisms of force generation in metastasis may lead to therapeutics that target force generation during metastasis.

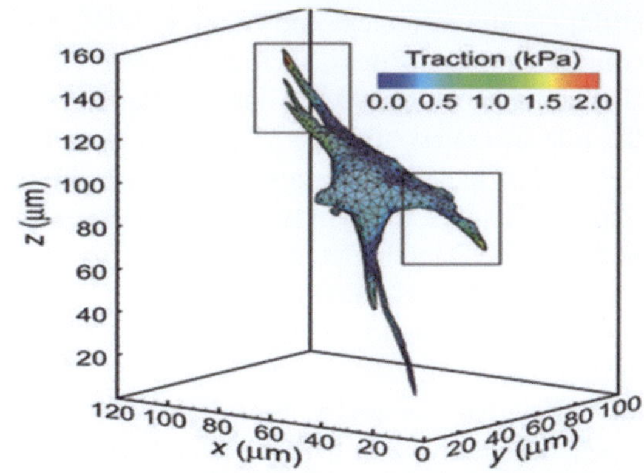

Fig. 4.12 3D traction force microscopy. Contour plot of the tractions exerted by the cell (From Legant et al. 2010)

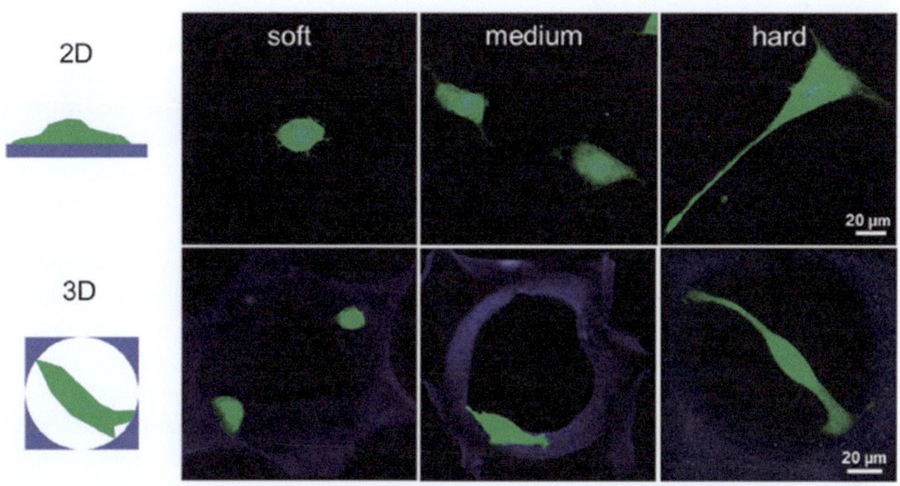

Fig. 4.13 Elongation of fibroblasts on stiff substrates (Adapted from Lee et al. 2013) Micrographs of fibroblasts on 2D flat substrates (*top*) and on 3D spherical pores of 100 mm (*bottom*). Micrographs in the same column are made of the same AC/BIS (5/0.2 %, 7/0.2 %, and 12/0.3 % for soft, medium, and hard gels, respectively). The cell body is shown in *green* based on expression of whole-body GFP. Fibronectin on the pore surface is labeled with Cy5 (*purple*) to visualize the pore

Mechanotransduction in Cancer: Tumor Response to Imposed Forces

Mechanotransduction is a broad term that describes the transduction of external mechanical cues into chemical signals within the cell. These cues can include (but are not limited to) changes in the mechanical properties of the extracellular matrix, compressive pressures, and tensional forces.

Most types of solid tumors are stiffer than normal tissue. There has been extensive work investigating the effects of matrix stiffness on cell behavior, some of which date back to 1990 when it was first noted that cell spreading increases on stiffer substrates (Keese and Giaever 1991). With the advent of tractable systems to control matrix stiffness, including the development of polyacrylamide substrates (Pelham and Wang 1998; Wang and Pelham 1998), several studies have pointed to the role of matrix stiffness in promoting malignancy (Kraning-Rush et al. 2012a; Levental et al. 2009; Paszek et al. 2005). Integrins and focal adhesion have been implicated as the mechanosensors of matrix stiffness. Additionally, the laboratories of both Dr. Daniel Fletcher (author, Chap. 7) and Dr. Atef Asnacios from the University of Paris, Diderot, France (site report, Appendix B), have used modified cantilever systems to investigate cellular force response to active changes in stiffness (Crow et al. 2012; Mitrossilis et al. 2010). Interestingly, Dr. Ming-Jer Tang's lab at Tunghai University, Taiwan, has shown that transformed cells display altered mechanosensitivity, resisting soft substrate-induced apoptosis (Wang et al. 2007; site report, Appendix C).

Tumors are also subjected to pressure and compressive stresses when confined during growth. Recent work from Dr. Lance Munn's group (author, Chap. 5) suggests that these forces enhance invasion by increasing cell-matrix adhesion (Tse et al. 2012). However, data from Dr. Jean-François Joanny's group at the Institute Curie, France (site report, Appendix B), suggests that imposed mechanical stress can inhibit tumor growth by inhibiting cell proliferation (Fig. 4.14). Clearly, there is a significant need to continue work in this field to better understand how mechanical stresses affect tumor growth.

As is the case in most fields investigating mechanotransduction in various physiological systems, there is growing interest in identifying and characterizing molecular mechanotransducers within the cell. Given the important role of cadherins in maintaining tissue structure and their location at cell-cell junctions, there is significant research ongoing to understand the role of cadherins in mechanosensing. Using a magnetic twisting cytometry system, Johan de Rooij's lab at the Hubrecht Institute,

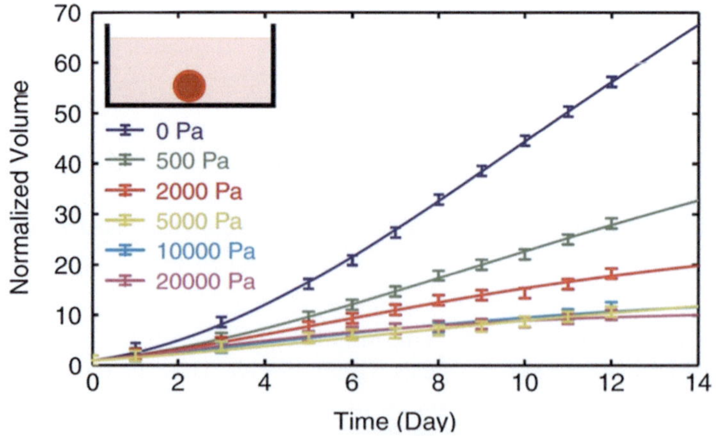

Fig. 4.14 Mechanical stress inhibits tumor growth. Spheroids were subjected to imposed pressures and their growth monitored over 2 weeks (From Montel et al. 2011)

Netherlands (site report, Appendix B), has found that E-cadherin is a mechanotransducer that causes cell stiffening when under tension (le Duc et al. 2010). This response is mediated by vinculin, a protein commonly associated with focal adhesions that is also known to localize to cell-cell junctions in structures termed focal adherens junctions. François Gallet's laboratory at the University of Paris, Diderot, France (site report, Appendix B), has also investigated cadherin mechanics, specifically the cross-talk between cadherins and integrins (Al-Kilani et al. 2011). Interestingly, their data suggest a negative feedback loop between cadherin binding and integrin binding. A similar relationship has been shown for endothelial cells, where integrin-matrix adhesion is altered by matrix stiffness and VE-cadherin engagement is disrupted (Huynh et al. 2011). There is work being done in the United States that is also investigating cadherin-binding mechanics, including a collaboration between Deborah Leckband's group at the University of Illinois at Urbana-Champaign and Johan de Rooij at the Hubrecht (site report, Appendix B; Leckband et al. 2011; le Duc et al. 2010). Together, these data suggest that changes in matrix stiffness could facilitate cancer cell invasion by enhancing integrin adhesion and disrupting cell-cell adhesion. These results are important in our understanding of the microenvironmental cues that stimulate metastasis because they indicate that increased matrix binding may actively lead to decreased cell-cell adhesion.

Mechanotransductive cues in cancer are not limited to pressure and increased stiffness. In lung cancer, for example, cells are exposed to periodic cyclic, tensile stresses due to lung expansion during breathing. Recent work from Chau-Hwang Lee's laboratory at the Academia Sinica has shown that cyclic stress reduces myofibroblast activation of cancer cell migration (Huang et al. 2013; site report, Appendix C). This work was enabled by the development of a novel device that allows for co-culture of fibroblasts and cancer cells, optical observation of cell behavior, and the application of well-defined tensile forces (discussed in Chap. 7).

Mechanoactivation of cells has also been a focus of G.V. Shivashankar's lab in the Mechanobiology Institute in Singapore. Specifically, they have an interest in understanding how physical forces alter transcriptional activity. Using magnetic particles attached to the surface of the cell membrane, they imposed well-calibrated forces and investigated the subsequent effects on actin polymerization, chromatin remodeling and nuclear transport (Fig. 4.15). Forces imposed on the membrane can result in changes in actin polymerization and chromatin reorganization. The altered F/G actin ratio can alter nuclear transport of transcription factors to the nucleus. This work was enabled by a device that could both impose forces and image fluorescence anisotropy. In fact, most mechanotransduction experiments require the integration of a device with cells, and approaches to create tractable devices will help transition mechanotransduction experiments into cell biology labs.

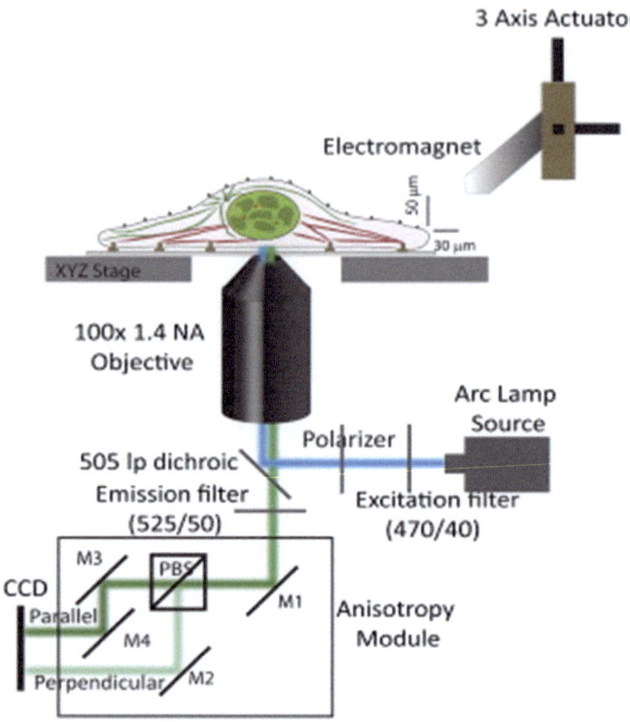

Fig. 4.15 Schematic of experimental setup application of force on single living cells (From Iyers et al. 2012)

Discussion

Efforts in the United States, Europe, and Asia have demonstrated that cell mechanics is a critical component to our understanding of cancer progression. Mechanics cannot be viewed as a complement to molecular biology, but rather as a partner. The biophysical properties of cells and the response of cells to mechanical forces is intimately involved in tumor growth and the metastatic process. Efforts to train scientists at the interface of biology and mechanics, such as those being pioneered at the Mechanobiology Institute in Singapore, will produce an entirely new breed of investigator that is able to work at the interface of the physical sciences and oncology. Close connections with clinicians, like those found at the University of Leipzig, are essential to translating cell mechanics studies to patient diagnosis and treatment. There are many questions that have not yet been answered, but we are now well-positioned to tackle them with the advent of new, tractable methodologies to probe cell mechanics. Connecting the biophysical properties of cells to the invasive behaviors of cells is a critical step in bringing cell mechanics work closer to therapeutic treatment.

Challenges to the Field of Cell Mechanics

Assessing the state-of-the-art research throughout Europe, Asia, and the United States has revealed several universal challenges in the field of cell mechanics.

Adapting Cell Mechanics Concepts to Biology Labs and the Clinic

While the importance of mechanics is increasingly appreciated, most of the current oncology studies continue to focus exclusively on molecular biomarkers, signaling pathways, and small molecule inhibitors. Significant technology development has occurred in the molecular biomarker field, producing more user-friendly, high-throughput methods for analyzing the molecular signature of cells. These methodologies are readily adaptable by scientific labs and have found their way into clinical assays. In contrast, there has been less done to make measurements of cell mechanics tractable to scientists outside of the mechanics field. As a result, the impact of cell mechanics has been more limited as fewer research labs have adopted methods or become familiar with the conceptual framework of cell mechanics.

Likewise, unlike molecular properties, mechanical properties are difficult to manipulate. There is no analogous technique to siRNA or knockdown in mechanics. Intervention in pathways related to mechanics often alters multiple signaling pathways. Therefore, it is difficult to test whether a specific cell behavior (e.g., invasion, migration, or proliferation) is specifically due to mechanical changes.

High-Throughput Screening of Single Cells

Cell mechanical testing is traditionally done by probing cells on an individual basis and has not been widely translated into high throughput methods. Most mechanics assays—both the actual experimental testing and the analysis required to convert the measurements into meaningful values—are typically very time consuming because each cell is tested and analyzed individually. Analysis of populations becomes difficult simply due to the time required to collect the amount of data necessary to analyze statistical differences. In contrast, many molecular techniques such as Western blotting and PCR are designed to test cell populations and are relatively high-throughput compared to mechanical testing techniques. For mechanics to be widely adopted, higher throughput methods will be beneficial.

The Need for Interdisciplinary Training

Cell mechanics and mechanobiology differ from fields like molecular biology in pedagogy as well. Mechanobiology, while an old field, is not offered as a unique degree program unlike many other biology sub-disciplines. While there are a plethora of biology textbooks available, there are few that could be considered the "authority" on mechanobiology. The Mechanobiology Institute in Singapore has made significant strides in this area in the creation of MBInfo (mechanobio.info), a multi-media resource, containing chapters describing mechanobiology across scales. The information is written and reviewed by expert scientists in the field. Its wiki-based format allows for information to be continually updated, a necessary element in a constantly changing field. Training students and scientists in both biology and mechanics is essential to making significant strides in tying oncology to the physical sciences.

Perceptions of the Integration of Mechanical Measurements with Molecular Biology

A potential limitation to the efforts being made in the field of cell mechanics is related to the perceptions of the larger biological community about cell mechanical measurements relative to what is already known. Molecular biology approaches and molecular biomarkers are often the dominant concepts in biomedical research. As such, there is a tendency for researchers to try to link mechanical changes to specific genetic and molecular changes. While it is logical that physical changes have their roots in genetic changes, it is possible that this may not always be the case. For instance, any number of different genetic or signaling changes can result in the same mechanical phenotype. Therefore, screening a population could result in the identification of a certain mechanical phenotype without finding a unifying underlying molecular biomarker. Additionally, the mechanical traits of a cell are transient—changing in time as a function of migration and protrusive activity, cell cycle state, matrix properties, and other factors. Therefore, it may not be possible to capture individual molecular or signaling changes that are responsible for mechanical properties. Given the lack of success in identifying universal molecular biomarkers of cancer progression and the recent surge of data showing that metastatic cells are more deformable and exert stronger forces, it is possible that mechanical biomarkers may be a promising avenue for diagnosing and treating cancer. A greater understanding of cell behavior may come from viewing mechanics as complementary to, but not necessarily rooted in, molecular biology.

Acknowledgement The author would like to thank Denis Wirtz and Paul Janmey for helpful discussions.

References

Al-Kilani, A., de Freitas, O., Dufour, S., Gallet, F.: Negative feedback from integrins to cadherins: a micromechanical study. Biophys. J. **101**, 336–344 (2011)

Baker, E.L., Bonnecaze, R.T., Zaman, M.H.: Extracellular matrix stiffness and architecture govern intracellular rheology in cancer. Biophys. J. **97**(4), 1013–1021 (2009)

Balaban, N.Q., Schwarz, U.S., Riveline, D., Goichberg, P., Tzur, G., Sabanay, I., Mahalu, D., Safran, S., Bershadsky, A., Addadi, L., Geiger, B.: Force and focal adhesion assembly: a close relationship studied using elastic micropatterned substrates. Nat. Cell Biol. **3**, 466–472 (2001)

Bao, G., Suresh, S.: Cell and molecular mechanics of biological materials. Nat. Mater. **2**, 715–725 (2003)

Beerling, E., Ritsma, L., Vrisekoop, N., Derksen, P.W., van Rheenen, J.: Intravital microscopy: new insights into metastasis of tumors. J. Cell Sci. **124**, 299–310 (2011)

Carey, S.P., Starchenko, A., McGregor, A.L., Reinhart-King, C.A.: Leading malignant cells initiate collective epithelial cell invasion in a three-dimensional heterotypic tumor spheroid model. Clin. Exp. Metastasis **30**(5), 615–630 (2013)

Cross, S.E., Jin, Y.S., Rao, J., Gimzewski, J.K.: Nanomechanical analysis of cells from cancer patients. Nat. Nanotechnol. **2**, 780–783 (2007)

Crow, A., Webster, K.D., Hohlfeld, E., Ng, W.P., Geissler, P., Fletcher, D.A.: Contractile equilibration of single cells to step changes in extracellular stiffness. Biophys. J. **102**, 443–451 (2012)

De, R., Zemel, A., Safran, S.A.: Do cells sense stress or strain? Measurement of cellular orientation can provide a clue. Biophys. J. **94**, L29–L31 (2008)

De, R., Zemel, A., Safran, S.A.: Theoretical concepts and models of cellular mechanosensing. Methods Cell Biol. **98**, 143–175 (2010)

Dembo, M., Wang, Y.L.: Stresses at the cell-to-substrate interface during locomotion of fibroblasts. Biophys. J. **76**, 2307–2316 (1999)

du Roure, O., Saez, A., Buguin, A., Austin, R.H., Chavrier, P., Silberzan, P., Ladoux, B.: Force mapping in epithelial cell migration. Proc. Natl. Acad. Sci. U. S. A. **102**, 2390–2395 (2005)

Friedl, P., Alexander, S.: Cancer invasion and the microenvironment: plasticity and reciprocity. Cell **147**, 992–1009 (2011)

Friedl, P., Wolf, K.: Tumour-cell invasion and migration: diversity and escape mechanisms. Nat. Rev. Cancer **3**, 362–374 (2003)

Friedl, P., Wolf, K., Lammerding, J.: Nuclear mechanics during cell migration. Curr. Opin. Cell Biol. **23**, 55–64 (2011)

Guck, J., Ananthakrishnan, R., Mahmood, H., Moon, T.J., Cunningham, C.C., Käs, J.: The optical stretcher: a novel laser tool to micromanipulate cells. Biophys. J. **81**, 767–784 (2001)

Guck, J., Schinkinger, S., Lincoln, B., Wottawah, F., Ebert, S., Romeyke, M., Lenz, D., Erickson, H.M., Ananthakrishnan, R., Mitchell, D., Käs, J., Ulvick, S., Bilby, C.: Optical deformability as an inherent cell marker for testing malignant transformation and metastatic competence. Biophys. J. **88**, 3689–3698 (2005)

Gupta, S., Marcel, N., Sarin, A., Shivashankar, G.V.: Role of actin-dependent nuclear deformation in regulating early gene expression. PLoS One **7**, e53031 (2012)

Harris, A.K., Wild, P., Stopak, D.: Silicone rubber substrata: a new wrinkle in the study of cell locomotion. Science **208**, 177–179 (1980)

Heilbrunn, L.V.: Protoplasmic viscosity changes during mitosis. J. Exp. Zool. **34**, 417–447 (1921)

Hironaka, K., Morishita, Y.: Encoding and decoding of positional information in morphogen-dependent patterning. Curr. Opin. Genet. Dev. **22**, 553–561 (2012)

Huang, J.W., Pan, H.J., Yao, W.Y., Tsao, Y.W., Liao, W.Y., Wu, C.W., Tung, Y.C., Lee, C.H.: Interaction between lung cancer cell and myofibroblast influenced by cyclic tensile strain. Lab Chip **13**, 1114 (2013)

Huynh, J., Nishimura, N., Rana, K., Peloquin, J.M., Califano, J.P., Montague, C.R., King, M.R., Schaffer, C.B., Reinhart-King, C.A.: Age-related intimal stiffening enhances endothelial permeability and leukocyte transmigration. Sci. Trans. Med. **3**, 112–122 (2011)

Icard-Arcizet, D., Cardoso, O., Richert, A., Henon, S.: Cell stiffening in response to external stress is correlated to actin recruitment. Biophys. J. **94**, 2906–2913 (2008)

Isermann, P., Lammerding, J.: Nuclear mechanics and mechanotransduction in health and disease. Curr. Biol. **23**(24), R1113–R1121 (2013)

Iyer, K.V., Pulford, S., Mogilner, A., Shivashankar, G.V.: Mechanical activation of cells induces chromatin remodeling preceding MKL nuclear transport. Biophys. J. **103**(7), 1416–28 (2012)

Jain, N., Iyer, K.V., Kumar, A., Shivashankar, G.V.: Cell geometric constraints induce modular gene-expression patterns via redistribution of HDAC3 regulated by actomyosin contractility. Proc. Natl. Acad. Sci. U. S. A. **110**, 11349–11354 (2013)

Keese, C.R., Giaever, I.: Substrate mechanics and cell spreading. Exp. Cell Res. **195**, 528–532 (1991)

Khalil, A.A., Friedl, P.: Determinants of leader cells in collective cell migration. Integr. Biol. (Camb.) **2**, 568–574 (2010)

Khatau, S.B., Bloom, R.J., Bajpai, S., Razafsky, D., Zang, S., Giri, A., Wu, P.H., Marchand, J., Celedon, A., Hale, C.M., Sun, S.X., Hodzic, D., Wirtz, D.: The distinct roles of the nucleus and nucleus-cytoskeleton connections in three-dimensional cell migration. Sci. Rep. **2**, 488 (2012)

Koch, T.M., Munster, S., Bonakdar, N., Butler, J.P., Fabry, B.: 3D Traction forces in cancer cell invasion. PLoS One **7**, e33476 (2012)

Kraning-Rush, C.M., Carey, S.P., Califano, J.P., Smith, B.N., Reinhart-King, C.A.: The role of the cytoskeleton in cellular force generation in 2D and 3D environments. Phys. Biol. **8**, 015009 (2011)

Kraning-Rush, C.M., Califano, J.P., Reinhart-King, C.A.: Cellular traction stresses increase with increasing metastatic potential. PLoS One **7**, e32572 (2012a)

Kraning-Rush, C.M., Carey, S.P., Califano, J.P., Reinhart-King, C.A.: Quantifying traction stresses in adherent cells. Methods Cell Biol. **110**, 139–178 (2012b)

Lam, W.A., Rosenbluth, M.J., Fletcher, D.A.: Chemotherapy exposure increases leukemia cell stiffness. Blood **109**, 3505–3508 (2007)

le Duc, Q., Shi, Q., Blonk, I., Sonnenberg, A., Wang, N., Leckband, D., de Rooij, J.: Vinculin potentiates E-cadherin mechanosensing and is recruited to actin-anchored sites within adherens junctions in a myosin II-dependent manner. J. Cell Biol. **189**, 1107–1115 (2010)

Leckband, D.E., le Duc, Q., Wang, N., de Rooij, J.: Mechanotransduction at cadherin-mediated adhesions. Curr. Opin. Cell Biol. **23**, 523–530 (2011)

Lee, M.H., Wu, P.H., Staunton, J.R., Ros, R., Longmore, G.D., Wirtz, D.: Mismatch in mechanical and adhesive properties induces pulsating cancer cell migration in epithelial monolayer. Biophys. J. **102**, 2731–2741 (2012)

Lee, Y.H., Huang, J.R., Wang, Y.K., Lin, K.H.: Three-dimensional fibroblast morphology on compliant substrates of controlled negative curvature. Integr. Biol. **5**, 1447–1455 (2013)

Legant, W.R., Miller, J.S., Blakely, B.L., Cohen, D.M., Genin, G.M., Chen, C.S.: Measurement of mechanical tractions exerted by cells in three-dimensional matrices. Nat. Methods **7**, 969–971 (2010)

Levental, K.R., Yu, H., Kass, L., Lakins, J.N., Egeblad, M., Erler, J.T., Fong, S.F., Csiszar, K., Giaccia, A., Weninger, W., Yamauchi, M., Gasser, D.L., Weaver, V.M.: Matrix crosslinking forces tumor progression by enhancing integrin signaling. Cell **139**, 891–906 (2009)

Liu, Z., Tan, J.L., Cohen, D.M., Yang, M.T., Sniadecki, N.J., Ruiz, S.A., Nelson, C.M., Chen, C.S.: Mechanical tugging force regulates the size of cell-cell junctions. Proc. Natl. Acad. Sci. U. S. A. **107**, 9944–9949 (2010)

Logothetis, N.K.: Francis Crick 1916–2004. Nat. Neurosci. **7**, 1027–1028 (2004)

Mitrossilis, D., Fouchard, J., Pereira, D., Postic, F., Richert, A., Saint-Jean, M., Asnacios, A.: Real-time single-cell response to stiffness. Proc. Natl. Acad. Sci. U. S. A. **107**, 16518–16523 (2010)

Montel, F., Delarue, M., Elgeti, J., Malaquin, L., Basan, M., Risler, T., Cabane, B., Vignjevic, D., Prost, J., Cappello, G., Joanny, J.F.: Stress clamp experiments on multicellular tumor spheroids. Phys. Rev. Lett. **107**, 188102 (2011)

Morishita, Y., Iwasa, Y.: Coding design of positional information for robust morphogenesis. Biophys. J. **101**, 2324–2335 (2011)

Okuda, S., Inoue, Y., Eiraku, M., Sasai, Y., Adachi, T.: Reversible network reconnection model for simulating large deformation in dynamic tissue morphogenesis. Biomech. Model. Mechanobiol. **12**, 627–644 (2013)

Paszek, M.J., Zahir, N., Johnson, K.R., Lakins, J.N., Rozenberg, G.I., Gefen, A., Reinhart-King, C.A., Margulies, S.S., Dembo, M., Boettiger, D., Hammer, D., Weaver, V.: Tensional homeostasis and the malignant phenotype. Cancer Cell **8**, 241–254 (2005)

Pelham Jr., R.J., Wang, Y.L.: Cell locomotion and focal adhesions are regulated by the mechanical properties of the substrate. Biol. Bull. **194**, 348–349 (1998); discussion 349–350

Pelling, A.E., Horton, M.A.: An historical perspective on cell mechanics. Pflugers Arch. **456**, 3–12 (2008)

Roca-Cusachs, P., del Rio, A., Puklin-Faucher, E., Gauthier, N.C., Biais, N., Sheetz, M.P.: Integrin-dependent force transmission to the extracellular matrix by α-actinin triggers adhesion maturation. Proc. Natl. Acad. Sci. U. S. A. **110**(15), 1361–1370 (2013)

Rowat, A.C., Jaalouk, D.E., Zwerger, M., Ung, W.L., Eydelnant, I.A., Olins, D.E., Olins, A.L., Herrmann, H., Weitz, D.A., Lammerding, J.: Nuclear envelope composition determines the ability of neutrophil-type cells to passage through micron-scale constrictions. J. Biol. Chem. **288**(12), 8610–8618 (2013)

Safran, S.A., De, R.: Nonlinear dynamics of cell orientation. Phys. Rev. E Stat. Nonlin. Soft Matter Phys. **80**, 060901 (2009)

Schwarz, U.S., Balaban, N.Q., Riveline, D., Bershadsky, A., Geiger, B., Safran, S.A.: Calculation of forces at focal adhesions from elastic substrate data: the effect of localized force and the need for regularization. Biophys. J. **83**, 1380–1394 (2002)

Schwarz, U.S., Erdmann, T., Bischofs, I.B.: Focal adhesions as mechanosensors: the two-spring model. Biosystems **83**, 225–232 (2006)

Seifriz, W.: An elastic value of protoplasm, with further observations on the viscosity of protoplasm. J. Exp. Biol. **2**, 1–11 (1924)

Stricker, J., Sabass, B., Schwarz, U.S., Gardel, M.L.: Optimization of traction force microscopy for micron-sized focal adhesions. J. Phys. Condens. Matter **22**, 194104 (2010)

Tambe, D.T., Hardin, C.C., Angelini, T.E., Rajendran, K., Park, C.Y., Serra-Picamal, X., Zhou, E.H., Zaman, M.H., Butler, J.P., Weitz, D.A., Fredberg, J.J., Trepat, X.: Collective cell guidance by cooperative intercellular forces. Nat. Mater. **10**, 469–475 (2011)

Tan, J.L., Tien, J., Pirone, D.M., Gray, D.S., Bhadriraju, K., Chen, C.S.: Cells lying on a bed of microneedles: an approach to isolate mechanical force. Proc. Natl. Acad. Sci. U. S. A. **100**, 1484–1489 (2003)

Trepat, X., Fredberg, J.J.: Plithotaxis and emergent dynamics in collective cellular migration. Trends Cell Biol. **21**, 638–646 (2011)

Tse, J.M., Cheng, G., Tyrrell, J.A., Wilcox-Adelman, S.A., Boucher, Y., Jain, R.K., Munn, L.L.: Mechanical compression drives cancer cells toward invasive phenotype. Proc. Natl. Acad. Sci. U. S. A. **109**, 911–916 (2012)

Vedula, S.K., Hirata, H., Nai, M.H., Brugués, A., Toyama, Y., Trepat, X., Liu, C.T., Ladoux, B.: Epithelial bridges maintain tissue integrity during collective cell migration. Nat. Mater. **13**(1), 87–96 (2014)

Wang, Y.L., Pelham Jr., R.J.: Preparation of a flexible, porous polyacrylamide substrate for mechanical studies of cultured cells. Methods Enzymol. **298**, 489–496 (1998)

Wang, Y.H., Chiu, W.T., Wang, Y.K., Wu, C.C., Chen, T.L., Teng, C.F., Chang, W.T., Chang, H.C., Tang, M.J.: Deregulation of AP-1 proteins in collagen gel-induced epithelial cell apoptosis mediated by low substratum rigidity. J. Biol. Chem. **282**, 752–763 (2007)

Wilson, L.V.H.A.W.L.: A rational approach to the problem of cancer chemotherapy. Biol. Bull. **113**, 388–396 (1957)

Wirtz, D.: Particle-tracking microrheology of living cells: principles and applications. Annu. Rev. Biophys. **38**, 301–326 (2009)

Wirtz, D., Konstantopoulos, K., Searson, P.C.: The physics of cancer: the role of physical interactions and mechanical forces in metastasis. Nat. Rev. Cancer **11**, 512–515 (2011)

Wu, P.H., Hale, C.M., Chen, W.C., Lee, J.S., Tseng, Y., Wirtz, D.: High-throughput ballistic injection nanorheology to measure cell mechanics. Nat. Protoc. **7**(1), 155–170 (2012)

Yu, C.H., Law, J.B., Suryana, M., Low, H.Y., Sheetz, M.P.: Early integrin binding to Arg-Gly-Asp peptide activates actin polymerization and contractile movement that stimulates outward translocation. Proc. Natl. Acad. Sci. U. S. A. **108**(51), 20585–20590 (2011)

Zemel, A., Rehfeldt, F., Brown, A.E., Discher, D.E., Safran, S.A.: Cell shape, spreading symmetry and the polarization of stress-fibers in cells. J. Phys. Condens. Matter **22**(19), 194110 (2010)

Chapter 5
Fluid Mechanics and Transport in Tumors

Lance L. Munn

Introduction

The physical structure of a tumor is determined by "solid," structural components such as actin and collagen, but most of its volume is fluid. Fluid flows through the blood vessels and bathes the extravascular tissue. The structural microenvironment of a solid tumor plays a role in tumor physiology by compartmentalizing tissues and providing a dynamic substrate to support cell migration and differentiation (see Chaps. 3 and 4), but the fluid phase is equally important.

Within the scope of tumor fluid mechanics are a number of interesting processes relevant to progression and treatment (Fig. 5.1). Blood flowing through the vasculature carries nutrients necessary for growth, or drugs designed to kill a tumor. Blood also carries circulating tumor cells important for metastasis, and DNA, the recent Holy Grail for cancer diagnosis and biomarker development. Flowing blood also exerts forces on the cells lining the blood vessels. This can affect their behavior by mediating local diameter adjustments for flow optimization. Because vessels are permeable to water—and those in tumors are especially leaky—fluid can leave the blood vessels and flow past cells in the extravascular space before leaving via another blood or lymphatic vessel. This flow provides cues to stromal, immune, and cancer cells as well as cells in the blood vessel wall. In addition, this convecting fluid can skew gradients of growth factors or cytokines produced locally by cells in the tissue. By following the resulting asymmetric gradient, cells can sense flow direction during migration.

This chapter provides an overview of the major issues in tumor fluid mechanics and dynamics, focusing on current research in the United States, Europe, and Asia.

L.L. Munn (✉)
Massachusetts General Hospital/Harvard Medical School, Charlestown, MA, USA
e-mail: lance@steele.mgh.harvard.edu

P. Janmey et al. (eds.), *Physical Sciences and Engineering Advances in Life Sciences and Oncology*, Science Policy Reports,
DOI 10.1007/978-3-319-17930-8_5

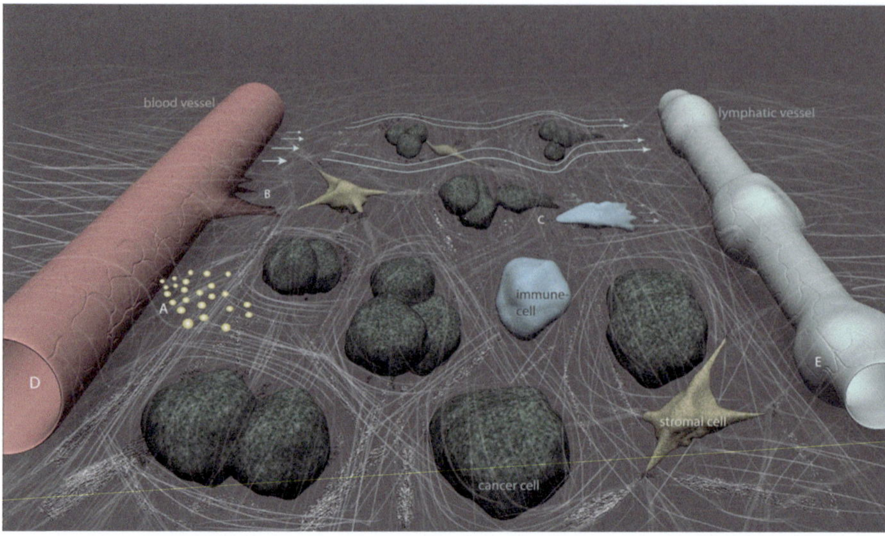

Fig. 5.1 The fluid microenvironment in and around tumors (Courtesy of L. Munn) (**a**) Fluid movement between the blood and lymphatic systems is important for drug delivery and (**b**) exposes endothelial cells to fluid forces that can guide angiogenesis and (**c**) migration of cancer or immune cells. (**d**) Hemodynamic forces within blood vessels control vascular tone and network remodeling, and circulation of cells and antigens in the (**e**) lymphatic system is critical for immunosurveillance

Research

Delivery of Nutrients and Drugs to Tumors

The primary function of blood vessels is to carry cells and biomolecules to tissues. In normal tissue, this is an optimized process; in tumors, on the other hand, there are many barriers that make delivery difficult (Jain 1994).

Many of the problems stem from the fact that tumor vasculature is topologically abnormal and not well-regulated (Jain 1996). This leads to non-uniform blood flow, with some vessels having fast flow, and others with little or no perfusion. Blood flow is also temporally non-uniform, so that different regions of the tumor are perfused at different times, again preventing uniform delivery.

Another important determinant of tumor physiology is vascular permeability. High permeability can disrupt normal flow patterns, leading to extravascular shunts and high interstitial fluid pressure (IFP) (Boucher et al. 1996; Yuan et al. 1994b). These problems are evident in images showing nonuniform delivery of drugs in tumor tissue (Yuan et al. 1994a), with high concentrations near some vessels but little penetration from other vessels (Fig. 5.2).

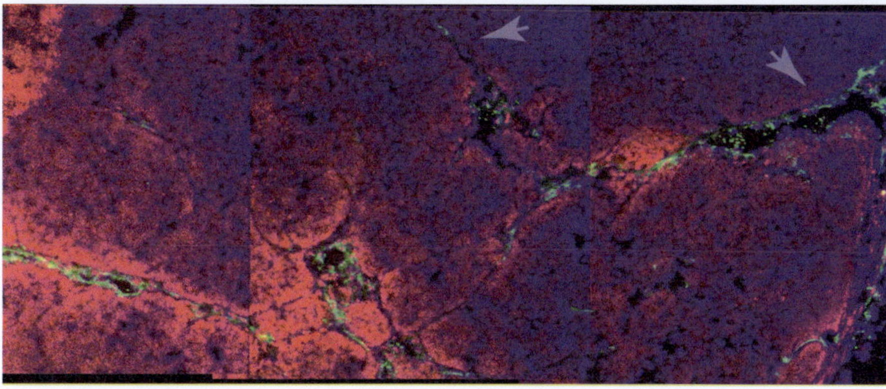

Fig. 5.2 Heterogeneous delivery of doxorubicin in an orthotopic MCaIV mammary carcinoma grown in a SCID mouse (Unpublished data from L. Munn)
The doxorubicin (15 mg/kg) was administered 4 h before harvesting the tumor for analysis. Note the high concentrations around some vessels, but lack of delivery from other vessels (*arrows*) *Blue*: dapi; *red*: doxorubicin; *green*: blood vessel marker

Vascular Normalization as a Means to Improve Drug Delivery

Injected drugs only reach some of the tumor areas because of the mentioned problems with the vasculature and blood flow. In the late 1990s, research in the Steele Lab at Massachusetts General Hospital highlighted the fact that anti-angiogenic drugs were not killing blood vessels as originally intended; instead, they were decreasing vessel diameters, reducing leaks, pruning inefficient segments, and increasing overall blood flow. This led Dr. Rakesh Jain to hypothesize that anti-angiogenic therapies might be used to normalize tumor blood vessels—increasing blood flow and restoring uniformity—to enhance delivery of subsequently-injected chemotherapeutics (Jain 2001). Preclinical work and recent clinical trial data support this strategy. For example, in a recent trial for newly-diagnosed glioblastoma patients, those with increased blood flow after treatment with Cediranib (a VEGF tyrosine kinase inhibitor) had the best response to chemotherapy and radiation (Pinho et al. 2012).

Transport of Cancer Cells in the Blood Stream

In addition to carrying nutrients and drugs, the blood stream is a conduit for metastasizing cancer cells. In recent years, significant effort has gone into developing methodologies for isolating these circulating tumor cells (CTCs) from whole blood drawn from patients, with the goal of using the cells as biomarkers to guide treatment strategies or study them *ex vivo*. In general, several milliliters of blood will contain on the order of 10–100 cancer cells, so the task of extracting ~1010 red blood cells is daunting.

Most strategies attempt to capture CTCs from whole blood directly, with little pre-processing. One example is the commercially-available Veridex Cell Search™ system, which is a semi-automated sample preparation system that enriches the sample based on the expression of epithelial-cell adhesion molecule using antibody-coated magnetic beads; it also labels the cells with a fluorescent nucleic acid dye for counting (https://www.cellsearchctc.com/). Another proven method, developed by Dr. Toner and colleagues at the Harvard Medical School, extracts cancer cells from whole blood using microarrays of PDMS posts coated with specific antibodies against cancer antigens (Nagrath et al. 2007). Channel geometry plays an important role in determining how the CTCs transit through these devices, and can be adjusted to enhance collisions with antibody-coated surfaces or to independently enrich the nucleated cell population (Kirby et al 2012; Jain and Munn 2009).

It is also possible to expose the whole blood to hypotonic solution, lysing the red blood cells (RBCs), which are more sensitive to osmotic shock than nucleated cells. Centrifugation then removes the lysed RBC debris. The resulting solution, composed mainly of CTCs and leukocytes, can then be plated and analyzed using high-throughput imaging to identify and quantify the cancer cells (Nieva et al. 2012; Wendel et al. 2012). Though they are showing some promise in the clinic (Budd et al. 2006; Hayes et al. 2006; Wendel et al. 2012), the existing technologies are tedious and difficult to generalize to multiple tumor types because the requirement for specific antigen binding for collection and identification (not all tumors have known antigens). Therefore, researchers are now trying to find robust and efficient ways to collect these clinically relevant cells. One example of this approach is work in the lab of Ying-chih Chang at the Genomic Research Center, Academia Sinica, Taiwan, in which antibody-functionalized supported lipid bilayer are being developed as non-fouling membrane coating using antibodies specific for CTCs (Wu et al. 2013). The physical and chemical features of surfaces made from phospholipids more closely mimic the endothelial surface in the vasculature and have the potential to increase specificity of cell capture and viability of cells, compared to methods using standard materials such as PDMS or plastic.

Autologous Chemotaxis

In addition to carrying cells, biomolecules, and drugs to and from tumors, fluid flow can also directly affect cancer cells by changing or establishing local chemotactic gradients (Polacheck et al. 2011). Melody Swartz and colleagues at the École Polytechnique Fédérale de Lausanne (site report, Appendix B) have shown that production of a chemokine, such as CCL21, by cancer cells can lead to a skewed gradient in the direction of flow (Fig. 5.3, Shields et al. 2007). Because much of the fluid percolating through a tumor eventually enters the peritumor lymphatic vessels, the process—called autologous chemotaxis—allows the cancer cells to find lymphatic vessels, facilitating metastasis. CCR7 is the necessary receptor for CCL21 in this system. Research by Swartz and colleagues has shown that fibroblasts can use similar mechanisms via autologous TGF beta signaling and matrix rearrangement to create paths for invading cancer cells (Shieh et al. 2011).

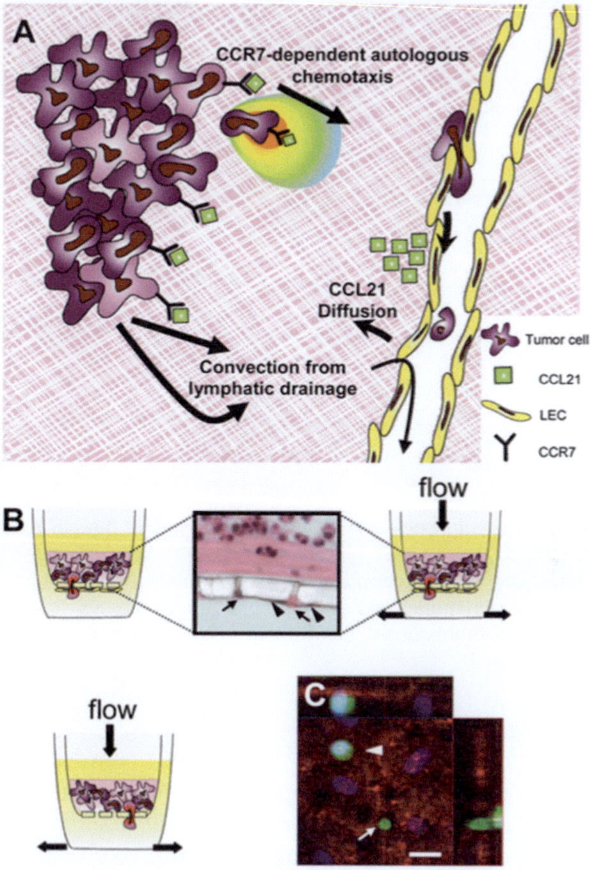

Fig. 5.3 Cancer cells producing CCL21 chemokine detect the direction of interstitial flow because of the shift in the local chemical gradient. In this process, known as autologous chemotaxis, the cells follow the flow to nearby lymphatic vessels (From Shields et al. 2007)

Flow-Guided Morphogenesis

Lymphangiogenesis

Flowing fluids also exert forces on cells transmitting mechanical signals during formation of the blood or lymphatic vasculature. Although we have known for decades that fluid shear stresses and pressures can affect blood vessel wall development, it was only recently that interjunctional and interstitial flows have been established as directors of blood and lymph vessel angiogenesis.

This was first elegantly demonstrated by Boardman and Swartz in a model of lymphangiogenesis in the mouse tail (Boardman and Swartz 2003). After creating an annular wound around the tail to remove the existing lymphatic network, a collagen gel was used to fill the wound (Fig. 5.4). By injecting a fluorescent tracer in the tip of the tail, fluid flow was forced through the collagen gel. Interestingly, this flow accelerated reformation of the damaged lymphatic network, and the lymphangiogenesis was initiated in the direction of the flow.

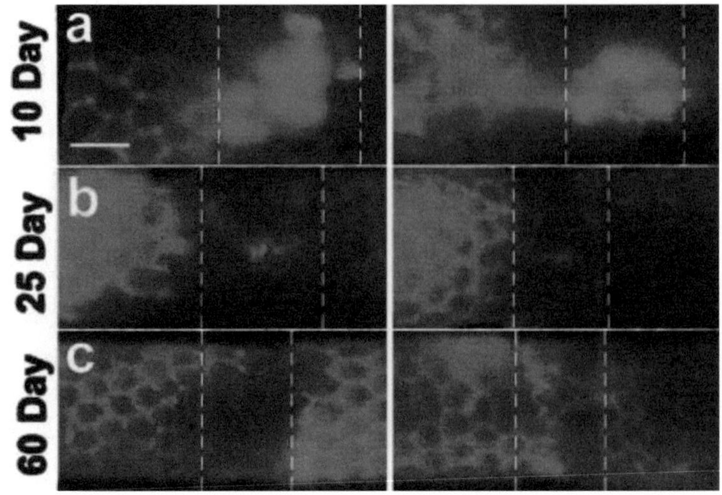

Fig. 5.4 Lymphangiogenesis is directed by fluid flow in the mouse tail (From Boardman and Swartz 2003)
After removing the dermis and existing lymphatic network in a localized region (*dashed lines*), collagen gel is implanted and fluid flow forced through from the distal end of the tail (*left side*). With flow, reformation of the lymphatic network is accelerated, and is initiated at the upstream side of the wound

Angiogenesis

More recently, several groups have developed microfluidic devices that accurately reproduce the process of sprouting to study angiogenesis *in vitro* (Song and Munn 2011; Jeong et al. 2011; Song et al. 2012; Yeon et al. 2012). In general, the endothelial cells exist in a monolayer that lines predefined channels adjacent to a collagen gel. Flow can be applied in the channels and/or directed through the gel (Fig. 5.5). Under the controlled conditions provided by the microfluidics, it is possible to study growth factor gradients and interstitial flow independently, and more precisely determine how they cooperate to effect morphogenesis.

Findings from these devices add to our understanding of flow-induced morphogenesis. Interestingly, the direction of flow is important for the initiation of blood vessels sprouts. Flow out of a vessel (abluminal-basal; a-b) induces dilation of vessels, while flow into a vessel (basal-abluminal; b-a) encourages sprouting (Song and Munn 2011; Jeong et al. 2011). This has important implications for angiogenesis in inflammation and cancer—both are cases of dynamically leaky vasculature.

It is not clear what mechanisms endothelial cells use to sense the flow and determine its direction, but they likely involve strains in cell-cell or cell-substrate anchoring structures caused by flow. This idea is supported by recent work from Roger Kamm's group at the Massachusetts Institute of Technology showing that the actin cytoskeleton reorganizes differently depending on flow direction (Vickerman and Kamm 2012). They showed that application of b-a, but not a-b flow, induced capillary morphogenesis, and redistribution of VE-cadherin and FAK phosphorylation patterns.

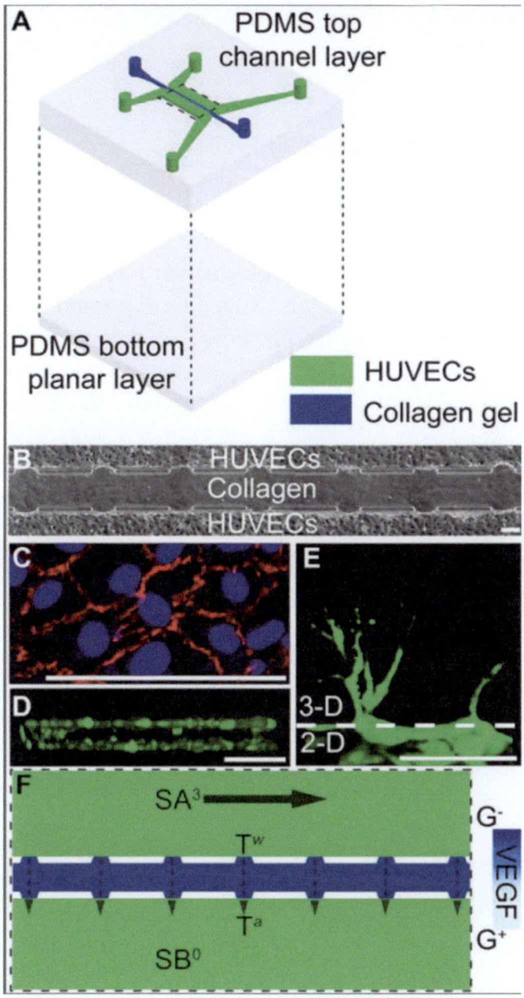

Fig. 5.5 Microfluidic device with localized 3D extracellular matrix for fluid force-mediated angiogenic sprouting and morphogenesis (From Song and Munn 2011)
(**a**) Multilayer fabrication of the polydimethylsiloxane (PDMS) microfluidic device. (**b**) HUVECs seeded into two channels separated by collagen gel visualized under phase microscopy. (**c**) VE-cadherin expression (*red*, lines) in the HUVEC monolayer. (**d**) Cross-section view of one of the HUVEC channels. (**e**) HUVEC-GFP cells sprouting into the 3D collagen gel migrate into the bulk of the gel rather than along the *top* or *bottom* surface. (**f**) Close-up view of boxed area in (**a**) showing seven apertures that allow contact between the HUVEC cells (*green*, four-armed structure) and the collagen barrier (*blue*, two-node line). Each HUVEC channel has independent input and outlet ports, allowing strict control over flow in both channels and across the collagen matrix

These results support the hypothesis that flow-mediated responses involve signaling at cell-cell and cell-matrix interfaces.

Flow-Guided Embryonic Development

Flow-guided morphogenesis and migration could also be important in development. It has been shown that flow patterns help drive embryonic tissue formation and compartmentalization. For example, Dr. Vincent Fleury and colleagues at the University of Paris, Diderot (site report, Appendix B) found that embryonic tissue behaves as a viscous fluid, and that formation of complex flow patterns during development follow the laws of hydrodynamics (Boryskina et al. 2011). They describe tissue formation as an initial broken symmetry that is up-scaled by a slow, continuous, viscoelastic flow that mirrors the animal's body plan. Similar correlations are seen during chicken chorioallontoic membrane circulation system development. Tissue flow patterns appear to pre-determine the shape of the forming vasculature (le Noble et al. 2004).

Convection and diffusion also play essential roles in pattern formation during neuronal development, as studied in the context of eye development by Prof. Yoshiki Sasai and colleagues at the RIKEN Center for Developmental Biology in Kobe, Japan. A current emphasis in this lab is on understanding how the self-organization of biological systems occurs through mechanisms involving reaction-diffusion processes, convection, and other physical phenomena, with the goal of using such data to improve methods for *in vitro* organogenesis (Sasai 2013; Sasai et al. 2012).

Angioadaptation

Blood flow within vessels is central to vascular development and maintenance, imposing shear forces that direct angioadaptation. Angioadaptation is the process by which vessel networks remodel to provide efficient, optimized networks for delivering blood to tissues. One of the reasons that tumor vasculature and blood flow are abnormal is impaired angioadaptation.

This process is driven, in part, by fluid mechanics. Shear forces exerted by blood flow are detected by the endothelium, and if the shear is too high or low, the vessel dilates or contracts appropriately. These local changes in diameter have the effect of equalizing the shear stress throughout the network and minimizing flow resistance. Other, longer-range signals are transmitted along the vessel wall or advected downstream with the flow to control overall flow into and out of the capillary beds (Fig. 5.6, Pries et al. 2011).

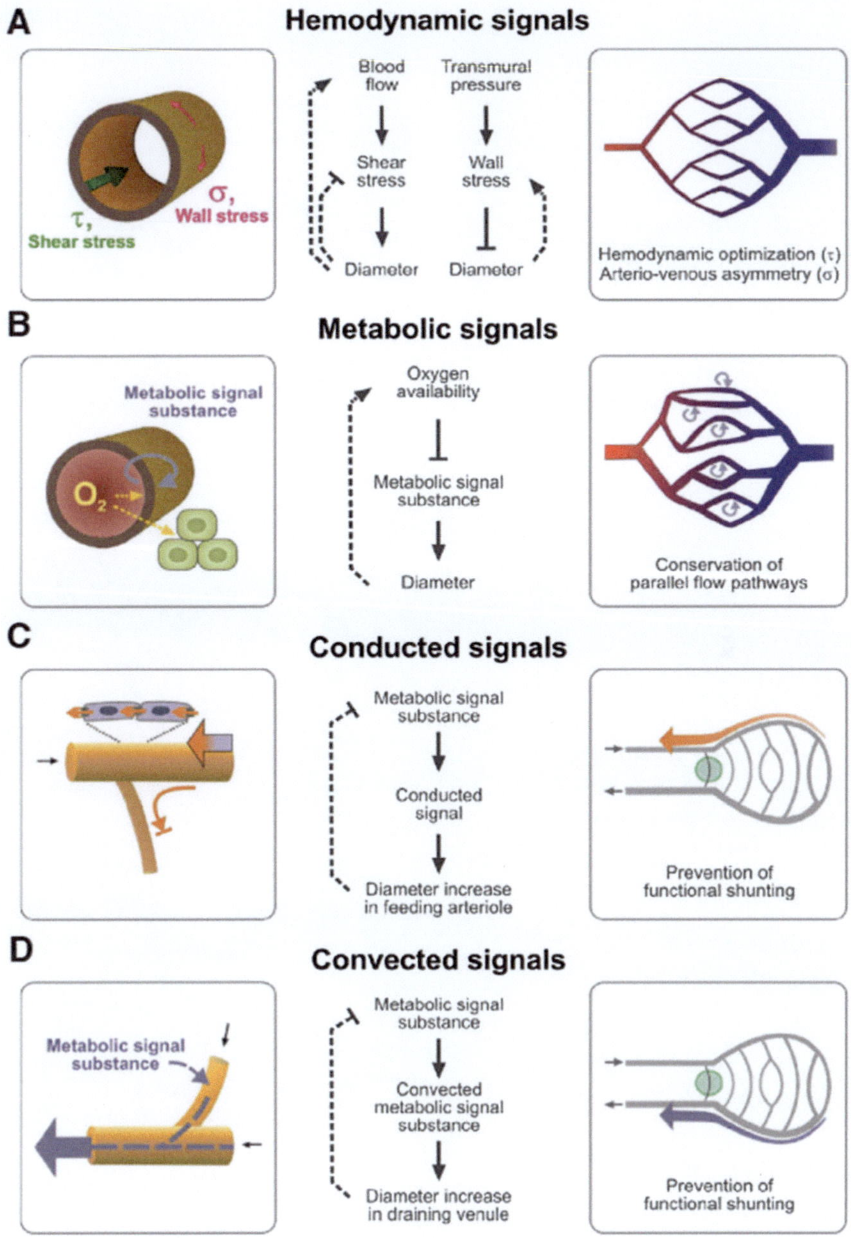

Fig. 5.6 Angioadaptation in normal and tumor vasculatures (From Pries et al. 2011)
Vessel diameters can be modulated in response to many signals including: **(a)** local hemodynamic shear and pressure forces, **(b)** metabolic signals from the tissue or vessel wall, and **(c)** longer range signals conducted up the vessel wall or **(d)** convected downstream in the flow. When operating correctly, these mechanisms cooperate to optimize flow through the capillary beds. **(e)** Tumor vasculature differs in topology from that in the normal mesentery (*top row*). However, mesenteric vasculature remodeled using tumor-derived angioadaptation rules resembles tumor vasculature (*bottom right panel*), while tumor vasculature exposed to normal remodeling rules closely resembles normal vasculature (*bottom left*). This suggests that angioadaptation mechanisms are defective in tumors (Pries et al. 2009)

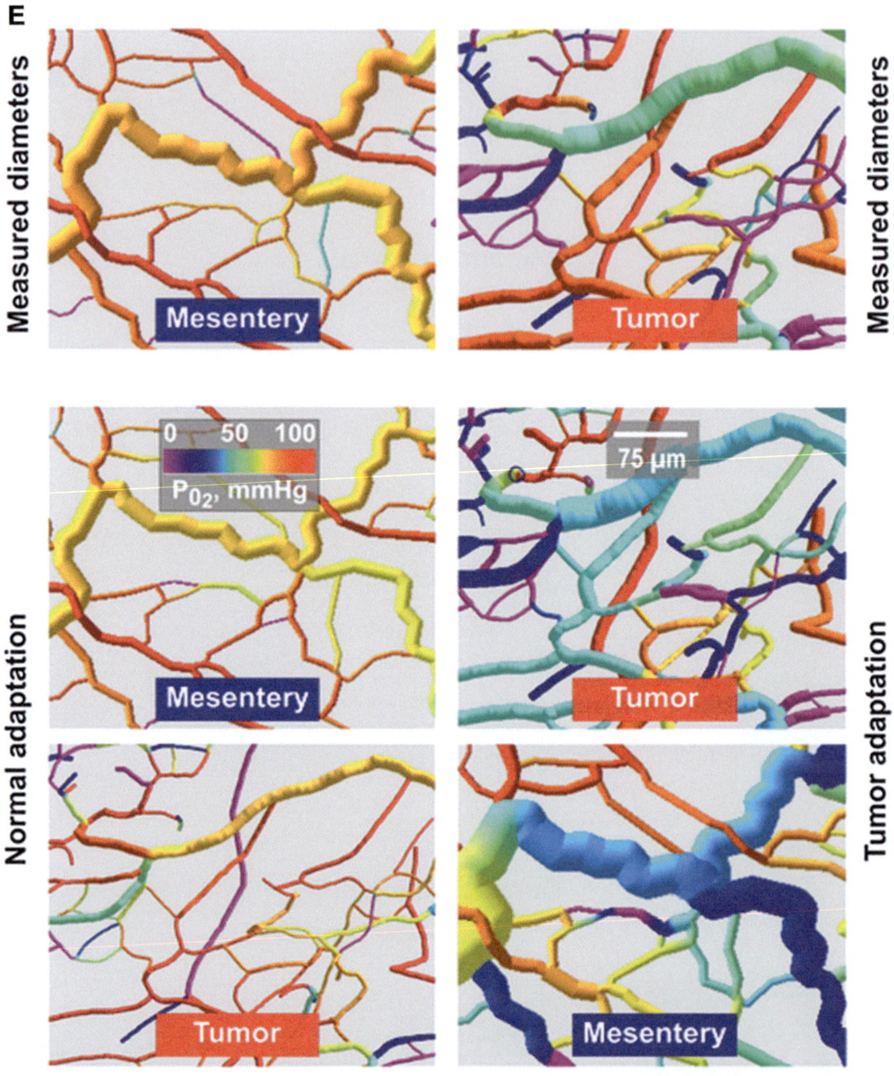

Fig. 5.6 (continued)

Malfunction of these control mechanisms could have important implications for tumor physiology and blood flow. By extracting the adaptation rules for tumor and normal vessel networks, it is possible to predict how a normal network would change if exposed to tumor rule and vice versa. The networks can then be analyzed to compare their structures and functions: specifically, diameter mismatch at bifurcations and oxygen delivery. Pries and colleagues found that the remodeling rules were sufficient to determine the efficiency of the network. They did this by performing the simulations for tumor and mesentery networks with the adapted tumor net-

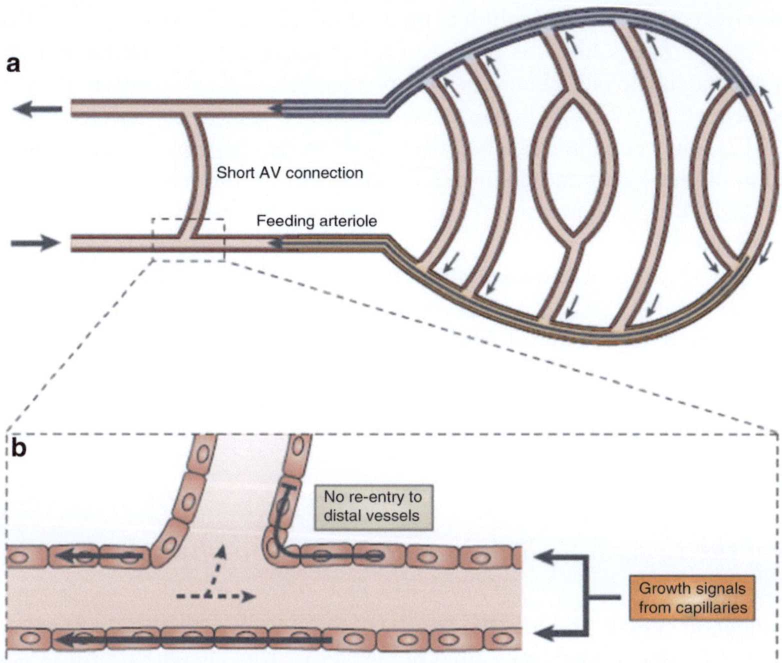

Fig. 5.7 Upstream and downstream signals are critical for distributing flow correctly between AV shunts and capillary beds. It is possible that these signals are absent in tumors (From Pries et al. 2010)

work functioning as well as the normal mesentery network, and the mesentery network adapted with tumor rules functioning as poorly as the original tumor network (Fig. 5.6e, Pries et al. 2009).

A corollary to this work actually relates to the poor distribution of blood flow in tumors, and focuses on potential "shunts" that, when malfunctioning, can prevent proper distribution of flow within tumors. The mechanisms primarily responsible for this are upstream conducted signals and downstream convected signals. If these are not operating, then the distribution of flow at bifurcations is not regulated and can shunt within the tumor rather than perfuse the capillary beds effectively (Fig. 5.7, Pries et al. 2010).

It is not known how the endothelial cells and vessel walls sense and respond to the blood flow or long-range and metabolic signals to adjust vessel diameter chronically. Acutely, vascular endothelial growth factor (VEGF) and nitric oxide (NO) are potential candidates for local control, and long-range signals are thought to propagate through gap junctions in the vessel wall. It is possible that chronic changes in endothelial arrangement can also induce longer-term remodeling. By growing endothelial cells on patches of matrix with defined aspect ratios, colleagues from Dr. Sylvain Gabriele's lab at the Université de Mons, Belgium (site report, Appendix B) were able to force alignment of the endothelial cells and elongation/deformation of the nuclei. By using a combination of micromanipulation tools, they found that tension in central stress fibers is produced by anisotropic force contraction dipoles, which expand as the cell elongates and spreads. Increased elongation of the nuclei was associated with more chromatin condensation and cell proliferation (Versaevel et al. 2012). They conclude that large-scale cell shape changes induce a drastic condensation of chromatin and dramatically affect cell proliferation. This could translate *in vivo* into proliferation of endothelial cells in vessels that are chronically dilated by shear.

Mechanical and geometrical control of cell function is also extensively studied at the Mechanobiology Institute of the National University of Singapore, directed by Prof. Michael Sheetz. Here for example, work by the labs of Chwee Teck Lim and Benoit Ladoux has used micropatterning and force measurements to show how cell-cell adhesions affect collective movements that enable tissue organization (Vedula et al. 2012).

Role of the Lymphatic System in Tumor Pathophysiology

The lymphatic system provides the major conduit for interstitial fluid to reenter the circulation, thereby maintaining fluid homeostasis. It is also the transit system for immune cells and contains the lymph nodes where T and B cells expand in response to antigen. To function properly, interstitial fluid with a potential antigen filters through the tissue and drains into initial draining lymphatic vessels. It then passes to collecting lymphatics, which contain one-way valves and can actively pump the lymph fluid via smooth muscle dilation and contraction. However, lymphatic vessels are only present outside the periphery of tumors (Padera et al. 2002), limiting antigen entry and dendritic cells (DCs) into the draining lymph nodes (LNs) and altering homeostasis and immune response.

Many factors influence lymph drainage. In normal physiology, lymph flow is achieved by both active and passive processes in collecting lymphatic vessels (Schmid-Schonbein 1990). Collecting lymphatic vessel pumping is due to autonomous lymphatic contraction driven by NO, a process that is altered by extravascular NO from the tumor tissue (Fig. 5.8, Liao et al. 2011). Proper flow in peritumor lymphatics may also be hindered by valve malformations that do not enforce unidirectional flow during pumping (Fig. 5.9, Hagendoorn et al. 2006).

While impaired lymph drainage can cause high interstitial fluid pressure and prevent immune cell trafficking, sufficient lymph flow can actually help tumors evade the immune response. Tumors often metastasize to lymph nodes. Because lymph nodes are specialized sites of immunomodulation, it is possible that exposure of cancer cells to the resident T and B cells induces immune-tolerance, protecting the tumor from host immunity (Swartz and Lund 2012). Lymph flow from tumors can also be increased, passing more antigen to the sentinel node and contributing to this induction of immunotolerance (Fig. 5.10). Future work is needed to determine

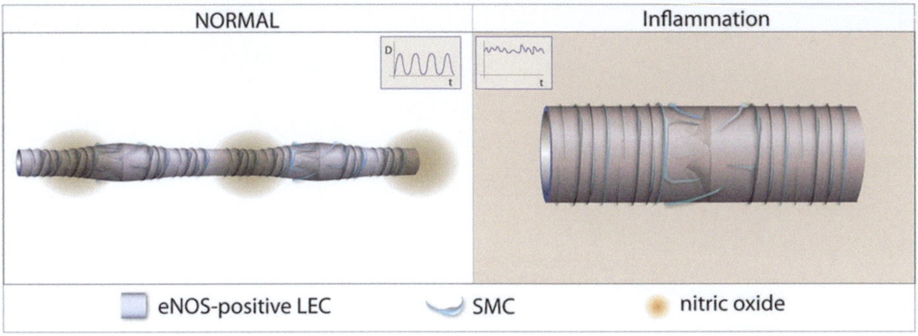

Fig. 5.8 Lymphatic vessel pumping dynamics (From Liao et al. 2011)
In normal tissue, cycles of shear stress and nitric oxide induce vessel dilation, which propagates along collecting lymphatic vessels to pump fluid. In inflammation or around tumors, extravascular nitric oxide overwhelms the system, causing chronic dilation and little pumping. This can have important implications for trafficking of immune cells and antigens

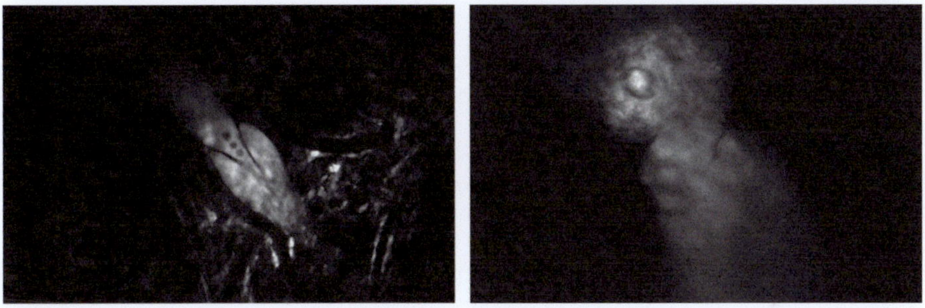

Fig. 5.9 Valve structures in collecting lymphatic vessels in normal (*left*) and peritumor (*right*) tissue. The valve leaflets in the peritumor lymphatic are malformed and unable to close to prevent back flow (From Hagendoorn et al. 2006)

whether these tumor-specific issues of lymphatic drainage and flow-induced mechanotransduction in the stroma allow tumors to escape the immune response by hijacking lymphatic mechanisms of peripheral tolerance.

Discussion

In summary, fluid dynamics in and around tumors plays an important role in disease progression and treatment. Similar to mechanochemical signals provided by matrix components, fluid flow through tissue can trigger morphogenesis and guide cell migration, facilitating immune response as well as angiogenesis and metastasis. In normal vasculature, fluid forces inside the blood vessels help optimize blood flow;

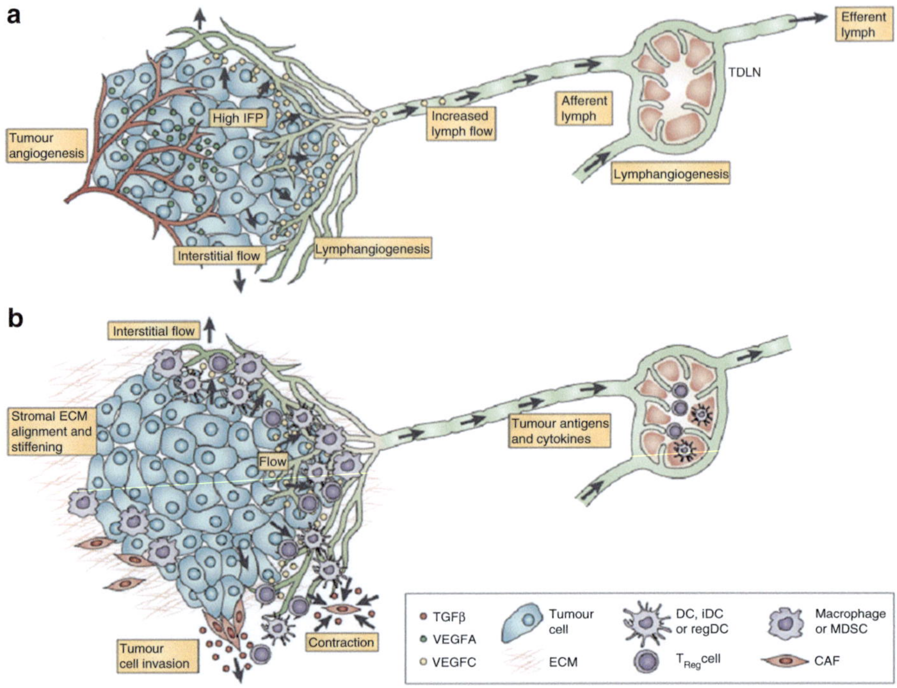

Fig. 5.10 Fluid pathways in the tumor microenvironment (From Swartz and Lund 2012) (**a**) Leaky tumor increases IFP within tumors, driving flow through tissue to draining lymphatics at the tumor margin. The lymph travels to sentinel or tumor-draining lymph nodes (TDLNs). (**b**) Increased interstitial and lymphatic flows in and around tumors affect the immune response and tumor stroma

it remains to be seen whether these mechanisms are active in tumors, or whether their malfunction contributes to the poor perfusion of tumors.

The lymphatic system plays an important role in fluid homeostasis and immunosurveillance and may be complicit in immune response escape by tumors. Much research in the United States and Europe (especially the Swartz lab) is aimed at inducing the endogenous immune response against tumors. Recent studies suggest that the lymphatic system and the way it helps modulate the natural process of inflammation may help tumors escape the immune response. Further elucidation of these mechanisms may lead to therapies that restore immune response in tumors.

Because of the diverse areas in which fluid mechanics affects biology, there is potential for new and unique targets for cancer therapy. Identification of the fluid mechanosensors for flow-induced angiogenesis and invasion would be a major step in this direction. Similarly, determination of the mechanisms for blood flow regulation and vascular adaptation in tumors would allow modulation of tumor blood flow to enhance drug delivery or accomplish tumor starvation.

Unfortunately, relatively few research groups are focused on fluid dynamics in tumors, both in the United States and Europe. This is likely due to: (1) a lack of appreciation for the importance of fluid flow in tumors (both intravascular and extravascular) in the physics and engineering communities; (2) the absence of flow-activated gene pathways accessible to molecular biologists; and (3) the historical difficulty in studying fluid mechanobiology in tissues or *in vitro*. Recent advances in microfluidics and tissue analogs are providing solutions for (3); these new methodologies are already producing exciting and intriguing data that should provide a more firm foundation for others to enter this area, thus eventually solving problems (1) and (2).

References

Boardman, K.C., Swartz, M.A.: Interstitial flow as a guide for lymphangiogenesis. Circ. Res. **92**, 801–808 (2003)

Boryskina, O.P., Al-Kilani, A., Fleury, V.: Body plan in tetrapods: is it patterned by a hyperbolic tissue flow? Cell Cycle **10**, 3801–3802 (2011)

Boucher, Y., Leunig, M., Jain, R.K.: Tumor angiogenesis and interstitial hypertension. Cancer Res. **56**, 4264–4266 (1996)

Budd, G.T., Cristofanilli, M., Ellis, M.J., Stopeck, A., Borden, E., Miller, M.C., Matera, J., Repollet, M., Doyle, G.V., Terstappen, L.W.M.M., Hayes, D.: Circulating tumor cells versus imaging—predicting overall survival in metastatic breast cancer. Clin. Cancer Res. **12**, 6403–6409 (2006)

Hagendoorn, J., Tong, R., Fukumura, D., Lin, Q., Lobo, J., Padera, T.P., Xu, L., Kucherlapati, R., Jain, R.K.: Onset of abnormal blood and lymphatic vessel function and interstitial hypertension in early stages of carcinogenesis. Cancer Res. **66**, 3360–3364 (2006)

Hayes, D.F., Cristofanilli, M., Budd, G.T., Ellis, M.J., Stopeck, A., Miller, M.C., Matera, J., Allard, W.J., Doyle, G.V., Terstappen, L.W.M.M.: Circulating tumor cells at each follow-up time point during therapy of metastatic breast cancer patients predict progression-free and overall survival. Clin. Cancer Res. **12**, 4218–4224 (2006)

Jain, R.K.: Barriers to drug delivery in solid tumors. Sci. Am. **271**, 58–65 (1994)

Jain, R.K.: Delivery of molecular medicine to solid tumors. Science **271**, 1079–1080 (1996)

Jain, R.K.: Normalizing tumor vasculature with anti-angiogenic therapy: a new paradigm for combination therapy. Nat. Med. **7**, 987–989 (2001)

Jain, A., Munn, L.L.: Determinants of leukocyte margination in rectangular microchannels. PLoS One **4**, e7104 (2009)

Jeong, G.S., Han, S., Shin, Y., Kwon, G.H., Kamm, R.D., Lee, S.H., Chung, S.: Sprouting angiogenesis under a chemical gradient regulated by interactions with an endothelial monolayer in a microfluidic platform. Anal. Chem. **83**, 8454–8459 (2011)

Kirby, B.J., Jodari, M., Loftus, M.S., Gakhar, G., Pratt, E.D.: Functional characterization of circulating tumor cells with a prostate-cancer-specific microfluidic device. PLoS One **7**, e35976 (2012)

le Noble, F., Moyon, D., Pardanaud, L., Yuan, L., Djonov, V., Matthijsen, R., Breant, C., Fleury, V., Eichmann, A.: Flow regulates arterial-venous differentiation in the chick embryo yolk sac. Development **131**, 361–375 (2004)

Liao, S., Cheng, G., Conner, D.A., Huang, Y., Kucherlapati, R.S., Munn, L.L., Ruddle, N.H., Jain, R.K., Fukumura, D., Padera, T.P.: Impaired lymphatic contraction associated with immunosuppression. Proc. Natl. Acad. Sci. U. S. A. **108**, 18784–18789 (2011)

Nagrath, S., Sequist, L.V., Maheswaran, S., Bell, D.W., Irimia, D., Ulkus, L., Smith, M.R., Kwak, E.L., Digumarthy, S., Muzikansky, A., Ryan, P., Balis, U.J., Tompkins, R.G., Haber, D.A., Toner, M.: Isolation of rare circulating tumour cells in cancer patients by microchip technology. Nature **450**, 1235–1239 (2007)

Nieva, J., Wendel, M., Luttgen, M.S., Marrinucci, D., Bazhenova, L., Kolatkar, A., Santala, R., Whittenberger, B., Burke, J., Torrey, M., Bethel, K., Kuhn, P.: High-definition imaging of circulating tumor cells and associated cellular events in non-small cell lung cancer patients: a longitudinal analysis. Phys. Biol. **9**, 016004 (2012)

Padera, T.P., Kadambi, A., di Tomaso, E., Carreira, C.M., Brown, E.B., Boucher, Y., Choi, N.C., Mathisen, D., Wain, J., Mark, E.J., Munn, L.L., Jain, R.K.: Lymphatic metastasis in the absence of functional intratumor lymphatics. Science **296**, 1883–1886 (2002)

Pinho, M.C., Polaskova, P., Jennings, D., Wen, P.Y., Jain, R.K., Sorensen, A.G., Gerstner, E.R., Batchelor, T.: Impact of adjuvant anti-VEGF therapy on treatment-related pseudoprogression in patients with newly diagnosed glioblastoma receiving chemoradiation with or without anti-VEGF therapy. Paper presented at: 2012 ASCO annual meeting (2012)

Polacheck, W.J., Charest, J.L., Kamm, R.D.: Interstitial flow influences direction of tumor cell migration through competing mechanisms. Proc. Natl. Acad. Sci. U. S. A. **108**, 11115–11120 (2011)

Pries, A.R., Cornelissen, A.J., Sloot, A.A., Hinkeldey, M., Dreher, M.R., Hopfner, M., Dewhirst, M.W., Secomb, T.W.: Structural adaptation and heterogeneity of normal and tumor microvascular networks. PLoS Comput. Biol. **5**, e1000394 (2009)

Pries, A.R., Hopfner, M., le Noble, F., Dewhirst, M.W., Secomb, T.W.: The shunt problem: control of functional shunting in normal and tumour vasculature. Nat. Rev. Cancer **10**, 587–593 (2010)

Pries, A.R., Reglin, B., Secomb, T.W.: Modeling of angioadaptation: insights for vascular development. Int. J. Dev. Biol. **55**, 399–405 (2011)

Sasai, Y.: Cytosystems dynamics in self-organization of tissue architecture. Nature **493**, 318–326 (2013)

Sasai, Y., Eiraku, M., Suga, H.: *In vitro* organogenesis in three dimensions: self-organising stem cells. Development **139**, 4111–4121 (2012)

Schmid-Schonbein, G.W.: Mechanisms causing initial lymphatics to expand and compress to promote lymph flow. Arch. Histol. Cytol. **53**(Suppl), 107–114 (1990)

Shieh, A.C., Rozansky, H.A., Hinz, B., Swartz, M.A.: Tumor cell invasion is promoted by interstitial flow-induced matrix priming by stromal fibroblasts. Cancer Res. **71**, 790–800 (2011)

Shields, J.D., Fleury, M.E., Yong, C., Tomei, A.A., Randolph, G.J., Swartz, M.A.: Autologous chemotaxis as a mechanism of tumor cell homing to lymphatics via interstitial flow and autocrine CCR7 signaling. Cancer Cell **11**, 526–538 (2007)

Song, J.W., Munn, L.L.: Fluid forces control endothelial sprouting. Proc. Natl. Acad. Sci. U. S. A. **108**, 15342–15347 (2011)

Song, J.W., Bazou, D., Munn, L.L.: Anastomosis of endothelial sprouts forms new vessels in a tissue analogue of angiogenesis. Integr. Biol. (Camb.) **4**(8), 857–862 (2012)

Swartz, M.A., Lund, A.W.: Lymphatic and interstitial flow in the tumour microenvironment: linking mechanobiology with immunity. Nat. Rev. Cancer **12**, 210–219 (2012)

Vedula, S.R., Leong, M.C., Lai, T.L., Hersen, P., Kabla, A.J., Lim, C.T., Ladoux, B.: Emerging modes of collective cell migration induced by geometrical constraints. Proc. Natl. Acad. Sci. U. S. A. **109**, 12974–12979 (2012)

Versaevel, M., Grevesse, T., Gabriele, S.: Spatial coordination between cell and nuclear shape within micropatterned endothelial cells. Nat. Commun. **3**, 671 (2012)

Vickerman, V., Kamm, R.D.: Mechanism of a flow-gated angiogenesis switch: early signaling events at cell-matrix and cell-cell junctions. Integr. Biol. (Camb.) **4**, 863–874 (2012)

Wendel, M., Bazhenova, L., Boshuizen, R., Kolatkar, A., Honnatti, M., Cho, E.H., Marrinucci, D., Sandhu, A., Perricone, A., Thistlethwaite, P., Bethel, K., Nieva, J., Heuvel, M.V., Kuhn, P.: Fluid biopsy for circulating tumor cell identification in patients with early-and late-stage non-small cell lung cancer: a glimpse into lung cancer biology. Phys. Biol. **9**, 016005 (2012)

Wu, J.C., Tseng, P.Y., Tsai, W.S., Liao, M.Y., Lu, S.H., Frank, C.W., Chen, J.S., Wu, H.C., Chang, Y.C.: Antibody conjugated supported lipid bilayer for capturing and purification of viable tumor cells in blood for subsequent cell culture. Biomaterials **34**, 5191–5199 (2013)

Yeon, J.H., Ryu, H.R., Chung, M., Hu, Q.P., Jeon, N.L.: *In vitro* formation and characterization of a perfusable three-dimensional tubular capillary network in microfluidic devices. Lab Chip **12**, 2815–2822 (2012)

Yuan, F., Leunig, M., Huang, S.K., Berk, D.A., Papahadjopoulos, D., Jain, R.K.: Microvascular permeability and interstitial penetration of sterically stabilized (stealth) liposomes in a human tumor xenograft. Cancer Res. **54**, 3352–3356 (1994a)

Yuan, F., Salehi, H.A., Boucher, Y., Vasthare, U.S., Tuma, R.F., Jain, R.K.: Vascular permeability and microcirculation of gliomas and mammary carcinomas transplanted in rat and mouse cranial windows. Cancer Res. **54**, 4564–4568 (1994b)

Chapter 6
The Dynamics of Cell Motility

Owen McCarty

Introduction

Cell-matrix adhesions function as structural anchor points for the organization of the actin cytoskeleton. Integrin engagement regulates the activity of several members of the Rho family of small guanosine triphosphatases (GTPases), which mediate rapid changes in cytoskeletal dynamics. The activity of Rho GTPases is principally controlled by the balance between the binding of guanine nucleotide exchange factors (GEFs) or GTPase-activating proteins (GAPs). Three family members—Rac, Rho, and Cdc42—are well-studied regulators of cell motility (Burridge and Wennerberg 2004; Hall 1998; Machesky and Insall 1999). Rac induces the assembly of focal complexes and actin polymerization during the formation of lamellipodia. Rho induces the formation of stress fibers, whereas Cdc42 induces actin polymerization for the formation of filopodia (Fig. 6.1).

Rho GTPases act as molecular switches that cycle between GTP (guanosine triphosphate)- and GDP (guanosine diphosphate)-bound states (Fig. 6.2). When cells are stimulated by growth factors, bound GDP is exchanged for GTP. The GTPases are active in the GTP-bound state and interact with specific downstream effectors, leading to the translocation and activation of the effector and induction of reorganization of the actin cytoskeleton. Thus, the Rho family of small GTPases conveys unique biological effects through distinct effector proteins that act through different signaling pathways. These GTPases also play pivotal roles in reorganization of the actin cytoskeleton and cell movement in invading cancer cells.

Rho GTPases have been shown to have oncogenic activity and promote cancer cell invasion (Heasman and Ridley 2008; Vega and Ridley 2008). An increased

O. McCarty (✉)
Department of Biomedical Engineering, Oregon Health & Science University,
3303 SW Bond Ave, Portland, OR 97239, USA
e-mail: mccartyo@ohsu.edu

P. Janmey et al. (eds.), *Physical Sciences and Engineering Advances in Life Sciences and Oncology*, Science Policy Reports,
DOI 10.1007/978-3-319-17930-8_6

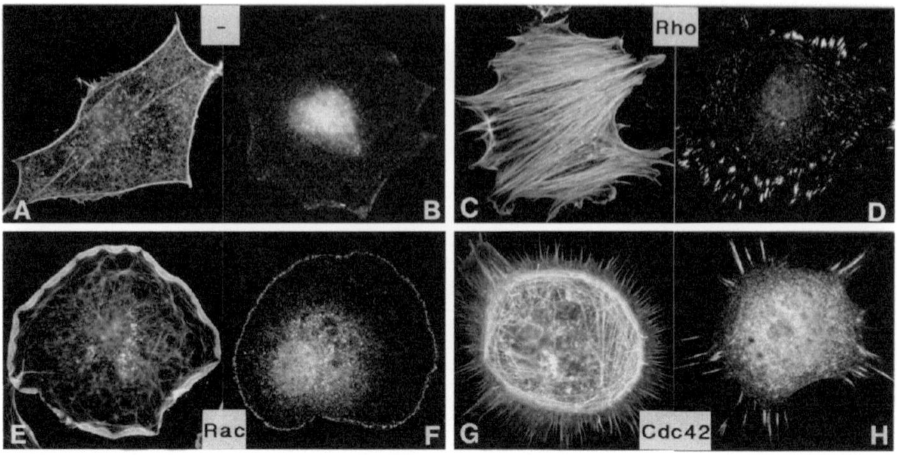

Fig. 6.1 Rho, Rac, and Cdc42 control the assembly and organization of the actin cytoskeleton (From Hall 1998)
Quiescent, serum-starved Swiss 3 T3 fibroblasts (–) contain very few organized actin filaments (**a**) or vinculin-containing integrin adhesion complexes (**b**). The effects of Rho, Rac, or Cdc42 activation in these cells can be observed in several different ways such as with the addition of extracellular growth factors, microinjection of activated GTPases, or microinjection of guanosine diphosphate (GDP)–guanosine triphosphate (GTP) exchange factors. Addition of the growth factor lysophosphatidic acid activates Rho, which leads to stress fiber (**c**) and focal adhesion formation (**d**). Microinjection of constitutively active Rac induces lamellipodia (**e**) and associated adhesion complexes (**f**). Microinjection of FGD1, an exchange factor for Cdc42, leads to formation of filopodia (**g**) and the associated adhesion complexes (**h**). Cdc42 activates Rac; hence, filopodia are intimately associated with lamellipodia, as shown in (**g**). In (**a**), (**c**), (**e**), and (**g**), actin filaments were visualized with rhodamine phalloidin; in (**b**), (**d**), (**f**), and (**h**), the adhesion complexes were visualized with an antibody to vinculin. Scale: 1 cm = 25 μm

expression and activity of Rho GTPases have been reported in a variety of cancers (Mardilovich et al. 2012). In contrast, other Rho GTPase family members appear to act as tumor suppressors and are mutated or downregulated in some cancers. The knowledge that Ras proteins are mutated in 30 % of human cancers of different origins suggests that the same might hold true for the Rho family of small GTPases (Lauth 2011). However, to date, no oncogenic mutations have been found in Rho proteins. Only one member of the Rho family of small GTPases has been reported to be genetically altered in human cancer (Mulloy et al. 2010). Apparently, mutational activation or inactivation of Rho proteins is not favorable for the initiation or progression of tumors. This has led to the hypothesis that disregulation of Rho GTPase signaling in cancer occurs at the level of expression or activation of Rho GTPases. A possible mechanism of GTPase deregulation in cancer is the altered expression and/or activity of their regulatory proteins, GEFs that promote formation of the active GTP-bound state and GAPs that return the GTPase to its GDP-bound inactive state.

Perhaps Rho GTPases play a role in regulating the ability of the physical microenvironment and mechanical forces to drive cancer? This chapter provides an

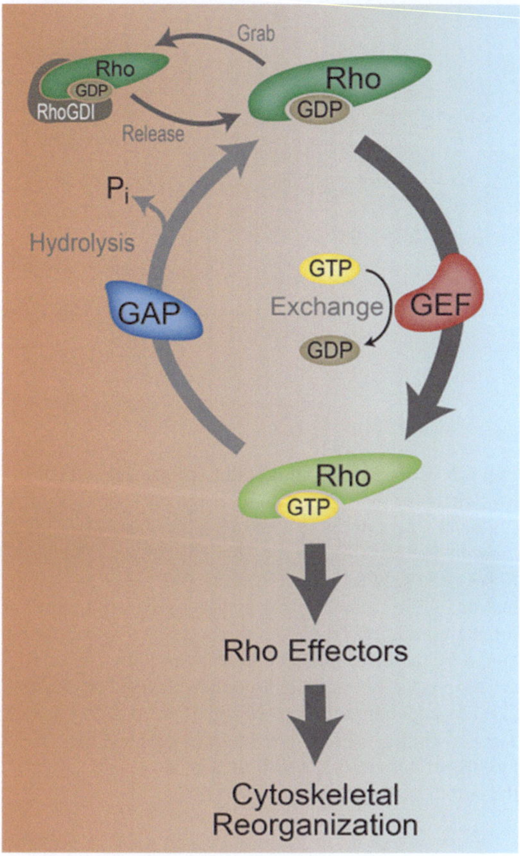

Fig. 6.2 Rho GTPase activation and regulation (From Aslan and McCarty 2013)
In their GTP-bound states, the Rho GTPases, including RhoA, Cdc42, and Rac1, associate with specific downstream effectors to regulate cytoskeletal remodeling events and platelet function. RhoA, Cdc42, and Rac1 are cyclically regulated by specific GEF and GAP proteins. GEFs promote GDP to GTP exchange to activate Rho GTPases. GAP proteins accelerate the hydrolysis of Rho-GTP to GDP and effectively inhibit Rho GTPase activation. Rho proteins may also be regulated by RhoGDI proteins, which "grab" and "release" Rho GTPases to sequester their activities

overview of the research being done to characterize the role that the physical and genetic drivers of cell motility play in cancer development and progression.

Research

The Cell Biology of Migration

Directed cell migration is a critical feature of various physiological and pathological processes, including development, wound healing, immunity, angiogenesis, and metastasis. Cell-matrix adhesions function as structural anchor points for the organization of the actin cytoskeleton, and are coupled to components of the actin-assembly machinery, such as actin-related protein 2/3 (Arp2/3) (Machesky and Gould 1999). In addition, integrin engagement regulates the activity of several members of the Rho family of small GTPases, which control the growth or contraction of filamentous actin (F-actin) fibers through Arp2/3 and myosin (Jaffe and Hall

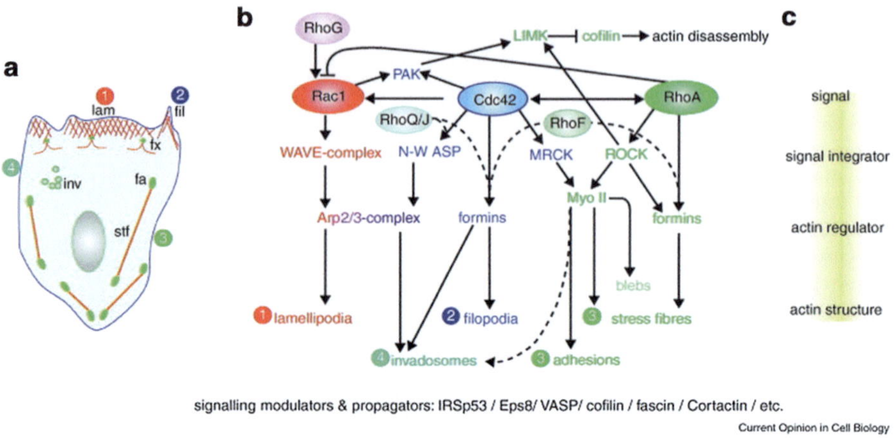

Fig. 6.3 Rho GTPase signaling to protrusion and adhesion (From Rottner and Stradal 2011) (**a**) Schematic cell migrating in a mesenchymal fashion while forming lamellipodia (1) filopodia (2), actin stress fibers (3), and invadosomes (4). (**b**) Actin cytoskeletal reorganization is regulated through the Rho family of small GTPases. The best studied members are Rac1, Cdc42, and RhoA. Rac-GTPases such as Rac1 (*red pathway*) are prominently involved in lamellipodium formation, while Cdc42 (*blue pathway*) drives filopodia formation. RhoA (*green pathway*) is crucial for the regulation of contractile actin arrays and stress fiber formation. Actin turnover is maintained by the actin filament severing/depolymerization activity of cofilin. Cofilin is constitutively active in cells, but inhibited, perhaps locally, by phosphorylation downstream of Rac1, RhoA, and Cdc42. (**c**) Schematic steps of signal propagation from active Rho-GTPase to the output structure. The active Rho-GTPase (signal) becomes engaged with a specific pathway by binding to an effector (signal integrator), which then binds to and activates a given actin regulator generating in turn the output actin structure

2005; Ridley 2001). Members of the Src family of tyrosine kinases (SFKs) also localize in cell-matrix adhesions (Huveneers and Danen 2009). In addition to regulating protein-protein interactions and augmenting cell-matrix adhesion turnover, SFKs that are downstream of integrins control GEFs and GAPs that act on the Rho-family small GTPases: Rac, Rho, and Cdc42 (Fig. 6.3, Ridley 2001).

Rac and Cdc42 induce protrusions in most cells by activating the WASp homologue (WH) domain-containing proteins neural WASp (N-WASp) and WAVE, which in turn induce actin polymerization by directly activating the Arp2/3 complex (Insall and Machesky 2009). Rac and Cdc42 also bind and activate the PAK Ser/Thr kinases (PAKs) (Bishop and Hall 2000). As a downstream effector of Rac, PAKs function in the regulation of the actin cytoskeleton, motility, and invasion. They are subdivided into PAK1-like (group I) or PAK4-like (group II) kinases. The team led by Dr. Ed Manser of A-Star, Singapore (site report, Appendix C) has recently shown that in contrast to group I kinases (which are inhibited in *trans* and requires new phosphorylation of Ser144 and Thr423 for activation), prototype group II kinase PAK4 is strongly auto-inhibited by an auto-inhibitory domain, and is activated by the binding of Cdc42, as illustrated in Fig. 6.4 (Baskaran et al. 2012). PAKs have multiple cytoskeletal targets, including LIM kinase, which is activated by PAK and

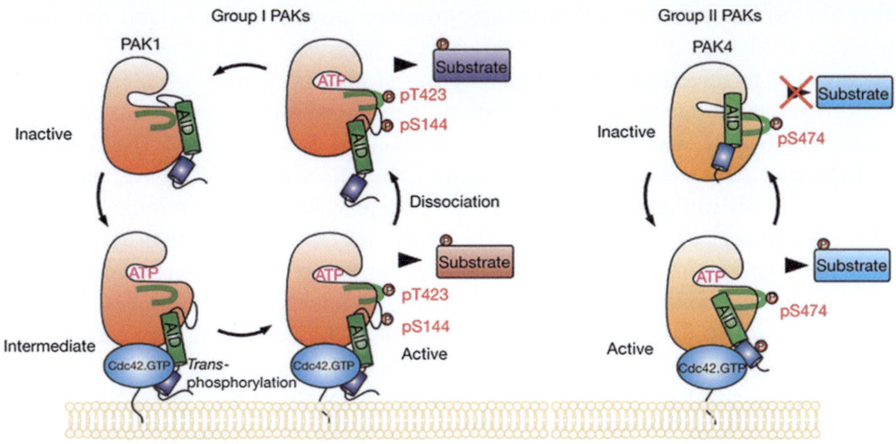

Fig. 6.4 Model of the alternate activation mechanisms for group I and II PAKs (From Baskaran et al. 2012)
Cdc42 binding in both cases results in relief from auto-inhibition, but through different underlying mechanisms. In the case of PAK1, this requires new phosphorylation of Ser 144 and Thr 423. In contrast, PAK4 is constitutively phosphorylated on Ser 474, and requires Cdc42.GTP to induce and sustain an active kinase conformation that allows substrate interaction

enhances actin polymerization by inactivating cofilin—a protein that disassembles actin filaments (Breitsprecher et al. 2011). PAK also activates myosin II by phosphorylating its regulatory light chains (RLC) (Parsons et al. 2010). PAKs have been shown to be overexpressed and/or hyperactivated in several human tumors, where altered cytoskeletal reorganization may promote growth, invasion, and metastasis (Chan and Manser 2012). Several efforts are underway to develop PAK-targeted therapeutics to selectively target certain types of tumors.

Rac and Cdc42 induce protrusions in most cells by activating the WASp homologue (WH) domain-containing proteins neural WASp (N-WASp) and WAVE, which in turn induce actin polymerization by directly activating the Arp2/3 complex (Insall and Machesky 2009). Rac and Cdc42 also bind and activate the PAK Ser/Thr kinases (PAKs; Bishop and Hall 2000). As a downstream effector of Rac, PAKs function in the regulation of the actin cytoskeleton, motility, and invasion. They are subdivided into PAK1-like (group I) or PAK4-like (group II) kinases. The team led by Dr. Ed Manser of A-Star, Singapore (site report, Appendix C) has recently shown that in contrast to group I kinases (which are inhibited in *trans* and requires new phosphorylation of Ser144 and Thr423 for activation), prototype group II kinase PAK4 is strongly auto-inhibited by an auto-inhibitory domain, and is activated by the binding of Cdc42, as illustrated in Fig. 6.4 (Baskaran et al. 2012). PAKs have multiple cytoskeletal targets, including LIM kinase, which is activated by PAK and enhances actin polymerization by inactivating cofilin—a protein that disassembles actin filaments (Breitsprecher et al. 2011). PAK also activates myosin II by phosphorylating its regulatory light chains (RLC) (Parsons et al. 2010). PAKs have been shown to be overexpressed and/or hyperactivated in several human tumors, where altered cytoskeletal reorganization may promote growth, invasion, and metastasis (Chan and Manser 2012). Several efforts are underway to develop PAK-targeted therapeutics to selectively target certain types of tumors.

Rho activation leads to the maturation of focal adhesions through its ability to activate myosin II, which promotes adhesion maturation and stability (Huttenlocher and Horwitz 2011). Rho activates myosin through ROCK1 and ROCK2, which act mainly by inactivating a subunit of myosin phosphatases, thus sustaining myosin II RLC phosphorylation (Kaibuchi et al. 1999). Together, these pathways regulate actin polymerization, bundling, and adhesion formation. While a system of tyrosine kinases and effectors immediately downstream of integrins and receptors has been extensively investigated (Liu et al. 2009), little is known about how more distal signals interact and communicate with one another to regulate the spatial and temporal activation of Rho GTPases as well as their respective GEFs or GAPs. Current studies in the field are focused on defining how the extracellular matrix (ECM) microenvironment regulates these signaling networks to control actin polymerization and cytoskeletal reorganization and migration.

The Dynamics of Mechanobiology

Physical interactions between cells and the ECM are predominantly mediated by integrins. Integrins are heterodimeric αβ transmembrane receptors that connect the ECM to the cytoskeleton (Yamada and Geiger 1997). In mammals, 18 α-subunits and 8 β-subunits assemble into 24 different integrins, which bind collagens, laminins, or RGD (arginine-glycine-aspartic acid)-containing proteins, such as fibronectin (Ross 2004). Many integrins are known to adopt low-affinity, intermediate-affinity, and high-affinity conformations and exist in a dynamic equilibrium with one another (Cary et al. 1999). An increase in the proportion of heterodimers adopting high-affinity conformations is termed integrin activation, and can be induced either by cytoplasmic events (inside-out activation), or by extracellular factors (outside-in activation). Ligand-binding triggers integrin clustering (avidity), integrin connection to the cytoskeleton, and the formation of adhesion complexes (Cary et al. 1999). Moreover, integrin-ligand interactions induce a multifaceted cascade of ‘outside-in’ events such as cell spreading and migration, ECM assembly, and the activation of signal transduction pathways that regulate proliferation, survival, and gene expression (Aplin et al. 1998).

Upon binding, integrins gather together in membrane-specific regions and recruit several coupling proteins through their intracellular domains, which form the focal adhesion complex (Fig. 6.5) (Huttenlocher and Horwitz 2011). Among these proteins, focal adhesion kinase (FAK), a non-receptor tyrosine kinase, has a central role in mediating integrin-dependent mechanotransduction through focal adhesion contact. FAK and Src, another non-receptor tyrosine kinase, control the recruitment of adaptor proteins that then behave as molecular hubs for signal transduction (Etienne-Manneville and Hall 2002). Through these proteins, integrin engagement

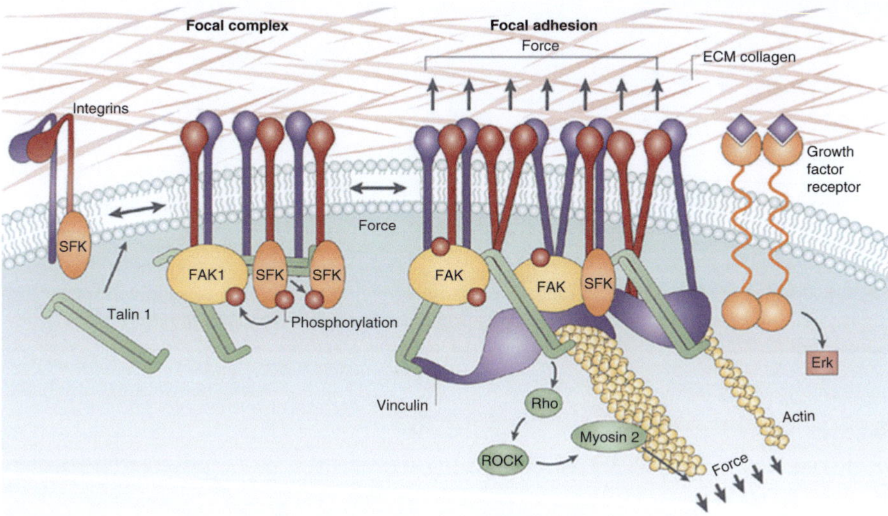

Fig. 6.5 Integrins and extracellular matrix in mechanotransduction (Butcher et al. 2009) Integrin activation leads to receptor clustering, signaling downstream of Src family kinases (*SFK*), formation of focal adhesions, and stimulation of Rho GTPase-dependent actin assembly

results in the activation of a vast spectrum of signaling pathways involving protein tyrosine kinases, small GTPases of the Rho family, as well as the activation of JNK, ERK, and PI3K/Akt modules (Jean et al. 2011). Strong evidence exists supporting the notion that not only FAK, but also integrin adaptors, have a role in cell transformation, tumor progression, and chemoresistance (Schaller 2010). Current studies are focused on defining the mechanisms by which integrin signaling pathways regulate the spatial-temporal dynamics of cellular mass, volume, and density in response to the ECM microenvironment.

The global adhesive forces of cells are regulated by a series of individual bonds, each which may exhibit exceedingly high bond strengths (Ingber 2003). While this explains the process of cell adhesion, it is unclear whether the dissociation of receptor-ligand interactions required for cell motility is chemical or physical by nature. Dr. Ana-Sunčana Smith from the University of Nurnberg-Erlangen (site report, Appendix B) recently modeled the effective binding affinity of a cell membrane (Fig. 6.6) (Reister-Gottfried et al. 2008). This model predicts that adhesion is a dynamic transition from nucleation of receptors in a free membrane to thermodynamically-regulated bond formation followed by saturation of receptor-ligand bonds. These simulations have been verified experimentally to show that intrinsically strong bonds can exhibit ultraweak adhesion mediated by transiently bound domains and can undergo a transition to a stable strong adhesion by locally increasing bond density (Fig. 6.7). These results and simulations provide a physical sciences perspective on the mechanism by which the cell migration through the extracellular matrix is thermodynamically favored, despite the fact that the bond strength of each individual receptor-ligand interaction is exceedingly high.

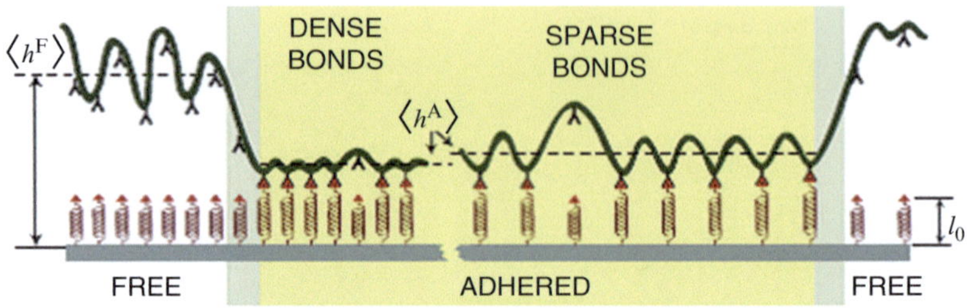

Fig. 6.6 Model of adhering membrane. The effective binding affinity of a membrane for a surface is governed by the receptor density, (*h*) separation distance, and (*l*) receptor length (From Reister-Gottfried et al. 2008)

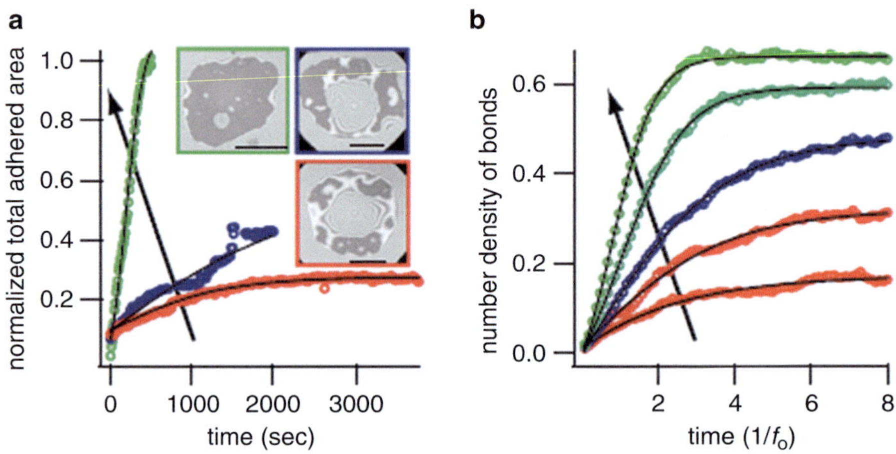

Fig. 6.7 Normalized total adhered area and number density of bonds (From Reister-Gottfried et al. 2008)
(**a**) Adhered area (normalized by the equilibrium contact zone area) versus time for vesicles on substrates with high, middle, and low E-selectin densities. For vesicles of comparable size halving the concentration of E-selectin approximately doubles the equilibration time. Scale bar: 10 μm. (**b**) Average number of bonds in time in sets of 200 simulation runs for $\exp(\upsilon_0/\kappa_B T) = 3.0, 3.25, 3.5, 3.9$, and 4.5. The directions of growing E-selectin density and υ_0 are shown with arrows

The group led by Dr. Joachim Spatz of the University of Heidelberg (site report, Appendix B) has developed an innovative strategy to experimentally characterize the role of spatial organization of the receptor-ligand interactions on cell morphology (Cavalcanti-Adam et al. 2006). This technique relied on the ability to control the spacing of RGD peptides on a 2D surface in order to control integrin-mediated cell adhesion at the single-receptor level (Fig. 6.8). Their work demonstrated that degree of cell adhesion and spreading is dependent on the physical spacing of the RGD ligands (Fig. 6.9). Moreover, the size and density of integrin clusters is decreased in cells plated on RGD surfaces at distances larger than 110 nm. This approach is being used to characterize the role that spatial organization of the extracellular environment plays in regulating cell-ECM interactions.

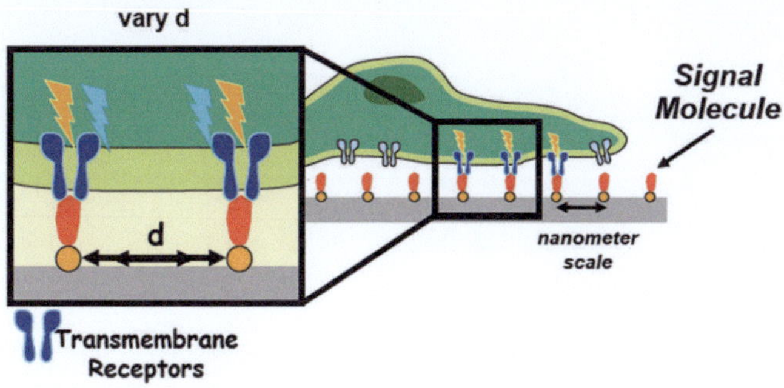

Fig. 6.8 Experimental design to characterize spatially and temporally coordinated migration (From Cavalcanti-Adam et al. 2006)

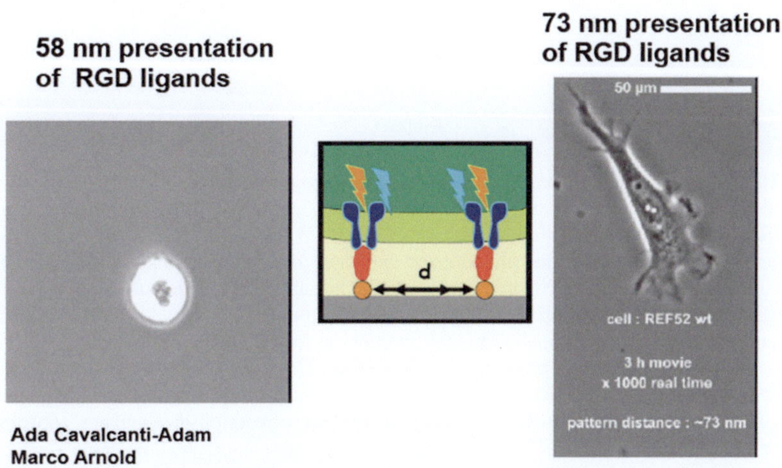

Fig. 6.9 Lateral spacing of integrin ligands regulates cell spreading and focal adhesion assembly (From Cavalcanti-Adam et al. 2006)

The Physics of Actin Dynamics

Actin filaments in cells are assembled into extended structures such as branched networks and bundles (Fig. 6.10). Branched networks often occur in lamellipodia, which are broad, thin protrusions extending at the front of a cell. The branched networks have a polarized structure with barbed ends preferentially pointing toward the membrane. The filament length is typically a fraction of a micron. Assembly of branched networks is aided by the Arp2/3 complex, a seven-protein complex that binds to the sides of existing filaments and creates new daughter branches (Machesky and Gould 1999). Because the rate of new branch formation is proportional to the number of existing branches, filament nucleation by branching is an autocatalytic process. Additional mechanical stability is provided by cross-linking proteins. Polymerization occurs mainly near the membrane and depolymerization occurs

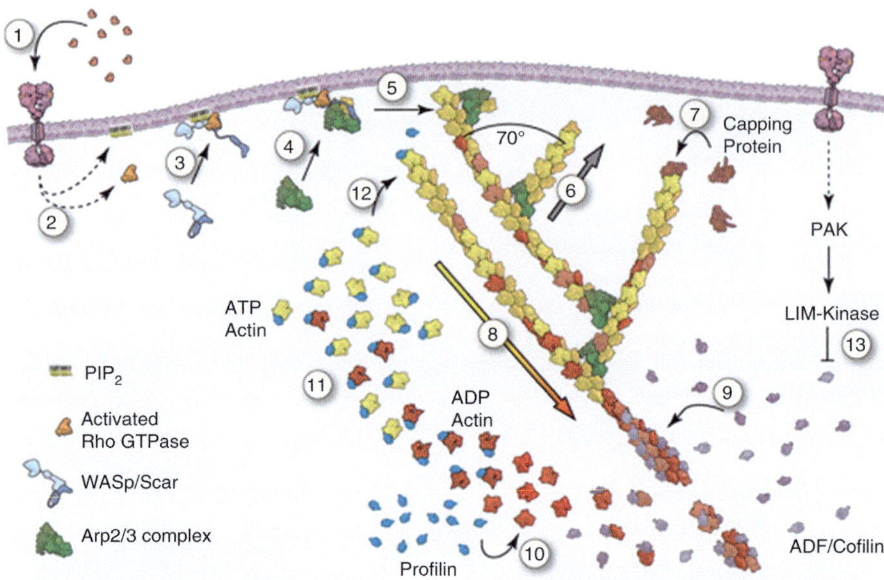

Fig. 6.10 Model of actin cytoskeleton rearrangement (From Pollard et al. 2000)
In the dendritic nucleation model, (*1*) several signaling pathways (*2*) converge to (*3*) activate WASp/Scar proteins, which in turn (*4*) activate the Arp2/3 complex. (*5*) The active Arp2/3 complex binds to the side of an existing filament and nucleates (*6*) new filament growth towards the cell membrane, while filaments age by hydrolysis of ATP bound to each actin subunit (white subunits turn yellow) followed by dissociation of the phosphate (subunits turn red) (*8*). The combined force from many growing filaments pushes the cell membrane forward, moving the cell. Actin binding proteins, including (*7*) capping protein, (*9*) ADF/cofilin, (*10*) profilin, tropomyosin, formins, and Ena/VASP modulate these events. While we know how many of these proteins modify actin filaments on their own, we do not fully understand how these reaction mechanisms coordinate to organize the actin network at the leading edge or how signaling molecules affect that organization

mainly away from the membrane. The assembly of branched actin networks is often signaled by upstream external agents such as growth factors (Rowinsky 2003). These activate receptors at the cell surface, which leads to activation of actin-binding proteins such as Arp2/3 complex. Disassembly of actin networks involves both severing of and depolymerization of actin filaments, but the balance between these two processes is unclear. Disassembly is typically viewed as occurring in the absence of active intervention. However, since proteins which sever actin filaments, like cofilin (Oser and Condeelis 2009), can be activated from the membrane, such active intervention is a possibility. Bundles are composed of tightly cross-linked filaments and occur in protrusions such as filopodia.

While the biophysical parameters of actin assembly have been well-established, it is unclear how the dynamics of actin assembly are coordinated to drive cell motility. The recent work led by Dr. Andreas Bausch of the Technical University of Munich, Germany (site report, Appendix B), has demonstrated a mechanism by which emergent properties of actin networks are coordinated to drive collective

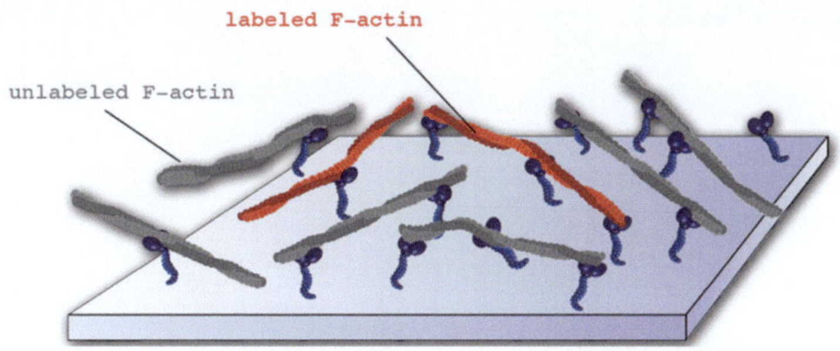

Fig. 6.11 Experimental model of the collective dynamics of actin cytoskeletal networks (Courtesy of A. Bausch, Technical University of Munich)

dynamics (Kohler et al. 2011). They show the emergence of collective motion in a high-density motility assay that consists of highly concentrated actin filaments propelled by immobilized molecular motors in a planar geometry. Above a critical density, the filaments self-organize to form coherently moving structures with persistent density modulations, such as clusters, swirls, and interconnected bands (Fig. 6.11). These polar nematic structures are long-lived and can span length scales that are orders of magnitudes larger than their constituents. The group's work highlighted the existence of absorbant states in natural systems. Specifically, they used a 2D system of actin filaments to demonstrate a combination of active directed motion and steric repulsion which caused the system to produce dynamic patterns in the form of density fluctuations and waves. These were short-lived structures that appear and disappear (Fig. 6.12). The system had similarities to other fluctuating nonequilibrium liquid states. When a critical amount of protein is reached, cross-linking between actin filaments drove the system to self-organize into a distinct moving state characterized by all the hallmarks of an absorbed or dynamically frozen state.

Measurement Science of Cell Migration in 3D

Receptors in the plasma membrane signaling to small GTPases such as Rho, Rac, and Cdc42 can trigger nucleation of new actin filaments via specific downstream pathways to generate either branched or linear filament arrays (Machesky and Gould 1999; Machesky and Insall 1999). It is now clear that the mechanotransduction of actin-based processes is multidimensional and that no simple linear relationship exists between activation of a GTPase and any single cellular structure output. While lamellipodia and filopodia are seen as the main types of protrusions that cells produce when moving on 2D surfaces *in vitro* (Arjonen et al. 2011), other structures are perhaps more important for motility *in vivo* (Pinner and Sahai 2008). For instance, in most tumors, several types of cell migration are observed

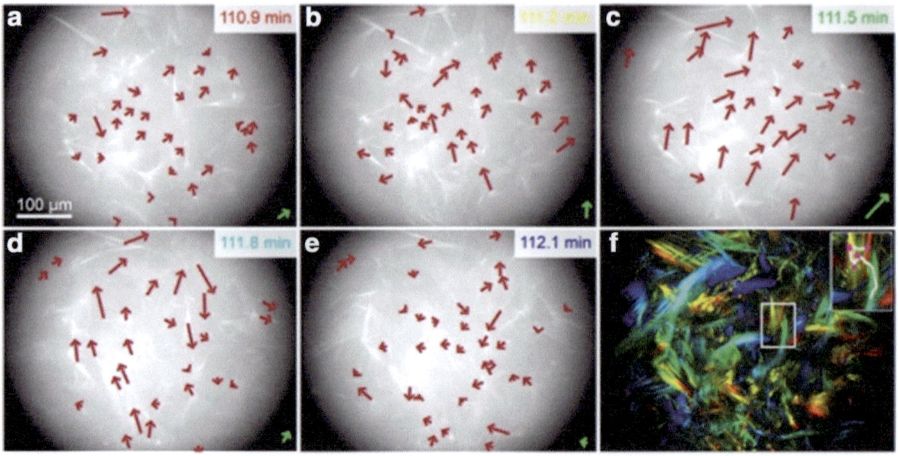

Fig. 6.12 Structure and dynamics of actin/fascin/myosin networks (From Kohler et al. 2011) (**a-e**) Fluorescence micrographs of active actin networks at indicated times after initiation of polymerization show the dynamic reorganization within the network. *Red arrows* (*dark arrows*) indicate the movement of individual points in the network with lengths 20-fold magnified and a time-average over three frames. *Green arrows* (*outside the circle*) show the resulting overall movement in the field of view (lengths are 40-fold magnified). (**f**) These long-range reorganization processes are summarized in the colored time overlay (*red* to *blue*). The connectivity between the structures is higher than what can be seen in these fluorescence micrographs. The trajectory of an individual structure exhibits persistent runs (inset, *white line*) intermitted by stall periods (inset, *magenta* (*dark*) *dots*). These run and stall phases are not only observed for individual structures, but also the whole network in the field of view exhibits pulsatile collective dynamics with movements being coordinated in time and direction: (**a**, **e**) In stall phases, individual structures move predominantly for only short periods and in random directions. The stall phases are followed by periods with high activity in the entire network. (**b**, **c**) During these run phases, most of the network shows long, persistent runs

simultaneously. In epithelial tumors, in which cell-cell junctions are present, cells move together in sheet-like structures (Gray et al. 2010; Khalil and Friedl 2010). This mode of migration is observed in epithelial cells during wound repair and in endothelial cells during angiogenesis (Trepat and Fredberg 2011). As dedifferentiation of epithelial cancer cells proceeds, the function of cadherin, a cell-cell junction protein, is suppressed and individual cells separate from the other cells and move individually. These modes of migration are called collective and individual cell migration, respectively. The transition from collective to individual migration is termed epithelial-mesenchymal transition and is a well-studied indicator of tumor progression (Thompson et al. 2005). Thus, cancer cells show distinct modes of cell migration according to differentiation states.

To elucidate the mechanisms by which tumor cells acquire an invasive phenotype, 3D *in vitro* assays have been developed that mimic the process of cancer cell invasion through a basement membrane or in the stroma. The group led by Dr. Laura Machesky at the Beatson Institute for Cancer Research, United Kingdom (site report, Appendix D) has developed a 3D circular invasion assay. They found that it provides a simple and amenable system to study cell invasion within a matrix in an environment that closely mimics 3D invasion (Fig. 6.13, Yu and Machesky 2012). The group led by Cynthia Reinhart-King at Cornell University in the United States has extended the use of 3D *in vitro* assays to demonstrate that the biophysical microarchitecture of the matrix regulates tumor cell behavior (Carey et al. 2012). In particular, they demonstrated that cell-scale gel microarchitecture is the predominate feature as compared to bulk-scale gel density in controlling specific cell behaviors. When studied in 3D, cancer cells typically show two types of morphologies. One is an elongated morphology similar to that of fibroblasts and the other is a rounded morphology. These two morphologies use different migratory mechanisms; elongated cells undergo mesenchymal migration and rounded cells undergo amoeboid migration (Parisi and Vidal 2011). At one extreme, mesenchymal cell migration is characterized by single cell motility and a multistep cycle of protrusion, adhesion formation, and stabilization at the leading edge followed by cell body translocation and release of adhesions and detachment of the cell's rear. Motile fibroblasts and some cancer cells show organized adhesion structures and can exert substantial contractile forces on the ECM (van Zijl et al. 2011). Mesenchymal migration in 3D tissues is associated with the degradation of ECM and regulated extracellular proteolysis. Amoeboid migration lies at the other extreme and is characterized by gliding and rapid migration (Harunaga and Yamada 2011); it is the primary mode of migration for highly motile cells including neutrophils, dendritic cells, and lymphocytes. These cells exert relatively weak integrin-mediated traction forces on the surrounding substrate and can even be integrin-independent. For example, work done at the Max Planck Institute in Germany by Lämmermann and colleagues showed that migration of dendritic cells in interstitial tissues does not require integrins (Lammermann et al. 2008), despite the fact that integrins are used for motility over some 2D surfaces. Although integrin-induced traction forces are generally weak during amoeboid cell motility, weak integrin-mediated adhesions likely modulate the fast gliding motility of amoeboid cells. An extreme kind of integrin-independent 3D migration can be shown by amoeboid cells, including dendritic cells (Lammermann et al. 2008). In this mode, the cells move via a blebbing type of migration driven by cortical actin cytoskeletal tension.

Coordination of Cell Motility by Cooperative Intracellular Forces

The migration of single cells is the best-studied mechanism of cell movement *in vitro* and is known to contribute to many physiological motility processes *in vivo*, such as development, immune surveillance, and cancer metastasis. Single cell

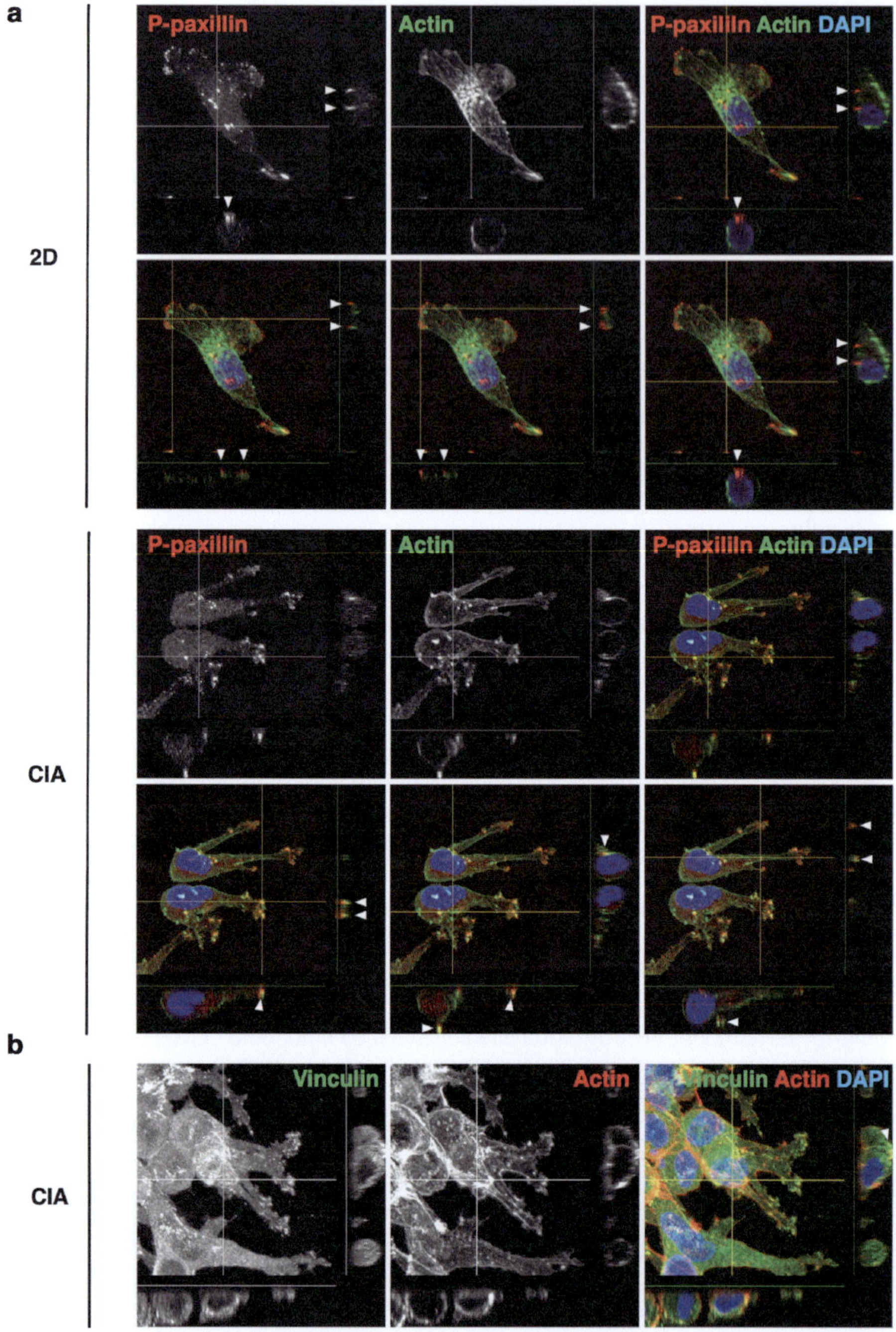

Fig. 6.13 Actin cytoskeletal and focal adhesion organization in MDA-MB-231 cells invading in modified circular invasion assay (CIA) (From Yu and Machesky 2012)
(**a**) Cells in wound healing assay without Matrigel on 2D surface and cells in CIA with Matrigel overlay were fixed and stained for actin (*green*, *outermost areas*), focal adhesion marker phospho-paxillin

migration allows cells to position themselves in tissues or secondary growths, as they do during morphogenesis, or to transiently pass through the tissue, as shown by immune cells. Collective migration is the second principal mode of cell movement. This mode differs from single cell migration in that cells remain connected as they move, which results in migrating cohorts and varying degrees of tissue organization. The molecular principles of actin turnover and polarized force generated by moving cell groups are similar to those in the migration of individual cells, but they are shared and coordinated between cells at different positions. The cortical actin network in the cell group shows supracellular organization, such that anterior protrusion activities and posterior retraction dynamics involve many cells working together. The mechanisms of supracellular cytoskeletal organization are not clear, but they probably reside in the combined actions of cadherin- and gap-junctional cell-cell coupling, as well as in the paracrine release of cytokines and growth factors.

The team of Drs. Yamao, Naoki, and Ishii from the Nara Institute of Science & Technology, Kyoto University, and RIKEN Computational Science Research Programs, respectively, in Japan (site report, Appendix C) have recently developed an *in silico* model of multi-cellular migration (Yamao et al. 2011). Their model predicts that collective cell migration is driven by two components: a deterministic force due to interactions between cells and a random force arising from Brownian motion (noise). Collective cell migration emerges when strong noise from migrating cells is coupled with weak forces from surrounding cells, whereas strong noise from non-migratory cells causes dispersive migration (Fig. 6.14).

The group led by Dr. Xavier Trepat at the University of Barcelona (site report, Appendix B) has developed a novel technique to characterize the mechanical stresses within the cell body and at cell-cell boundaries during migration (Tambe et al. 2011). This technique can generate high-resolution maps of stress components within an advancing monolayer sheet of cells (Trepat and Fredberg 2011). Their results demonstrate that local cellular trajectory follows local stress fields (Fig. 6.15) and that individual cells tend to migrate and remodel to maintain minimal intercellular shear stress. They have extended these studies to define the process termed "kenotaxis," whereby the cellular collective generates local traction forces pulling systematically and cooperatively towards unfilled spaces (Kim et al. 2013). This cell-patterning motif may be the underlying physiological principle that regulates collective cell migration during wound healing, development, or cancer cell migration and invasion. Future work is needed to define the role of Rho GTPases in regulating collective cell migration along orientations of minimal intercellular shear stress.

Fig. 6.13 (continued) (*red, extremity border*), and DNA (*blue, center*, nucleus). Z-stack confocal-images were captured and cell side views are shown to indicate positions of FA/FCs. White arrowheads indicate adhesion complexes. (**b**) Cells invading in CIA were fixed and stained with actin (*red, center panel*), focal adhesion marker vinculin (*green, left panel*), and DNA (*blue, right panel—center,* nucleus). Z-stack confocal images were captured and cell side views are shown to indicate positions of FA/FCs. White arrowheads indicate adhesion complexes

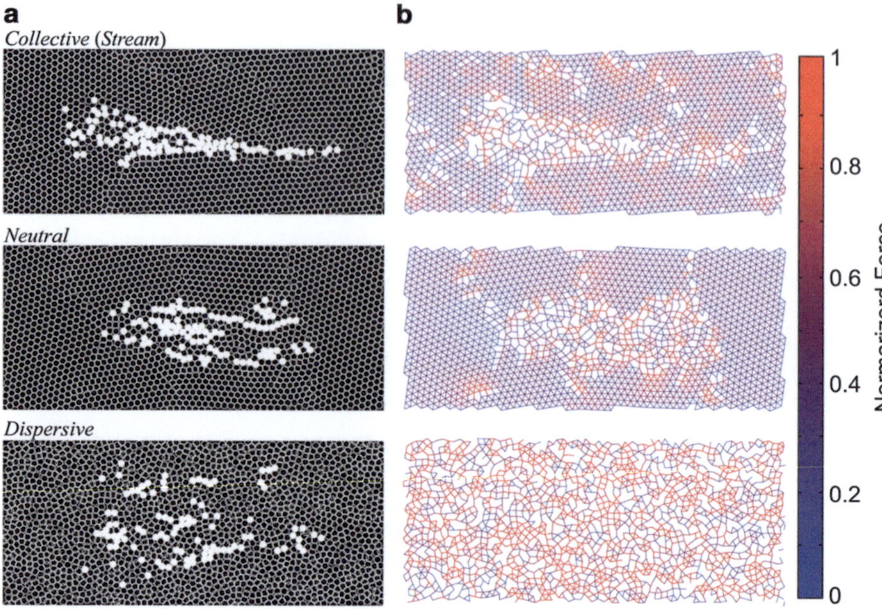

Fig. 6.14 Snapshots of migration patterns, cell contacts, and the migratory cell population (From Yamao et al. 2011)
(**a**) Migration patterns at a specific point in time. The *white* and *black circles* indicate migratory and non-migratory cells, respectively. (**b**) Cell contacts shown at the same time point as in (**a**). The links depict contacts between cells that interact by repulsive elastic forces, the strengths of which are indicated by their brightness (*red*), or darkness (*blue*)

Collective migration of cohesive cell groups *in vivo* is particularly prevalent during embryogenesis and drives the formation of many complex tissues and organs. The group led by Dr. Jerome Solon of the University of Barcelona (site report, Appendix B) has developed a novel approach to characterize collective cell migration *in vivo* during embryogenesis (Fig. 6.16) (Solon et al. 2009). They utilize the dorsal closure as a model system—a morphogenetic movement occurring at a late stage of *Drosophila* gastrulation. By using an integrative platform of high-resolution imaging, automated image processing, and physical modeling, they were able to characterize phenomena such as cell pulsing during embryogenesis. This showed analogs that tension-based dynamics and cell coupling control the force pulses that drive dorsal closure in the developing embryo.

With the recent development of photoactivatable analogue of Rac by Wang and colleagues in the United States at Johns Hopkins University (Wang et al. 2010), which allows for the rapid and reversible local activation or inactivation of Rac, the role of Rac in driving collective cell migration was recently shown in a *Drosophila* ovary model. This work demonstrated a key role for the spatial asymmetry of Rac

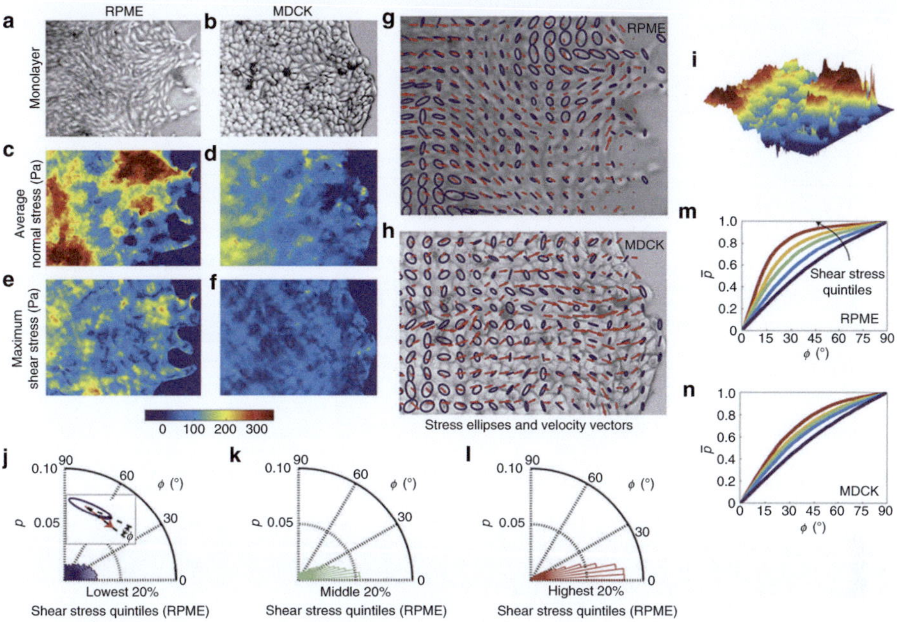

Fig. 6.15 Collective cell migration (From Tambe et al. 2011)
(**a**) Transmitted light image of the rat pulmonary microvascular endothelial (*RPME*) cell monolayer and (**b**) the Madin-Darby canine kidney (*MDCK*) cell monolayer. (**c, d**) Corresponding to these images are the maps of average normal stress, (**e, f**) maximum shear stress, and (**g, h**) principal stress ellipses (*blue circles and ellipses*) and cell velocity vectors (*red dashes and points*). (**i**) The map of average normal stress for the RPME cell monolayer is predominately tensile, but forms a rugged stress landscape. (**j-l**) The alignment angle, φ, between the major axis of the principal stress ellipse and the direction of the cellular motion (*j, inset*) shows that the greater the local maximum shear stress the narrower the distribution of φ. (**m**) The cumulative probability distribution varied strongly and systematically with stress anisotropy; curves, from blue (*bottom*) to red (*top*), are in the order of higher quintiles. (**n**) The cumulative probability distribution for the MDCK cell monolayer is also shown

activity for direction sensing, highlighting the role of physical parameters in regulating function.

PAK kinases have been implicated in the regulation of many of the hallmarks of cancer, including motility, growth, and apoptosis, making this family of serine/threonine protein kinases an attractive therapeutic target. The rationale for developing PAK inhibitors as cancer therapeutics is supported by studies documenting the correlation of increased cytoplasmic levels of PAK1 with recurrence rate and mortality of breast cancer (Bostner et al. 2007). Along these lines, the group led by Dr. Ching at the University of Hong Kong, China, demonstrated that PAK1 is frequently overexpressed in human hepatocellular carcinoma (HCC) and was that PAK1 overexpression was associated with more aggressive tumor behavior (Ching et al. 2007).

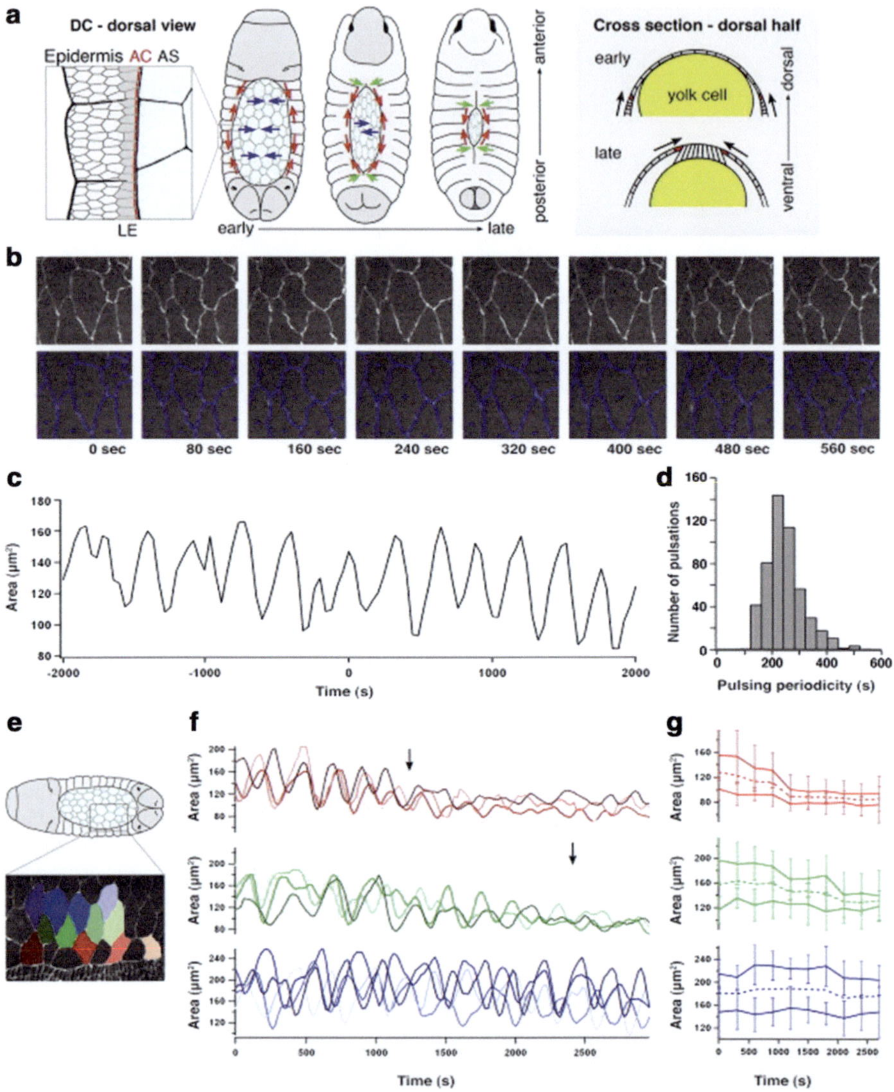

Fig. 6.16 Collective cell dynamics during dorsal closure (DC) (From Solon et al. 2009)
(**a**) Cartoon of DC embryos. *Colored arrows* depict forces produced by AS cells (*blue, two center images—central arrows*), AC (*red, arrows* following the curve of the ellipse), and zippering (*green, arrows* pointing to the median). *Black arrows* show the direction of LE movement. (**b**) Typical apical surface area pulsations of an AS cell in a GFP-Arm expressing embryo. The upper panel shows raw data, the lower panel shows the superimposed segmented image. (**c**) Apical cell surface fluctuations of an AS cell. Time point zero depicts the approximate onset of dorsal closure. (**d**) The period distribution of 505 pulses measured in 35 AS cells in two embryos is shown. The distribution is narrow and centered at 230 ± 76 s. (**e**) Image of a GFP-Arm-expressing embryo showing the epidermis (small cells at the *bottom*), the LE, and part of the AS tissue. AS cells are color-coded depending on their distance from the LE; the ventral-most cell row in *red*, the second row

The Ching group has recently translated this idea to demonstrate that an allosteric inhibitor of PAK1, IPA-3, is able to inhibit the growth and motility of HCC cells *in vitro* as well as suppress *in vivo* growth of HCC cells in a mouse xenograft model (Fig. 6.17) (Wong et al. 2013). Although IPA-3 is not stable or potent enough to be used in clinical trials, it is a promising lead for the development of effective inhibitors for Group I PAKs for the treatment of cancer.

Discussion

This chapter has given an overview of the research being done in Europe and Asia to define the role of physical parameters in the regulation of cell motility and migration. Understanding the mechanisms that govern collective cell migration in contrast to single cell motility may lead to the development of strategies to either suppress or enhance collective cell function in health and disease. Research is being done in Europe and Asia at length and time scales that span single molecule reactions to macroscopic cell migration *in vivo*. During our site visits, we witnessed the synergy between basic scientists and theorists and cell biologists that has led to the development of novel techniques to study the role of cell motility, cell-cell adhesion, signaling, and ECM remodeling to drive collective cell migration. In Europe, we observed a greater degree of multidisciplinary teamwork—and more importantly, infrastructure to house multidisciplinary teams in the same physical location—than is typically observed in the United States. A dynamic research environment exists in Asia, with multiple approaches to supporting team-based science. At the forefront of research in Asia is the development and utility of novel imaging and measurement science approaches to study basic cell biology questions in cancer research. Accordingly, the United States must continue to develop and support multidisciplinary teamwork in order to make scientific advances in understanding the pathophysiology of diseases such as cancer. A greater degree of emphasis on basic, mechanistic research was observed in Europe, as compared to the translational approach that is being emphasized in the United States. This has given the United States an advantage in the rational design and development of novel therapeutic approaches to treat and prevent disease such as cancer. Continued support of translational medicine in the United States is key to developing the next generation of therapeutics to produce better outcomes for cancer patients.

Fig. 6.16 (continued) in *green*, and the third row in *blue*. (**f**) Analysis of the apical surface fluctuations of the AS cells highlighted in (**e**). Cells sequentially cease pulsing (*indicated by arrows*): cells contacting the LE (*red*) are first (*top*), followed by the cells in the second row (*green, middle*). Cells in the third row (*blue*) continue pulsing throughout the analyzed time period (*bottom*). (**g**) The mean of the apical surface maxima and minima is shown for the different rows of AS cells in (**f**). The minima remain virtually constant over time while the maxima decrease sequentially. Consequently, the average surface (*dashed lines*) decreases mainly as function of the maxima reduction

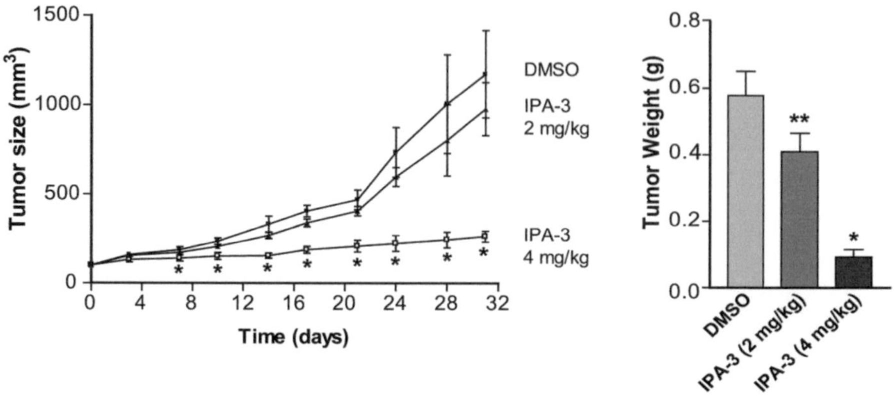

Fig. 6.17 The suppressive effect of PAK inhibition in a nude mouse xenograft model of hepatocellular carcinoma (From Wong et al. 2013)
MHCC97L cells were used for the xenograft model. Mice were treated three times weekly either with DMSO or the PAK1 inhibitor, IPA-3 (2 mg/kg or 4 mg/kg, i.p.). *P<0.001 (ANOVA) compared with the DMSO control group. Graph on the left demonstrates the tumor size, in mm^3, for the different treatments. Tumor weights were measured at the end of study, and shown in the graph on the right. *P<0.001, **P<0.01, (ANOVA) compared with the DMSO control group

References

Aplin, A.E., Howe, A., Alahari, S.K., Juliano, R.L.: Signal transduction and signal modulation by cell adhesion receptors: the role of integrins, cadherins, immunoglobulin-cell adhesion molecules, and selectins. Pharmacol. Rev. **50**, 197–263 (1998)

Arjonen, A., Kaukonen, R., Ivaska, J.: Filopodia and adhesion in cancer cell motility. Cell Adh. Migr. **5**, 421–430 (2011)

Aslan, J.E., McCarty, O.J.T.: Rho GTPases in platelet function. J. Thromb. Haemost. **11**, 35–46 (2013)

Baskaran, Y., Ng, Y.W., Selamat, W., Ling, F.T., Manser, E.: Group I and II mammalian PAKs have different modes of activation by Cdc42. EMBO Rep. **13**, 653–659 (2012)

Bishop, A.L., Hall, A.: Rho GTPases and their effector proteins. Biochem. J. **348**(Pt 2), 241–255 (2000)

Bostner, J., Ahnström Waltersson, M., Fornander, T., Skoog, L., Nordenskjöld, B., Stål, O.: Amplification of CCND1 and PAK1 as predictors of recurrence and tamoxifen resistance in postmenopausal breast cancer. Oncogene **26**, 6997–7005 (2007)

Breitsprecher, D., Koestler, S.A., Chizhov, I., Nemethova, M., Mueller, J., Goode, B.L., Small, J.V., Rottner, K., Faix, J.: Cofilin cooperates with fascin to disassemble filopodial actin filaments. J. Cell Sci. **124**, 3305–3318 (2011)

Burridge, K., Wennerberg, K.: Rho and Rac take center stage. Cell **116**, 167–179 (2004)

Butcher, D.T., Alliston, T., Weaver, V.M.: A tense situation: forcing tumour progression. Nat. Rev. Cancer **9**, 108–122 (2009)

Carey, S.P., Kraning-Rush, C.M., Williams, R.M., Reinhart-King, C.A.: Biophysical control of invasive tumor cell behavior by extracellular matrix microarchitecture. Biomaterials **33**, 4157–4165 (2012)

Cary, L.A., Han, D.C., Guan, J.L.: Integrin-mediated signal transduction pathways. Histol. Histopathol. **14**, 1001–1009 (1999)

Cavalcanti-Adam, E.A., Micoulet, A., Blummel, J., Auernheimer, J., Kessler, H., Spatz, J.P.: Lateral spacing of integrin ligands influences cell spreading and focal adhesion assembly. Eur. J. Cell Biol. **85**, 219–224 (2006)

Chan, P.M., Manser, E.: PAKs in human disease. Prog. Mol. Biol. Transl. Sci. **106**, 171–187 (2012)

Ching, Y.P., Leong, V.Y., Lee, M.F., Xu, H.T., Jin, D.Y., Ng, I.O.: P21-activated protein kinase is overexpressed in hepatocellular carcinoma and enhances cancer metastasis involving c-Jun NH2-terminal kinase activation and paxillin phosphorylation. Cancer Res. **67**, 3601–3608 (2007)

Etienne-Manneville, S., Hall, A.: Rho GTPases in cell biology. Nature **420**, 629–635 (2002)

Gray, R.S., Cheung, K.J., Ewald, A.J.: Cellular mechanisms regulating epithelial morphogenesis and cancer invasion. Curr. Opin. Cell Biol. **22**, 640–650 (2010)

Hall, A.: Rho GTPases and the actin cytoskeleton. Science **279**, 509–514 (1998)

Harunaga, J.S., Yamada, K.M.: Cell-matrix adhesions in 3D. Matrix Biol. **30**, 363–368 (2011)

Heasman, S.J., Ridley, A.J.: Mammalian Rho GTPases: new insights into their functions from *in vivo* studies. Nat. Rev. Mol. Cell Biol. **9**, 690–701 (2008)

Huttenlocher, A., Horwitz, A.R.: Integrins in cell migration. Cold Spring Harb. Perspect. Biol. **3**, a005074 (2011)

Huveneers, S., Danen, E.H.: Adhesion signaling – crosstalk between integrins, Src and Rho. J. Cell Sci. **122**, 1059–1069 (2009)

Ingber, D.E.: Mechanosensation through integrins: cells act locally but think globally. Proc. Natl. Acad. Sci. U. S. A. **100**, 1472–1474 (2003)

Insall, R.H., Machesky, L.M.: Actin dynamics at the leading edge: from simple machinery to complex networks. Dev. Cell **17**, 310–322 (2009)

Jaffe, A.B., Hall, A.: Rho GTPases: biochemistry and biology. Annu. Rev. Cell Dev. Biol. **21**, 247–269 (2005)

Jean, C., Gravelle, P., Fournie, J.J., Laurent, G.: Influence of stress on extracellular matrix and integrin biology. Oncogene **30**, 2697–2706 (2011)

Kaibuchi, K., Kuroda, S., Amano, M.: Regulation of the cytoskeleton and cell adhesion by the Rho family GTPases in mammalian cells. Annu. Rev. Biochem. **68**, 459–486 (1999)

Khalil, A.A., Friedl, P.: Determinants of leader cells in collective cell migration. Integr. Biol. (Camb.) **2**, 568–574 (2010)

Kim, J.H., Serra-Picamal, X., Tambe, D.T., Zhou, E.H., Park, C.Y., Sadati, M., Park, J.A., Krishnan, R., Gweon, B., Millet, E., Butler, J.P., Trepat, X., Fredberg, J.J.: Propulsion and navigation within the advancing monolayer sheet. Nat. Mater. **12**, 856–863 (2013)

Kohler, S., Schaller, V., Bausch, A.R.: Collective dynamics of active cytoskeletal networks. PLoS One **6**, e23798 (2011)

Lammermann, T., Bader, B.L., Monkley, S.J., Worbs, T., Wedlich-Soldner, R., Hirsch, K., Keller, M., Forster, R., Critchley, D.R., Fassler, R., Sixt, M.: Rapid leukocyte migration by integrin-independent flowing and squeezing. Nature **453**, 51–55 (2008)

Lauth, M.: RAS and Hedgehog–partners in crime. Front. Biosci. **17**, 2259–2270 (2011)

Liu, P., Cheng, H., Roberts, T.M., Zhao, J.J.: Targeting the phosphoinositide 3-kinase pathway in cancer. Nat. Rev. Drug Discov. **8**, 627–644 (2009)

Machesky, L.M., Gould, K.L.: The Arp2/3 complex: a multifunctional actin organizer. Curr. Opin. Cell Biol. **11**, 117–121 (1999)

Machesky, L.M., Insall, R.H.: Signaling to actin dynamics. J. Cell Biol. **146**, 267–272 (1999)

Mardilovich, K., Olson, M.F., Baugh, M.: Targeting Rho GTPase signaling for cancer therapy. Future Oncol. **8**, 165–177 (2012)

Mulloy, J.C., Cancelas, J.A., Filippi, M.D., Kalfa, T.A., Guo, F., Zheng, Y.: Rho GTPases in hematopoiesis and hemopathies. Blood **115**, 936–947 (2010)

Oser, M., Condeelis, J.: The cofilin activity cycle in lamellipodia and invadopodia. J. Cell. Biochem. **108**, 1252–1262 (2009)

Parisi, F., Vidal, M.: Epithelial delamination and migration: lessons from *Drosophila*. Cell Adh. Migr. **5**, 366–372 (2011)

Parsons, J.T., Horwitz, A.R., Schwartz, M.A.: Cell adhesion: integrating cytoskeletal dynamics and cellular tension. Nat. Rev. Mol. Cell Biol. **11**, 633–643 (2010)

Pinner, S., Sahai, E.: Imaging amoeboid cancer cell motility *in vivo*. J. Microsc. **231**, 441–445 (2008)

Pollard, T.D., Blanchoin, L., Mullins, R.D.: Molecular mechanisms controlling actin filament dynamics in nonmuscle cells. Annu. Rev. Biophys. Biomol. Struct. **29**, 545–576 (2000)

Reister-Gottfried, E., Sengupta, K., Lorz, B., Sackmann, E., Seifert, U., Smith, A.S.: Dynamics of specific vesicle-substrate adhesion: from local events to global dynamics. Phys. Rev. Lett. **101**, 208103 (2008)

Ridley, A.J.: Rho family proteins: coordinating cell responses. Trends Cell Biol. **11**, 471–477 (2001)

Ross, R.S.: Molecular and mechanical synergy: cross-talk between integrins and growth factor receptors. Cardiovasc. Res. **63**, 381–390 (2004)

Rottner, K., Stradal, T.E.: Actin dynamics and turnover in cell motility. Curr. Opin. Cell Biol. **23**, 569–578 (2011)

Rowinsky, E.K.: Signal events: cell signal transduction and its inhibition in cancer. Oncologist **8**(Suppl 3), 5–17 (2003)

Schaller, M.D.: Cellular functions of FAK kinases: insight into molecular mechanisms and novel functions. J. Cell Sci. **123**, 1007–1013 (2010)

Solon, J., Kaya-Copur, A., Colombelli, J., Brunner, D.: Pulsed forces timed by a ratchet-like mechanism drive directed tissue movement during dorsal closure. Cell **137**, 1331–1342 (2009)

Tambe, D.T., Hardin, C.C., Angelini, T.E., Rajendran, K., Park, C.Y., Serra-Picamal, X., Zhou, E.H., Zaman, M.H., Butler, J.P., Weitz, D.A., Fredberg, J.J., Trepat, X.: Collective cell guidance by cooperative intercellular forces. Nat. Mater. **10**, 469–475 (2011)

Thompson, E.W., Newgreen, D.F., Tarin, D.: Carcinoma invasion and metastasis: a role for epithelial-mesenchymal transition? Cancer Res. **65**, 5991–5995 (2005). discussion 5995

Trepat, X., Fredberg, J.J.: Plithotaxis and emergent dynamics in collective cellular migration. Trends Cell Biol. **21**, 638–646 (2011)

van Zijl, F., Krupitza, G., Mikulits, W.: Initial steps of metastasis: cell invasion and endothelial transmigration. Mutat. Res. **728**, 23–34 (2011)

Vega, F.M., Ridley, A.J.: Rho GTPases in cancer cell biology. FEBS Lett. **582**, 2093–2101 (2008)

Wang, X., He, L., Wu, Y.I., Hahn, K.M., Montell, D.J.: Light-mediated activation reveals a key role for Rac in collective guidance of cell movement *in vivo*. Nat. Cell Biol. **12**, 591–597 (2010)

Wong, L.L., Lam, I.P., Wong, T.Y., Lai, W.L., Liu, H.F., Yeung, L.L., Ching, Y.P.: IPA-3 inhibits the growth of liver cancer cells by suppressing PAK1 and NF-κB activation. PLoS One **8**, e68843 (2013)

Yamada, K.M., Geiger, B.: Molecular interactions in cell adhesion complexes. Curr. Opin. Cell Biol. **9**, 76–85 (1997)

Yamao, M., Naoki, H., Ishii, S.: Multi-cellular logistics of collective cell migration. PLoS One **6**, e27950 (2011)

Yu, X., Machesky, L.M.: Cells assemble invadopodia-like structures and invade into matrigel in a matrix metalloprotease dependent manner in the circular invasion assay. PLoS One **7**, e30605 (2012)

Chapter 7
Devices and New Diagnostic Principles

Daniel A. Fletcher

Introduction

Physical scientists and engineers have long made significant contributions to medical and biological sciences through the development of devices and new diagnostic principles. Many of the screening, diagnostic, and therapeutic technologies used today have their origins in fundamental discoveries of physical scientists and laboratory instruments developed by engineers. Examples from the past 150 years include the discovery of X-rays and creation of X-ray machines, discovery of radioactivity and its use in radiotherapy, discovery of nuclear magnetic resonance and development of magnetic resonance imaging (MRI), and development of flow cytometry and its use in high-throughput screening.

What is different today? In addition to continued development of new technologies for visualizing, measuring, and treating cancer and other diseases, physical scientists and engineers have become more engaged in understanding the fundamental mechanisms at work in biological systems and in creating new strategies to combat disease based on those fundamental principles. There is a growing appreciation that biological systems must follow the same rules as other physical systems, and therefore must exhibit the same properties we expect of physical systems—stochasticity, elasticity, fluctuations, electric field susceptibility, etc. Rather than working only on measurement instruments, physical scientists and engineers are treating the biological systems—the tumors, the cancer stem cells, the extracellular matrix—as instruments themselves, instruments that can be understood through reverse engineering. These new approaches, supported most prominently by the network of Physical Sciences-Oncology Centers initiated by the National Institutes of Health's

D.A. Fletcher (✉)
Bioengineering, University of California, 608B Stanley Hall 1762, 94720 Berkeley, CA, USA
e-mail: fletch@berkeley.edu

P. Janmey et al. (eds.), *Physical Sciences and Engineering Advances in Life Sciences and Oncology*, Science Policy Reports,
DOI 10.1007/978-3-319-17930-8_7

National Cancer Institute, are exploring new approaches to age-old questions, just as cancer research faces a critical juncture.

Despite significant improvements in patient healthcare that have accompanied new clinical technologies, cancer remains a significant medical challenge. No cure exists for cancer, and the precise biological origins of tumors in different tissues remain a mystery. Therapeutic management continues to be a challenge for most forms of the disease. As average life expectancy increases, the likelihood that an individual will be faced with cancer at some point in his or her life also increases. Even the developing world, saddled with the scourge of infectious diseases, must now confront the fact that cancer and other non-communicable diseases are serious public health problems (Hotez and Daar 2008).

Can a revolutionary idea or technology be found that detects cancer early? Will a novel insight from outside of the medical field result in a new and effective therapy? Can the biological origins of cancer be understood and addressed from a physical sciences perspective? Recent research initiatives in the United States and abroad have focused attention on the intersection of physical sciences, biological sciences, and oncology, bringing researchers with fundamental and applied interests together with biologists and oncologists to pursue a new understanding of cancer and new approaches to cancer therapies. These efforts are producing a community poised to make significant contributions to the field.

This chapter will describe new ideas and innovations at the intersection of physical sciences and oncology that benefit both the lab and the clinic. It will conclude with a summary of drivers of innovation in Europe, Asia, and the United States.

Research

New Ideas and Innovations

Research institutions in the United States, Europe, and Asia have produced a remarkable breadth of new ideas and innovative approaches to the diagnosis and treatment of cancer. The specific examples discussed below are based primarily on the APHELION study trip to European and Asian institutions, together with some examples from the United States. The examples are broken into six clusters of innovation:

- Optical microscopy: Use of light to visualize tumors and tumor progression *in vivo*
- Force microscopy: Measurement of mechanical properties to identify cancer cells
- New materials: Materials that can control or interrogate cells in new ways
- Microdevices: Microfabricated systems that provide new measurement capabilities

- Medical imaging: Methods for imaging within the human body to study or track tumors
- Novel therapies: Innovative approaches to understanding and treating cancer

Optical Microscopy

Optical microscopy is a mainstay of cancer research and diagnosis. Traditional histological analysis combines tissue sample staining with microscopic imaging to determine patient diagnoses and treatment plans. In research, optical microscopy using fluorescently labeled molecules has led to a mechanistic understanding of basic cellular processes and the ability to study the molecular details of tissue development and function. Researchers in the United States have pioneered a multitude of fluorescent proteins and nanocrystal tags, including the variations on green fluorescent proteins by Dr. Roger Tsien at University of California, San Diego, and quantum dots by Dr. Paul Alivisatos at University of California, Berkeley. While fluorescence microscopy has become routine for localizing molecules, many properties that are harder to measure, such as the local concentration of reactive oxygen within a cell, are important measurements to understanding metabolism. Dr. Jerker Windengren at the Royal Institute of Technology (KTH) in Stockholm, Sweden, (site report, Appendix B) has advanced a new approach to measuring small molecules by using not the fluorescence emission of a fluorescent molecule but rather the population of its triplet state (Hevekerl et al. 2011). By carefully quantifying photon fluxes, Dr. Windengren can carry out triplet-state imaging that may be useful in detecting molecules, such as reactive oxygen, that alter the triplet state.

Fluorescence microscopy and its variants are useful for fundamental studies based on cells and tissues that can be genetically or otherwise manipulated to contain fluorescent labels on the molecules of interest. Label-free approaches are desirable if optical microscopy is to be applied to humans for diagnosis and therapy. Research into multiphoton interactions with biological molecules has revealed that high contrast can be achieved without labels by measuring 2nd and even 3rd harmonic interactions. Dr. Peter Friedl at the Nijmegen Centre for Molecular Life Sciences, Netherlands, (site report, Appendix B) is a pioneer of multiphoton imaging and has brought the technology to bear on questions of tumor progression and metastasis (Friedl et al. 2007; Ilina et al. 2011). Recent developments in his laboratory have enabled real-time and structure-specific imaging of live tissue with both 2nd harmonic and 3rd harmonic contrast without the need to introduce fluorescent labels or other contrast tabs (Fig. 7.1). This technology is already advancing the study of tumor progression in model animals.

Dr. Masaru Ishii of the Department of Immunology and Cell Biology in the Graduate School of Medicine at Osaka University, Japan, (site report, Appendix C) uses intravital two-photon microscopy to identify normal and aberrant behavior of osteoclasts in response to chemokines and other signals. His findings are pointing to novel regulators of motility that have implications for the study autoimmune dis-

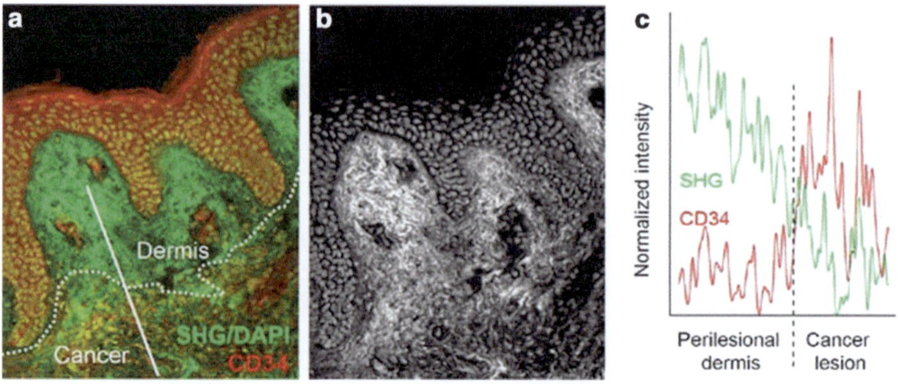

Fig. 7.1 Images of human dermatofibrosarcoma protuberans (From Friedl et al. 2007) (**a**) The tissue was stained for CD34 (*red*, *top* and *bottom* layer of tissue, showing cancer cells) and DAPI (*green*, dermis). The image also shows second harmonic generation from collagen (*green*, dermis), which is shown together with DAPI as a grayscale image in (**b**). (**c**) Intensity curves show that SHG (*top*) is reduced at the border with the cancer lesion

eases as well as cancer. In one important advance for understanding the underlying mechanism of rheumatoid arthritis, Dr. Ishii discovered that osteoclasts *in vitro* will exhibit directed motility toward sphigosine-1-phosphate (S1P), a signaling sphingolipid. His recent work has expanded on that finding to develop methods for studying and controlling osteoclast chemotaxis and bone resorption activity *in vivo*. This work takes full advantage of the ability of two-photon imaging to track single cells over time, allowing study of their movement speeds, paths, and morphology. For example, Dr. Ishii and his colleagues recently showed that the hormonally active form of vitamin D reduces bone resorption by modulating S1P-mediated motility (Fig. 7.2, Kikuta et al. 2013). Ongoing work has identified a role for cell cycle in the behavior of osteoclasts that may be a general mechanism modulating cell motility in rheumatoid arthritis and other situation. Importantly, the molecular mechanisms that have been found to modulate osteoclast activity may also be important in cancers such as colon cancer.

Force Microscopy

The old observation that solid tumors are stiffer than their surroundings has taken on new importance with the finding that mechanical properties of cells and their matrix are both indicators and drivers of disease. Studies have shown that cancer cell stiffness is increased in leukemia and decreased in breast cancer, suggesting that measuring these properties in hospitals (which do not currently use assays to quantify cellular or extracellular matrix stiffness) may be clinically useful. A variety of methods for measuring the mechanical properties of cells have been developed, including atomic force microscopy (AFM) and parallel plate rheometry (see Chap. 4).

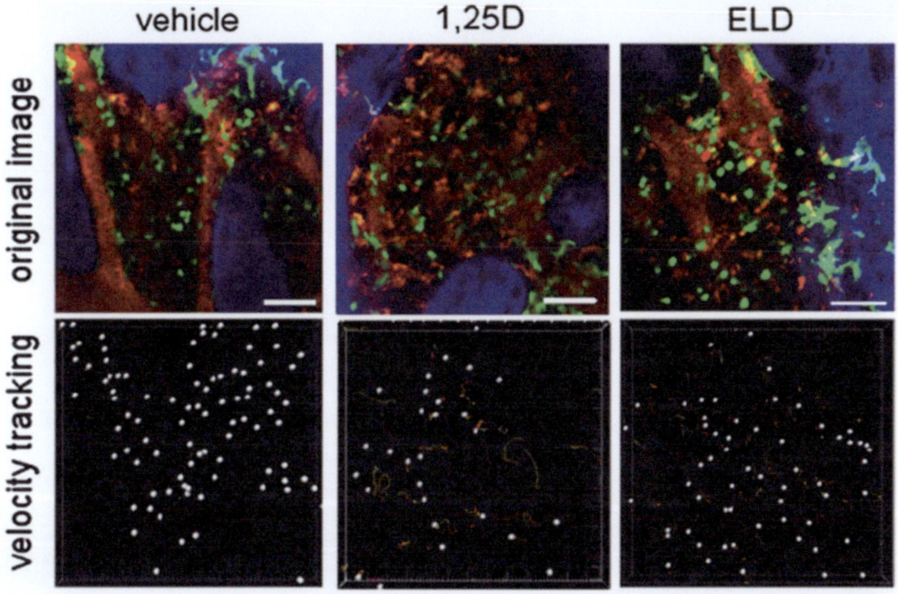

Fig. 7.2 Intravital two-photon imaging of S1PR2-mediated control of migration of osteoclast precursor monocytes (From Kikuta et al. 2013)
Imaging of the skull bone tissues of heterozygous CX3CR1-EGFP knock-in mice showing CX3CR1-EGFP–positive cells (*green*), microvasculature containing Texas Red-conjugated 70 kDa dextran (*red*), and bone surface (*blue*). The study revealed that the active form of vitamin D affects migration of osteoclast precursors to S1P. Scale bars = 50 μm

Researchers in the United States have been particularly active in the application of AFM and related indentation technologies for imaging and measurement of cells and their properties, including measurements of the correlation between cell stiffness and extracellular matrix stiffness by Dr. Paul Janmey of the University of Pennsylvania and analysis of bone structure by Dr. Paul Hansma at University of California, Santa Barbara. These methods have proven very useful for fundamental studies in which small numbers of samples may be needed to establish a point, but their throughput is woefully inadequate for a clinical setting in which large numbers of blood cells from many different patients would need to be assayed.

Dr. Josef Käs at the University of Leipzig, Germany, (site report, Appendix B) has developed a high-throughput method for quantifying mechanical properties of cells called the optical stretcher (Fig. 7.3; Guck et al. 2000). This technology takes advantage of refractive index differences between the cell and surrounding media and uses two counter-propagating lasers to actively deform single cells. Video imaging of the cell deformation and subsequent relaxation gives a measure of cell mechanical properties. A major advantage of this technology is that it has been automated to enable high-throughput analysis of cell populations. Since the mechanical properties of patient cells can be heterogeneous, and the cells of interest may represent only a small fraction of the population, assays like the optical stretcher

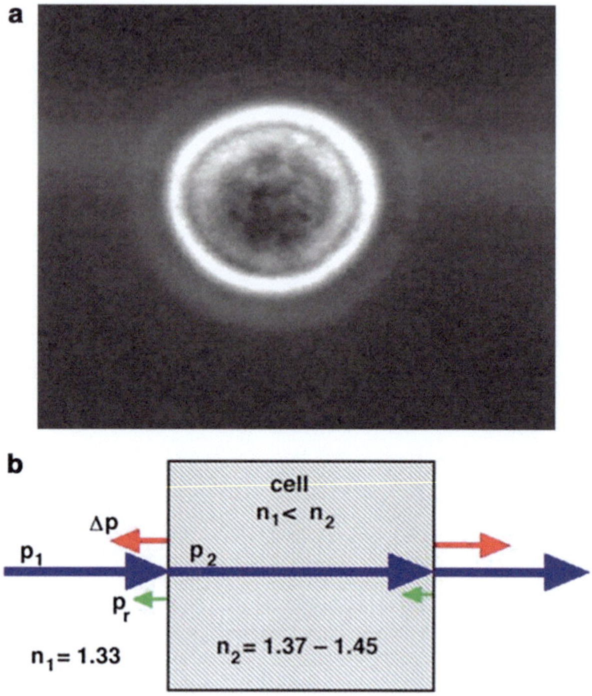

Fig. 7.3 The optical stretcher deforms cells with light, allowing non-contact measurement of mechanical properties after exposure (From Guck et al. 2000) **(a) A phase image of a stretched cell. (b) Schematic representation of the light path and refractive indices in the optical stretcher**

quantify properties of hundreds or thousands of cells to determine statistically reliable data that could be missed by evaluation of only tens of cells with traditional approaches (Fritsch et al. 2010).

New Materials

A particularly active field of research is the development of new materials for biological and biomedical applications. Advancements range from the creation of new polymeric gels to direct cell growth to the development of new nanoparticle probes for measuring cell behavior. Researchers in the United States are actively pursuing an understanding of cell-matrix interactions and development of new extracellular matrix materials, such as the alginate-based gels from Dr. David Mooney at Harvard University and the dynamic biomaterials from Dr. Kristi Anseth at the University of Colorado. Dr. Wilhelm Huck at the Nijmegen Centre for Molecular Life Sciences in the Netherlands (site report, Appendix B) has created polyacrylamide gels with crosslinked collagen that are able to alter cell activation (Trappmann et al. 2012). Dr. Pierre Nassoy at the Institute Curie in Paris, France, (site report, Appendix B) is developing a novel alginate encapsulation method for cell aggregates that will help to understand how cell growth responds to confinement. Dr. Clair Wilhelm at the University of Paris, Diderot, (site report, Appendix B) has produced magnetic

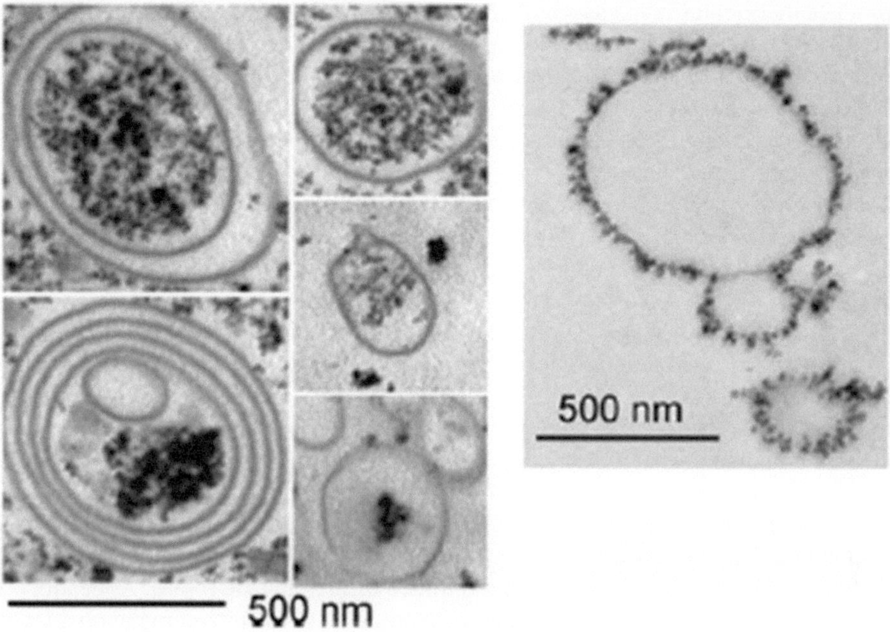

Fig. 7.4 Magnetic particles and biological vesicles (From Lesieur et al. 2011)
(*Left*) Vesicles derived from cells that have internalized the magnetic particles. (*Right*) Vesicles coated with magnetic particles after derivation from cells

nanoparticles and nanorods that are taken up by cells and can be externally manipulated to measure cytoplasmic properties of cells (Fig. 7.4). She is also developing technologies to form liposomes from cell membranes that have the potential to be more effective drug delivery materials (Lesieur et al. 2011).

Microdevices

Microfabrication technology originally developed for the semiconductor has allowed the creation of microdevices with diverse capabilities. Miniaturized versions of conventional laboratory equipment, such as cell counting or gel separation technologies, are now possible with microfabrication. Outstanding fabrication facilities have been created at U.S. universities that have fostered the development of an active microfabrication community in the United States, including the MicroLab at University of California, Berkeley, and the Nanofabrication Facility at Stanford University. A major advantage of microfabrication is the ability to extract data from small sample sizes and multiplex measurement methodologies within a single device. One example of this new technology is from Dr. Jean-Louis Viovy at the Institute Curie, France, (site report, Appendix B). He has designed a microdevice that automates fluorescence *in situ* hybridization, potentially making high-throughput analysis of patient samples less time-consuming. New fabrication technologies to produce thermoplastic devices by lamination are also under development (Miserere et al. 2012). Advances in microfluidic technologies will be further advanced by the formation of a new center in Paris associated with the de Gennes Foundation.

Research institutes in Asia have been particularly active in developing novel microdevices that increase the speed, sensitivity, and/or complexity of possible measurements. At Peking University in Beijing, China, (site report, Appendix C), Dr. Jianzhong (Jeff) Xi has developed a high throughput RNA screen platform based on self-assembly cell microarray (SAMCell) (Fig. 7.5). Dr. Xi is extending his screening of RNA to include micro RNA (miRNA), which play important roles in transcriptional and post-transcriptional gene regulation.

Dr. Fan Bai, also of Peking University, (site report, Appendix C) is using microdevices to investigate the role of circulating tumor cells (CTCs) in cancer progression. He has developed a combination technology for isolating CTCs that first uses a commercial system to obtain an initial round of CTC selection and then uses custom automated imaging to select CTCs with specific characteristics. With the highly selected population, single cell sequencing is applied to identify new properties, similarities, and differences in CTCs. The work of Dr. Bai is ongoing but has exciting clinical promise.

Several research groups at Academia Sinica in Taiwan (site report, Appendix C) are pioneering new uses of microdevices to study cell growth and metabolism. Dr.

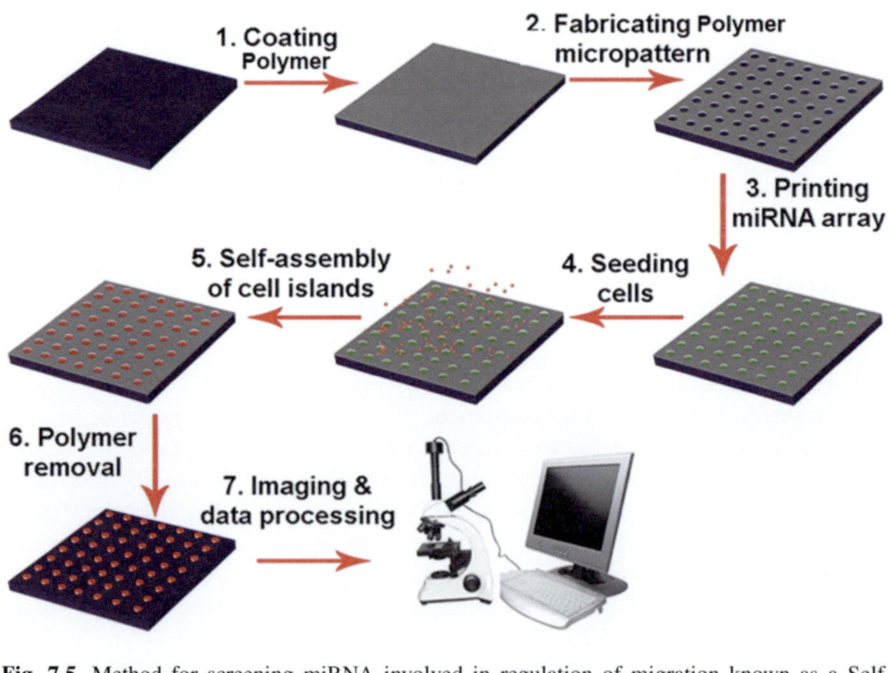

Fig. 7.5 Method for screening miRNA involved in regulation of migration known as a Self-Assembly Cell Microarray (SAMCell) developed by Dr. Jianzhong (Jeff) Xi (From Zhang et al. 2011)

Chau-Hwang Lee applies microfabrication to study cell behavior and its implications for cancer. Dr. Lee and his research group are motivated by the question of how the microenvironment of a tumor influences the behavior of tumor cells. It has become clear that mechanical properties and cyclic stresses are important modulators of cell behavior, and stromal cells are key players in the tumor microenvironment. To address the question of how stromal cells affect tumor cell behavior, Dr. Lee and colleagues have developed a microfluidic cell culture chip that can support the growth of multiple types of cells at a time, such as breast cancer cells, glioblastoma cells, smooth muscle cells, microvascular endothelial cells, and fibroblast cells. These cell culture chips allow control of solutions and substrate interactions while at the same time enabling high-resolution imaging of cell organization and behavior, including cell motility.

Fibroblasts are of increasing interest in cancer research as therapeutic targets. In particular, tumor-associated fibroblasts, which produce collagen type I that affects microenvironmental properties and cancer drug uptake, may play an important role in cancer biology and the development of new treatment strategies. Dr. Lee has recently developed a cell co-culture chip for lung cancer cells and lung fibroblasts (Fig. 7.6). By tracking the movements of each cell type, Dr. Lee and colleagues have found that the presence of fibroblasts in the culture affect the migration speeds of the cancer cells, presumably through a paracrine loop. This work suggests that preventing the fibroblasts from being activated could be essential for controlling cancer metastasis (Hsu et al. 2012). In related work, Chen-Hsien Liu is using dielectrophoresis-based technologies to create liver-like tissue constructs with heterogenous subpopulations (Ho et al. 2013). Shape, size, and cell placement can all be controlled.

Dr. Yi-Chung Tung, also at Academia Sinica, uses microdevices to control the fluid environments around cells and to analyze small numbers of cells. The microenvironment around cells influences their growth and behavior through soluble factors,

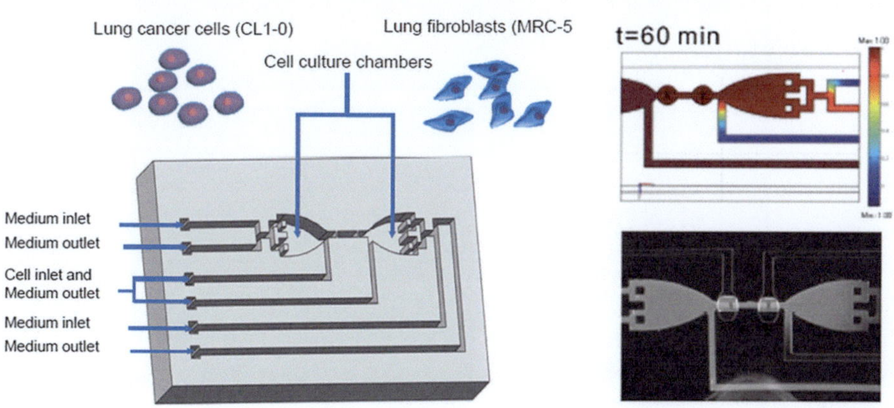

Fig. 7.6 Microfluidic co-culture chip for studying the effect of lung fibroblasts on lung cancer cell behavior. The co-culture chip was designed to ensure that all the cells sense similar conditioned medium concentration from the other chamber (Courtesy of C.-H. Lee)

physical cues, as well as dissolved gasses. It is well known that certain tissues and tumors are associated with changes in levels of dissolved gasses, but this property has been difficult to study *in vitro* due to the challenge of creating controlled gaseous microenvironments. Dr. Tung has addressed this issue by developing a microfluidic cell culture device with spatial control of dissolved O_2 (Fig. 7.7; Peng et al. 2013). The device works by confining oxygen scavenging chemical reactions underneath the cell culture chambers, causing the local oxygen levels to vary spatially.

Using the device, Dr. Tung and colleagues tested the effect of the anti-cancer drug triapazamine (TPZ) on an adenocarcinomic human alveolar basal epithelial cell line at three different oxygen levels. They found that the toxicity of TPZ is hypoxia induced, providing a new method for investigating the role of the gaseous microenvironment on the mechanism of this important anti-cancer drug. Dr. Tung is extending his group's ability to control the gaseous microenvironment by developing a device to control NO that is being used to test the role of NO levels and gradients on smooth muscle cells and other cells.

Microfabrication is also being used to study how spatial constraints on cell motility affect metastasis of cancer cells. Dr. Liyu Liu at the Institute of Physics at the Chinese Academy of Sciences (site report, Appendix C) has established a labora-

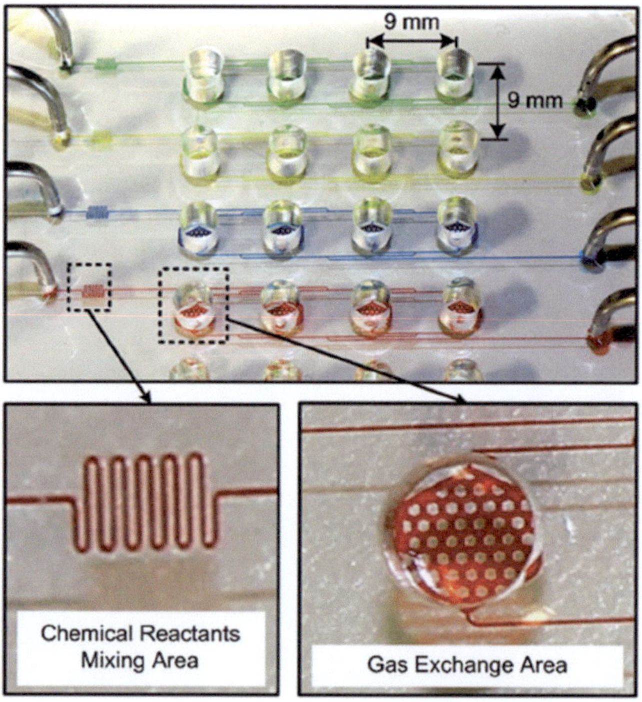

Fig. 7.7 Integrated microfluidic culture device with controlled gaseous microenvironments. The device enables variation in oxygen levels and establishment of oxygen gradients (From Peng et al. 2013)

tory that combines microfabrication, microfluidics, and cancer biology. In work while he was a research at Princeton University with Dr. Bob Austin, Dr. Liu explored the motility of prostate cancer cells within three-dimensional microfabricated structures (Liu et al. 2011). The mesa-like structures, referred to as Tepuis, have a 3:1 aspect ratio and are packed into arrays. By introducing highly metastatic PC-3 cells and non-invasive LNCaP cells into the Tepuis structures, Dr. Liu and colleagues found that the highly metastatic cells covered the tops of the Tepuis but non-invasive cells did not (Fig. 7.8). The difference in coverage of the microfabricated structures is proposed to be linked to the differences in contact inhibition of the two cells, which is reduced in the PC-3 cells.

Dr. Urs Frey at the RIKEN Quantitative Biology Center in Kobe, Japan, (site report, Appendix C) develops novel CMOS-based sensors for biology with applications in systems biology and systems neuroscience. The group is pioneering the design and use of microelectrode array technology that can simultaneously collect electrical signals from multiple positions, providing a method for tracking the

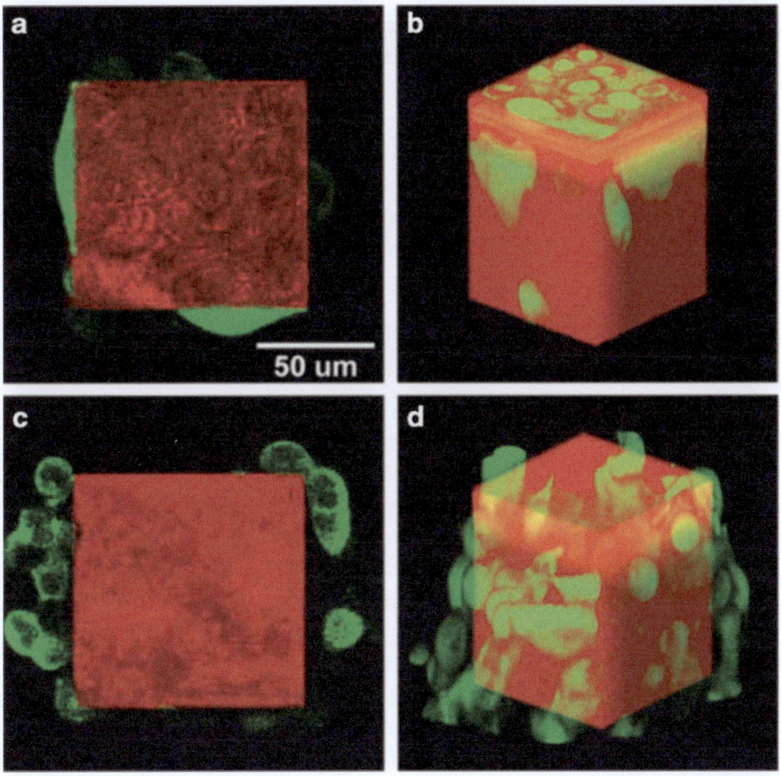

Fig. 7.8 Prostate cancer cells covering a microfabricated structure (From Liu et al. 2011) Highly metastatic cells (**a**, **b**) and non-invasive LNCaP cells (**c**, **d**), both expressing GFP, are introduced onto a microfabricated structure containing high aspect ratio pillars called Tepuis. The non-invasive cells are unable to cover the top of the Tepuis, while the invasive cells are

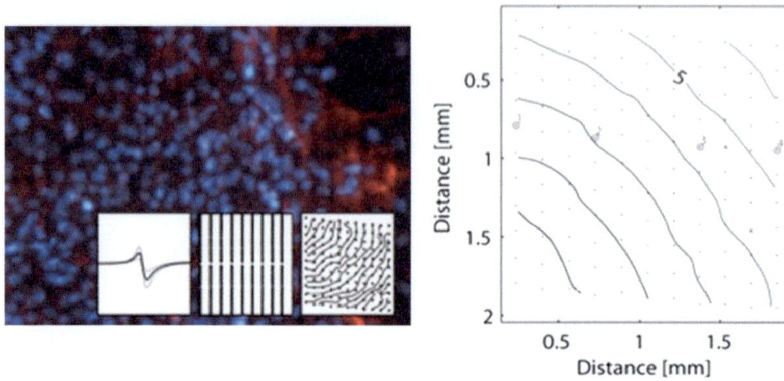

Fig. 7.9 Microelectrode array analysis of cardiomyocyte activity. Cultured cardiomyocytes exhibit propagating waves of electrical activity associated with contraction. Use of the microelectrode array device enables analysis of electrical activity over long distances (Courtesy of U. Frey)

propagation of signals and studying collective electrical behavior of groups of cells, including cancer cells. The bioelectric sensors developed by the Frey group have applications to a broad set of biological systems to better understand how they electrically communicate. Microelectrode arrays designed and fabricated to be compatible with cells enable recording of electrical signals with spatially high density, and the devices are being tested for use in several application areas. For example, propagation of a cardiomyocyte contraction wave (Fig. 7.9) and axonal action potentials (Bakkum et al. 2013) can be observed with a 2D microelectrode array.

Medical Imaging

A central goal of research at the intersection of physical sciences and cancer biology is to develop methods, insights, and understanding that improve patient outcomes. This requires active collaboration between engineers and physicians, which can be difficult to establish due to the busy schedules and differing demands on engineers and physicians. However, at some institutes, such as Peking University (site report, Appendix C), active collaboration between engineers and physicians is accelerating the process of technology development, particularly in the arena of medical imaging. Dr. Qiushi Ren of the Biomedical Engineering Department at Peking University is developing advancements in multi-modality molecular imaging systems (CT-PET-SPECT-FMT) that take advantage of each imaging technique to enhance data collection from individual patients. Dr. Hongbing Han has developed a new MRI method to study diffusion, such as that of a drug or other fluid, in the extracellular space in the brain (Fig. 7.10). Together with computational modeling to enable prediction of diffusion patterns, this work will aid in the design of better drug delivery to the brain. In a similar combination of MRI imaging and computer modeling, Dr. T. Christian Gasser at the Royal Institute of Technology, Sweden, (site report,

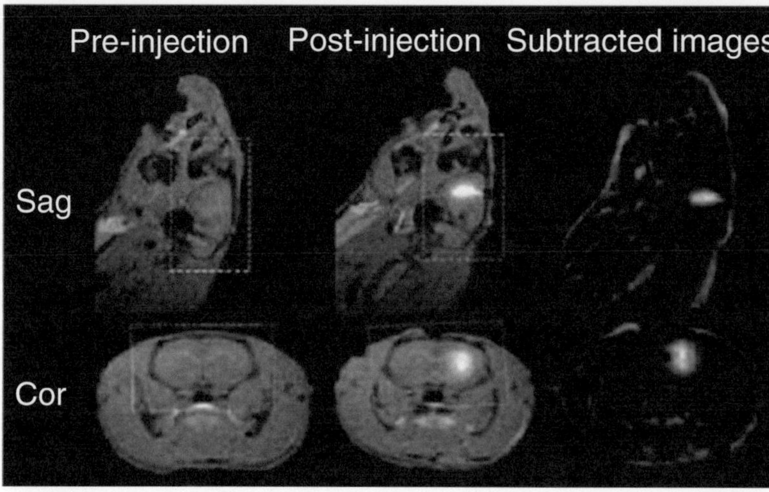

Fig. 7.10 Imaging of drug diffusion in the extracellular domain of a rat brain with MRI (Courtesy of H. Han)

Appendix B) has used computational modeling of blood vessel mechanics based on MRI images of patients to predict when surgical intervention is required (Martufi and Gasser 2011).

Novel Therapies

Many of the devices and diagnostic principles described so far are improvements on or refinement of existing technology. Sometimes, entirely new ideas and approaches are needed to solve problems. The United States has a long history of creating innovative medical devices, such as the artificial heart designed by Dr. Robert Jarvik. A more recent example is Dr. Yoram Palti at the Technion-Israel Institute of Technology and Novocure, Israel, (site report, Appendix B) who has developed a novel approach to cancer therapy based on high-frequency electric fields that are thought to disrupt microtubule organization in dividing cells, an approach known as tumor treating fields (Fig. 7.11, Kirson et al. 2007). Initial studies by Novocure, the company commercializing the technology, show promising results. These technologies demonstrate that fundamental and well-known physical principles have the potential to inspire creative researchers to develop new clinical therapies.

Novel therapies can have their origin not only in the development of new technologies but also in the development of new theoretical models. Dr. Chen-Hsiang Yeang in the Institute of Statistical Science at Academia Sinica (site report, Appendix C) creates algorithms, statistical methods, and models to identify drivers of tumor phenotypes by analyzing various types of molecular aberrations and single cell heterogeneities. Dr. Yeang identifies drivers of tumor phenotypes by developing mathematical models to capture aberrations at the molecular level. One approach

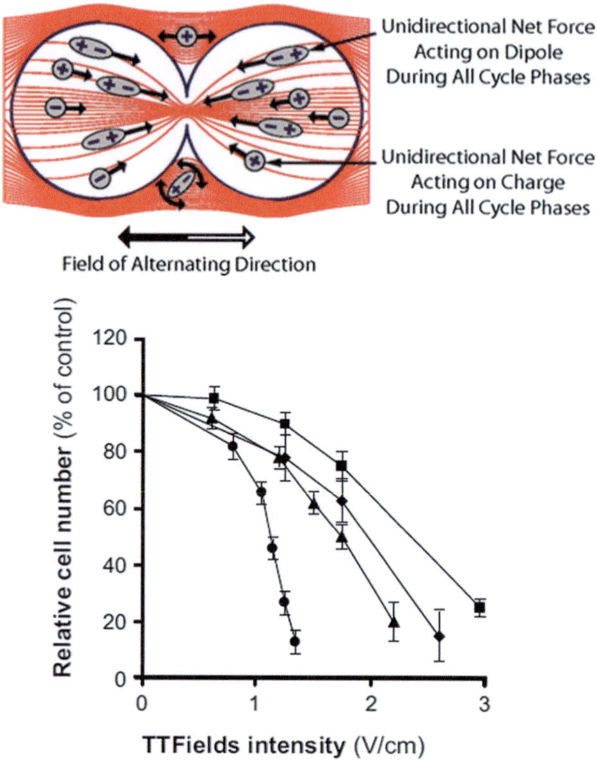

Fig. 7.11 Tumor treating fields (From Kirson et al. 2007) (*Top*) **Alternating current across a dividing cell applies forces that resist division.** (*Bottom*) **Decrease in cell number after 24 h of tumor treating fields for B16F1, MDA-MB-231, F-98, and H1299 cells**

employed by Dr. Yeang is to define association modules of the aberrations, such as mutation and copy number. The association modules are then integrated into a logistic regression model, and layered model selection is used to determine the associations. Dr. Yeang is in the process of applying this framework to glioblastoma multiform, making use of cancer genomic data from the Cancer Genome Atlas.

Dr. Yeang is also using mathematical modeling to understand the role that single-cell heterogeneity has on personalized medicine treatments. Tumors are notoriously heterogeneous at the single cell level, but most cancer therapies can't or don't take that heterogeneity into account. In a recent publication, Dr. Yeang and colleagues showed through modeling that accounting for this heterogeneity would motivate an alternate approach to personalized cancer therapy, which they termed 'nonstandard' personalized medicine (Fig. 7.12; Beckman et al. 2012). This work is another promising example of how statistical methods informed by cancer biology can lead to new treatment strategies. The statistical models developed by Dr. Yeang have important and direct implications for cancer therapies. His models, both those that are aimed at tracking association modules from cancer genomes and those that analyze the effects of single cell heterogeneity on cancer treatment, show that detailed analysis of data can help to guide thinking about cancer progression and therapy. With further clinical interactions, this work has the potential to guide new approaches to cancer treatment that would benefit many patients.

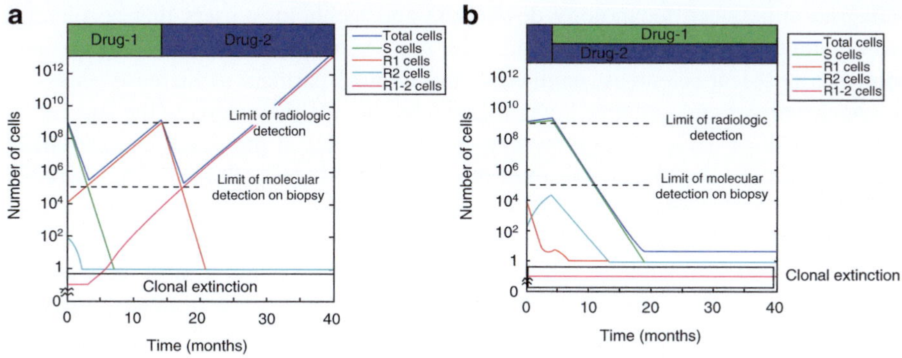

Fig. 7.12 Example of how cancer treatment with conventional personalized medicine (**a**) differs from the "nonstandard" personalized medicine (**b**), showing how treatments that incorporate single-cell heterogeneity have improved outcomes (From Beckman et al. 2012)

Dr. Ping Ao at Shanghai Jiao Tong University (site report, Appendix C) is tackling the challenge of developing predictive models of cancer progression, beginning with the formalisms of theoretical physics. Dr. Ao's efforts are aimed at creating the theoretical and computational tools that provide a framework for organizing, explaining, and predicting the behavior of tumors and other complex diseases. Based on experimental work that has identified oncogenes, tumor suppressors, and of other molecular components that are known to be linked to cancer progression, Dr. Ao and colleagues have put forward a concept they refer to as the endogenous molecular-cellular network hypothesis. Dr. Ao's hypothesis is based on interactions between sets of modules of activity in cells, such as cell cycle and cell division, that can operate autonomously but also interact with each other. According to Dr. Ao, "The interactions among these agents form an autonomous, nonlinear, stochastic, and collective dynamical network. We have tentatively named it as the endogenous molecular-cellular network." (Wang et al. 2012).

Discussion

Emerging Themes

Researchers in the United States, Europe, and Asia are actively creating new devices and diagnostic principles that show great promise for improving cancer diagnosis and therapy. While it is difficult to succinctly characterize research strengths and approaches at the national level, several generalizations can be made based our visits to research institutes, discussions with researchers, and comparison with our own academic experiences: (i) research in the United States is often distinguished by individuals who are able to introduce new principles or technologies; (ii) in contrast, research in Europe can be characterized as having more and better integration between technology developers and technology users that accelerates commercialization; and (iii) research in Asia benefits from active engagement between physical scientists, engineers, and clinical partners to increase access to clinical samples and patients. Several common themes characterized the successful development of devices and new diagnostic principles at the Asian and European institutions our team visited as part of the APHELION study. These themes not only advance the application of physical science concepts to problems in cancer, but also point to a highly collaborative research environment that integrates government funding, foundation support, industry engagement, and public health resources to improve cancer diagnosis and therapy. The four main themes also suggest areas for improvement in U.S.-supported efforts to drive technology innovation in medical devices and diagnostics.

Ease of Use by Non-specialists

Non-physicists and non-engineers should be able to easily use a new technology or device. This is critical for advancing its testing and adoption. The early stages of developing a novel device typically involve building a proof-of-concept system that only the expert researcher who built it (often a single graduate student) knows exactly how to run. That device satisfies the goal of testing the validity of a concept or measuring the sensitivity limits of a new diagnostic principle, but its value in spurring further use and testing can be limited. Additional efforts to improve the usability and reliability of the technology are often needed before it can be applied routinely to biological samples of interest. A new device can only generate a sufficient body of evidence to make a clinical impact when ease of use and availability are sufficient for non-specialists. Two examples from European institutions are the multiphoton imaging systems developed by Dr. Friedl and the optical stretcher developed by Dr. Käs. In both cases, the developers' significant efforts to advance the technologies beyond proof-of-concept have resulted in increasing popularity and growing evidence that these approaches can provide critical insight into disease processes.

Availability of Large-Scale Biobanks

Publically maintained biobanks of tissue samples and patient histories are critically important resources for evaluating new ideas for disease indicators and therapeutic strategies. While diagnostic principle development itself is the key step, testing a new approach to making a diagnosis or a new device requires access to patient samples. Such samples are even more useful if they are accompanied by detailed patient history and information about subsequent outcomes. Furthermore, serial samples from a single patient with history and outcome information can provide an exceptionally rich opportunity for retrospective studies of disease markers and signatures of disease progression. Europe has been very active in establishing and promoting biobanking. The European Strategy Forum on Research Infrastructures, funded by the European Commission, has established the Biobanking and Biomolecular Resources Research Infrastructure, which includes more than 225 associated organizations from over 30 countries (www.bbmri.eu). The goal of this infrastructure is to organize biobanking activities in Europe and strengthen the link between biological specimen collection and use in biological and medical research. Some individual countries have had long-established biobanking systems that now support extensive medical research. The Swedish National Biobank was established more than 20 years ago and has patient information allowing for both retrospective and prospective studies of a broad range of cancers (www.biobanks.se). Biobanks within Sweden now contain more than ten million pathology samples, 1.7 million plasma samples, and 24,000 fresh-frozen samples. In China, the active engagement of physicians in the development of new technologies ensures that the projects receive early feedback and access to clinical samples that guide further development.

Co-localization of Multiple Device Modalities

Diagnosis and treatment of cancer involves the use of many different devices and technologies. In order to understand how a new device or diagnostic principle compares with current or competing methods, data from several different measurement techniques on the same samples provides the most useful information. Indeed, obtaining the biological specimen, whether it is from a specially engineered mouse model or from a patient with a unique condition, can often be the limiting step in data collection. Therefore, maximizing data collection from an available sample is important. This can be achieved by combining multiple device modalities in a single location with open access to all instruments. Rather than have a separate facility for X-ray imaging, MRI, and a new imaging modality in the testing phase, combine these instruments into one imaging facility. This multipurpose facility will have easy access to commercial instruments as well as instruments under development, and it enables efficient use of samples and provides valuable comparative data. This is exactly what has been developed at the Preclinical Imaging Centre of the Radboud University Nijmegen Medical Centre in the Netherlands, where Dr. Friedl and colleagues in multiple departments have established an imaging center that offers MRI, microSPECT/CT, multiphoton microscopy, and other imaging modalities in a single facility (http://www.umcn.nl/Research/Departments/cdl/PRIME/Pages/MultiPhotonMicroscopy.aspx). This is also similar to the multi-model technologies being developed at Peking University. Access to multiple instruments in a central location enables researchers to easily interrogate tissue in multiple ways.

Close Industry Connections

It is necessary to commercialize technologies developed in research laboratories if they are to become broadly available for clinical use. The process of transitioning a technology from a research environment to an industrial environment is a complex one, often presenting significant barriers. Strong connections between academic researchers and industry can provide efficient routes for device commercialization as well as awareness of what problems need solving. Research institutions that are able to foster those strong connections have been successful in spawning start-up companies or partnering with established companies to further develop and commercialize technology. A successful example from our study trip is KTH in Sweden, where multiple technologies have gone from laboratory prototypes to commercial devices, including eXcillum which produces X-ray phase contrast equipment (www.excillum.com) and Adolesco which produces mobile 3D SPECT (www.adolesco.se). Two actions contribute to close collaborations between research and industry: (1) strong government support for collaborations with industry; and (2) a consensus among researchers that transitioning a technology to industry is part of a research project timeline. In Sweden, patent policy is also a contributing factor. Their policy states that the rights to an invention stay with the inventor rather than with the institution, which is counter to the United States policy. In part of Asia, academic-industry connections that would help to translate novel technologies into commercial products are still being developed.

Conclusions

The physical sciences have great potential to impact the diagnosis and treatment of cancer. Both European and Asian research institutions have created a supportive environment for the interaction of physical sciences and oncology that has fostered the development of novel devices and new diagnostic principles. This environment is characterized by highly interactive and interdisciplinary researchers spanning academic departments and medical institutions. The extraordinary support provided by the Chinese government to specific faculty and institutes in China has enabled the development of cutting-edge research facilities, and the continued support of RIKEN and associated institutions in Japan by the Japanese government has enabled the establishment and maintenance of world-renown research efforts. Collaborative funding initiatives from the European Union and country governments have contributed to Europe's highly creative and productive research, and an emphasis on industry partnerships has promoted commercialization of innovative new technologies. The Institute Curie in France, where internal funding is used to initiate cross-disciplinary projects, is a model of how basic researchers and clinical researchers can be located within a common research infrastructure and work together to advance fundamental principles and identify new therapeutic opportunities. The emerging themes identified above point to specific ways in which U.S. funding could further encourage the development of new devices and diagnostic principles for cancer.

References

Bakkum, D.J., Frey, U., Radivojevic, M., Russell, T.L., Muller, J., Fiscella, M., Takahashi, H., Hierlemann, A.: Tracking axonal action potential propagation on a high-density microelectrode array across hundreds of sites. Nat. Comm. **4**, 2181 (2013)

Beckman, R.A., Schemmann, G.S., Yeang, C.H.: Impact of genetic dynamics and single-cell heterogeneity on development of nonstandard personalized medicine strategies for cancer. Proc. Natl. Acad. Sci. U. S. A. **109**(36), 14586–14591 (2012)

Friedl, P., Wolf, K., von Andrian, U.H., Harms, G.: Biological second and third harmonic generation microscopy. Curr. Protoc. Cell. Biol. Chapter 4, Unit 4.15 (2007). doi:10.1002/0471143030.cb0415s34

Fritsch, A., Hockel, M., Kiessling, T., Nnetu, K.D., Wetzel, F., Zink, M., Käs, J.A.: Are biomechanical changes necessary for tumour progression? Nat. Phys. **6**, 730–732 (2010)

Guck, J., Ananthakrishnan, R., Moon, T.J., Cunningham, C.C., Käs, J.: Optical deformability of soft biological dielectrics. Phys. Rev. Lett. **84**, 5451–5454 (2000)

Hevekerl, H., Spielmann, T., Chmyrov, A., Widengren, J.: Forster resonance energy transfer beyond 10 nm: exploiting the triplet state kinetics of organic fluorophores. J. Phys. Chem. B **115**, 13360–13370 (2011)

Ho, C.T., Lin, R.Z., Chen, R.J., Chin, C.K., Gong, S.E., Chang, H.Y., Peng, H.L., Hsu, L., Yew, T.R., Chang, S.F., Liu, C.H.: Liver-cell patterning lab chip: mimicking the morphology of liver lobule tissue. Lab Chip **13**, 3578–3587 (2013)

Hotez, P.J., Daar, A.S.: The CNCDs and the NTDs: blurring the lines dividing noncommunicable and communicable chronic diseases. PLoS Negl. Trop. Dis. **2**, e312 (2008)

Hsu, T.-H., Kao, Y.-L., Lin, W.-L., Xiao, J.-L., Kuo, P.-L., Wu, C.-W., Liao, W.-Y., Lee, C.-H.: Migration speed of cancer cells influenced by macrophages and myofibroblasts co-cultured in a microfluidic chip. Int. Biol. **4**, 177 (2012)

Ilina, O., Bakker, G.J., Vasaturo, A., Hofmann, R.M., Friedl, P.: Two-photon laser-generated microtracks in 3D collagen lattices: principles of MMP-dependent and -independent collective cancer cell invasion. Phys. Biol. **8**, 015010 (2011)

Kikuta, J., Kawamura, S., Okiji, F., Shirazaki, M., Sakai, S., Saito, H., Ishii, M.: Sphingosine-1-phosphate-mediated osteoclast precursor monocyte migration is a critical point of control in antibone-resorptive action of active vitamin D. Proc. Natl. Acad. Sci. U. S. A. **110**(17), 7009–7013 (2013)

Kirson, E.D., Dbaly, V., Tovarys, F., Vymazal, J., Soustiel, J.F., Itzhaki, A., Mordechovich, D., Steinberg-Shapira, S., Gurvich, Z., Schneiderman, R., Wasserman, Y., Salzberg, M., Ryffel, B., Goldsher, D., Dekel, E., Palti, Y.: Alternating electric fields arrest cell proliferation in animal tumor models and human brain tumors. Proc. Natl. Acad. Sci. U. S. A. **104**, 10152–10157 (2007)

Lesieur, S., Gazeau, F., Luciani, N., Menager, C., Wilhelm, C.: Multifunctional nanovectors based on magnetic nanoparticles coupled with biological vesicles or synthetic liposomes. J. Mater. Chem. **21**, 14387–14393 (2011)

Liu, L., Sun, B., Pedersen, J.N., Aw Yong, K.M., Getzenberg, R.H., Stone, H.A., Austin, R.H.: Probing the invasiveness of prostate cancer cells in a 3D microfabricated landscape. Proc. Natl. Acad. Sci. U. S. A. **108**(17), 6853–6856 (2011). doi:10.1073/pnas.1102808108

Martufi, G., Gasser, T.C.: A constitutive model for vascular tissue that integrates fibril, fiber and continuum levels with application to the isotropic and passive properties of the infrarenal aorta. J. Biomech. **44**, 2544–2550 (2011)

Miserere, S., Mottet, G., Taniga, V., Descroix, S., Viovy, J.L., Malaquin, L.: Fabrication of thermoplastics chips through lamination based techniques. Lab Chip **12**, 1849–1856 (2012)

Peng, C.C., Liao, W.H., Chen, Y.H., Wu, C.Y., Tung, Y.C.: A microfluidic cell culture array with various oxygen tensions. Lab Chip **13**(16), 3239–3245 (2013)

Trappmann, B., Gautrot, J.E., Connelly, J.T., Strange, D.G., Li, Y., Oyen, M.L., Cohen Stuart, M.A., Boehm, H., Li, B., Vogel, V., Spatz, J.P., Watt, F.M., Huck, W.T.S.: Extracellular-matrix tethering regulates stem-cell fate. Nat. Mater. **11**, 642–649 (2012)

Wang, G.-W., Zhu, X.-M., Hood, L., Ao, P.: From Phage lambda to human cancer: endogenous molecular-cellular network hypothesis. Quant. Biol. **1**(1), 32–49 (2012). doi:10.1007/s40484-013-0007-1

Zhang, H., Hao, Y., Yang, J., Zhou, Y., Li, J., Yin, S., Sun, C., Ma, M., Huang, Y., Xi, J.J.: Genome-wide functional screening of miR-23b as a pleiotropic modulator suppressing cancer metastasis. Nat. Commun. **2**, 554 (2011)

Chapter 8
Clinical Perspective

Ross Levine

Introduction

There is growing support for the use of physics to study problems in cancer biology on a mechanistic and translational level. I am pleased to see multidisciplinary and translational efforts supported in the United States by the PS-OC program, which funds U.S. researchers conducting this work. I am also encouraged by the APHELION study, which reports on research into physical sciences applied to cancer research and treatment in Europe and Asia.

I have been asked to comment from a clinical perspective on the material covered in the previous six chapters. In my work as a physician-scientist at Memorial Loan-Kettering Cancer Center, I have had the opportunity to observe biological discoveries evolve from laboratory research projects to helpful clinical tools. It is encouraging to see such promise for multidisciplinary translational breakthroughs in the detection and treatment of cancer.

For a cancer biologist, there is one fundamental question for all of the studies presented in this report: How can this work be applied in clinical and translational contexts? In order to answer this question, we must determine if these efforts can be used to impact cancer biology and therapy in the near term, and, if so, what are the specific opportunities for these applications and how do we prioritize them? Finally, we need to look beyond the near term and identify the long-term efforts that are worth pursuing to link physical sciences and oncology.

I would like to address each chapter individually. The other authors of this report have answered this question in many different contexts, but also raised many

R. Levine, M.D. (✉)
Human Oncology & Pathogenesis Program, Memorial Sloan-Kettering Cancer Center, New York, NY, USA
e-mail: leviner@mskcc.org

P. Janmey et al. (eds.), *Physical Sciences and Engineering Advances in Life Sciences and Oncology*, Science Policy Reports,
DOI 10.1007/978-3-319-17930-8_8

important questions about how the physical sciences can help cancer research and treatment on a biological or translational level.

Research

Researchers in Asia, Europe, and the United States have all made tremendous efforts in developing better ways of studying living cells. One of the things we struggle with in cancer biology is that cells grow better *in vivo* than *ex vivo*. I firmly believe that the solution to working more effectively with cells lies in the fields of engineering or physics. Many of the methods discussed in the previous chapters will help us to keep cells alive longer and in a more physiologic context, and therefore conduct better-controlled experiments. Studying cells *ex vivo* is an important tool in the cell biologist's arsenal—it's not the tool that will give us the ultimate answer (a cure for cancer), but I believe it will give us the ability to find that answer. Another critical tool is the ability to study how cancer cells interact with non-cancer cells, i.e., studies of cell environment and heterogeneity using more precise, accurate experimental systems.

That said, the importance of studying cancer in its natural habitat should not be overlooked. It is gratifying to see how well researchers outside the United States are evolving from using cell lines and other archetypical models to real, living cancer models like genetically engineered mice and biopsied samples directly from the clinical setting. Techniques for using these real systems are getting more prevalent and better, but we need to be increasingly critical about their use.

Cancer biologists, clinicians, bioengineers, and biophysicists have spent a lot of time with data from The Cancer Genome Atlas (TCGA) trying to understand genomic information and infrastructure in order to solve problems and build tools. But there is a gap between cell biology and molecular biology that is impeding our progress. How do we analyze this data in the same way we analyze a genome or exome? We need to build that bridge.

A very important issue I would like to highlight is the use of tissue biopsies. Tissue samples are germane to everything we do in cancer biology. We cannot conduct studies of a patient's cancer without reliable methods to get robust biopsies of suitable quality and quantity for state-of-the-art studies. Ideally, tissue samples would be taken immediately before and after cancer therapies. However, there is often no clinical indication for a biopsy, and it is an expensive procedure to ask a surgeon or radiologist to do at a mercifully reduced rate. Our biologists are starved for this material, but there are incredible time and economic limitations here. Researchers in the United States could learn a lot from the countries that are doing biopsy collection well. If other countries are able to get these samples to their biologists in a more economical and time-sensitive way, and if they are able to encourage patient participation in sample collection, we need to bring their knowledge here. To me, this is probably the biggest priority.

Chapter 2: Parag Mallick, Complexity and Information

Parag Mallick's chapter addressed modeling complex systems and information—how to take complicated data from cancer systems and model them using mathematical approaches.

Most cancer biology and clinical studies have considered genes as a singular feature, not pieces of a complex network. This view is myopic and is starting to lose popularity in favor of a more complex systems point of view. From a clinical perspective, the first priority in modeling complex systems should be to develop tools to study the regulation of gene expression. We can apply this work to epigenetic analyses and protein expression studies. Surprisingly, we still do not understand—on a fundamental level—communication between DNA and RNA. Perhaps these tools will be better at predicting outcomes than traditional Western blots.

In many centers outside the United States, especially in Japan, there is an increasing focus on systems biology and development of approaches to describe complex systems. An example is the recent work from Hong Kong on the p53 pathway, which has already led to new insights (Chen et al. 2013). We have been studying p53 for 40 years. It is gratifying to see that we can learn new things about old systems using this new technology.

Chapter 3: Sharon Gerecht, Mimicking the Microenvironment

The cancer field has increasingly embraced the microenvironment and its role in cancer development and treatment response and resistance, but the models are not mature, either biologically or mathematically; and we need better tools to investigate the tumor microenvironment. The challenge here will not be convincing biologists that the microenvironment is important to look into. The challenge is that the systems to investigate it are not yet available, or if available are not yet tractable for most members of the cancer community. However, it is gratifying to see there is work going on to move them forward.

Researchers are using novel systems to study clues that the microenvironment uses to communicate with cancer cells. It will be key to use these systems in preclinical therapeutic studies. We think of drugs affecting cancer cells or affecting the microenvironment, but we tend not to think about situations where drugs affect multiple populations at once and how this contributes to therapeutic sensitivity and resistance.

Dr. Gerecht also reports an increase in the use of 3D systems that better capture the *in vivo* tumor environment. This is a very important advance. It is obvious that viewing the cancer and its behavior while alive in a 3-dimensional state is advantageous, but it is a technically complicated thing to do.

Information on the role of oxygen tension in cancer development and therapy will lead to better systems to manipulate oxygen levels in controlled cancer cell culture systems. Understanding the role of hypoxia is often discussed, but not studied broadly, in part because of technical challenges in controlling oxygen levels under experimental conditions.

Cancer biologists are also interested in modeling different types of cells (e.g., mesenchymal stem cells) and ligands, and their impact on cancer therapies. We need to know: What are the critical reagents for heterotypic interactions? What kinds of cells do cancer cells talk to? What ligands are most relevant? Cell modeling may help us to continue to learn how to "ping" the system for valuable data into how cell non-autonomous factors contribute to transformation and cancer therapeutic response.

Chapter 4: Cynthia Reinhart-King, Cancer Cell Mechanics

In Cynthia Reinhart-King's chapter, the important message for the biologist is the need to marry cell biology to molecular biology. These two areas have been largely overlooked in cancer biology and translational arenas. Cancer cells can look different in different patients, but no one seems to address why they look different and what this could mean to the biology of cancer.

There is clear relevance of mechanical studies to biology in the areas of metastasis, response to radiotherapy, and chemotherapy. It will be harder to find the link with epigenetics, but it is an important long-term question that needs to be addressed.

Another important mechanical question involves the process of cell adhesion and migration: How can these behaviors be linked to genetics and genomics, and further, how can it be linked to patient samples?

Integrins have become increasingly popular in cancer stem cell biology. We need to know their role in regulating cancer stem cell behavior. In the last 2 years, peer-reviewed literature has indicated that integrins have much more diverse roles than in regulating adhesion and migration.

Mechanical forces have known roles in many normal tissues (e.g., hematopoietic stem cells), but the relevance to different malignancies is not well studied.

Chapter 5: Lance Munn, Fluid Mechanics and Transport in Tumors

The hydrodynamics of fluid flow through blood cells and the lymphatic system is critical to maintaining tissue homeostasis, and as Lance Munn discusses, is frequently altered in growing tumors. Understanding solute and drug delivery by diffusion and convection in these systems is still hampered by lack of quantitative studies *in vivo* and in determining how fluid flows alter the biology of cells lining vessels or transported within them.

Current progress in imaging flow *in vivo* and in measuring interstitial pressures has great promise to alter or exploit the altered fluid flow within tumors to aid in targeted delivery of drugs or optimizing radiation treatments. At a more fundamental level, cell biologic studies are beginning to address how shear stresses and pressure gradients influence cancer and stromal cell biology.

An additional technical advance with relevance to cancer biology is the development of microfluidic methods and new biomimetic materials and surface coatings that improve methods to detect, capture, and grow circulating tumor cells, and to use these methods with the complex biological fluids relevant to patient diagnosis and potential treatment.

Chapter 6: Owen McCarty, Dynamics of Cell Motility

The link between genetics and cell biology of motility is well studied, especially in normal development and genomics. However, we do not understand a lot about how cell motility impacts human cancer biology.

Recent research shows an increasing realization that Rho proteins are mutated in cancer, but if you ask a cancer biologist why they are mutated, they have no idea where to start to get an answer. A critical event for us will be to answer how genomic alterations affect cell motility, and how this contributes to malignant transformation.

Many on/off targets of kinase inhibitors affect genes and pathways, but we also need to know if this is relevant to their therapeutic impact. For example, this would be useful in understanding unexpected toxicity in a therapy that has previously been harmless. This has been the case with the recent description of vascular/thrombotic complications with ponatinib, which were not expected and now require in depth mechanistic studies of the impact of cancer cells on thrombosis and vascular integrity.

Cancer biologists need better 3D systems to visualize motility in cancer cells and their interaction with the stroma and microenvironment. These data could be applied to information from metabolic measurements.

Having the ability to integrate these movement and interaction measurements with high-throughput genomic data will help us move beyond understanding cell lines and toward understanding patient cells, and how therapies like radiation affect these properties and affect cancer cell survival.

Chapter 7: Daniel Fletcher, Devices and New Diagnostic Principles

This may be the easiest section to see applications in the biologic or clinical field. We like and are comfortable using diagnostic tools. However, there are several areas for improvement.

There is a desire for new modalities to diagnose and follow cancer patients and to monitor response to therapy. We need to image the behavior of cells, not just the number of cells, in their *in situ* environment. This might initially be easier in model systems. Eventually, non-invasive microscopy methods could evolve to image cancer cells and the microenvironment, including secreted factors.

Microfabrication methods to study circulating cells and single cell studies is a rapidly moving field on the translational science front because it allows us to study cancer cells *ex vivo* very nicely.

Circulating tumor cell system technologies are also valuable, but most people do not die from having too many cancer cells in their blood. We need to understand the places where cancer cell proliferation leads directly to morbidity and mortality, and develop *ex vivo* ways of studying how it happens.

New imaging and diagnostic methods may give us more efficient ways of gathering data from biopsies. If it is easy and cheap to do and has direct translational impact, clinicians may be more inclined to make tissue collection a priority.

Discussion

The previous chapters are filled with new techniques and tools that have obvious applications in cancer biology. I believe that the biggest priority for researchers and clinicians like me is to develop better methods for tissue sampling. High-quality, frequent biopsies from patients would provide an incredible amount of information about the cancer as well as the response to therapy and inform the use of new physical sciences approaches. We can learn a lot from countries that already do this well—how they are able to collect the samples quickly and economically, as well as encouraging patient participation in tissue collection.

Another fascinating theme covered in this report is the development of microfabrication systems that allow us to study cancer cells *ex vivo* more thoroughly and marrying the results to single cell technology.

We need to understand how cancer cells travel to and impact different local sites, and develop *ex vivo* systems to investigate this process. There are a number of promising concepts in this report that include combining co-culture systems with imaging to study cell-cell interactions *in vivo* or *ex vivo*.

Finally, optimizing drug dosing and understanding how best to use cancer therapies including chemotherapy, radiation therapy, and targeted therapies is of great interest from a biological point of view. There are a million ways to give the therapies we have. If we have a way to visualize their impact and their behavior at different time points during treatment in a non-invasive way, we might develop better ways to use existing drugs in a more rapid fashion than through empiric, iterative clinical trials.

Reference

Chen, X., Chen, J., Gan, S., Guan, H., Zhou, Y., Ouyang, Q., Shi, J.: DNA damage strength modulates a bimodal switch of p53 dynamics for cell-fate control. BMC Biol. **11**, 73 (2013). doi:10.1186/1741-7007-11-73

Appendix A. Aphelion Study Panelists and Advisors

Panelists

Daniel Fletcher

Daniel Fletcher is professor of bioengineering and biophysics at the University of California, Berkeley; and a faculty scientist in the physical sciences at Larence Berkeley National Laboratory. He received a Ph.D. from Stanford University and a D.Phil. from Oxford University.

Dr. Fletcher's research interests include mechanics of leukemic cells, reconstruction of branched actin networks, measurement of platelet contraction, and mechanics of branching morphogenesis. He is currently exploring the mechanics of breast cancer cells, mechanical regulation of actin networks, and development of infectious disease diagnostics.

P. Janmey et al. (eds.), *Physical Sciences and Engineering Advances in Life Sciences and Oncology*, Science Policy Reports,
DOI 10.1007/978-3-319-17930-8

Sharon Gerecht

Sharon Gerecht is associate professor of chemical and biomolecular engineering at Johns Hopkins University. She received a Ph.D. in biotechnology from Technion Technical Institute of Israel and an M.S. in medical science from Tel Aviv University.

Dr. Gerecht's research interests include stem cell engineering, angiogenesis and vascular biology, and biomaterials. Her new research has branched into investigating microbioreactors, skin regeneration, and the blood/brain barrier.

Paul Janmey

Paul Janmey is professor of physiology, physics, and bioengineering at the Institute of Medicine and Engineering at the University of Pennsylvania. He received his Ph.D. in physical chemistry from the University of Wisconsin and completed his post-doc at the Hematology-Oncology Unit at Massachusetts General Hospital.

Dr. Janmey's research interests include the interaction between cytoskeletal and extracellular matrix stiffness, effects of substrate mechanics on cell structure and function, phosphoinositide signaling for actin assembly, fibrin-based materials for wound healing, and intermediate filament assembly and mechanics. His latest research explores signaling between integrins and HA receptors for cell proliferation and motility, and mechanosensing through cadherins.

Ross Levine

Ross L. Levine, MD is Laurence Joseph Dineen Chair in Leukemia Research, Human Oncology & Pathogenesis Program, at the Memorial Sloan-Kettering Cancer Center. Dr. Levine is a physician-scientist with a specific interest in the genetics and therapy of myeloid malignancies, including the myeloproliferative neoplasms (MPNs) polycythemia vera, essential thrombocytosis, and primary myelofibrosis. The focus of the Levine lab is to improve our understanding of the genetic basis of myeloid malignancies, with a specific focus on the role of oncogenic disease alleles in the pathogenesis of MPNs and acute myeloid leukemia.

Parag Mallick

Parag Mallick is assistant professor of radiology, Bio-X Program, at the Canary Center for Cancer Early Detection, Stanford University. He received his Ph.D. from the University of California, Los Angeles, and completed his post-doc in clinical proteomics and systems biology at the Institute for Systems Biology.

Dr. Mallick's current research interests include markers and mechanisms of therapeutic response to EGFR targeted therapies, models of tumor-to-circulation transmission, ProteoWizard software development, and systems models of cell-state. His work is evolving to include tumor microenvironment, cell biomechanics, and tumor evolution.

Owen McCarty

Owen McCarty is associate professor of biomedical engineering at the Oregon Health and Science University. He received his Ph.D. in chemical engineering at Johns Hopkins University and completed his post-doc in pharmacology at Oxford University.

Dr. McCarty's research has included characterization of the interaction of cancer cells with the blood microenvironment, development of anti-thrombotic strategies, and the role of Rho GTPases in platelet cell biology. He is now exploring the development of single cell imaging modalities and the identification of thrombotic risk factors in cancer patients.

Lance Munn

Lance Munn is associate professor at the Massachusetts General Hospital/Harvard Medical School. He received his Ph.D. in bioengineering at Rice University.

Dr. Munn's research projects include mechanisms of vascular remodeling during anti-angiogenic therapy, dynamics of vascular anastomosis, contribution of fluid forces to angiogenesis, collection of circulating tumor cells, and biomechanics of metastasis.

Cynthia Reinhart-King

Cynthia Reinhart-King is associate professor of biomedical engineering at Cornell University. She received her Ph.D. in bioengineering from the University of Pennsylvania.

Dr. Reinhart-King's current and past research includes topics in cell migration, cell-biomaterial interactions, cellular traction stresses, and cellular mechanotransduction. Her new research directions include microfabricated tissue structures, 3D microenvironments, and microfluidic devices for cellular studies.

Advisors

Antonio Tito Fojo

Antonio Tito Fojo is the head of the experimental therapeutics section at the National Cancer Institute, National Institutes of Health.

Dr. Fojo was born in Havana, Cuba. He moved to the United States with his family in 1960, and became a United States citizen in 1970. He received his M.D. and

Ph.D. from the University of Miami. He completed 3 years of training in internal medicine at Washington University/Barnes Hospital in St. Louis, and after a year as chief resident came to the NCI as a clinical associate in the Medicine Branch, now the Cancer Therapeutics Branch. After three years with Drs. Ira Pastan and Michael Gottesman, he assumed the position of senior investigator in the cancer therapeutics branch.

Denis Wirtz

Denis Wirtz is the Theophilus H. Smoot Professor in the department of chemical and biomolecular engineering and materials science in the Whiting School of Engineering and a member of the oncology department at the Johns Hopkins School of Medicine.

Dr. Wirtz is a recognized expert in cell and molecular biophysics and in the development of new methods grounded in physical principles, including statistical mechanics and polymer physics, to probe and establish the physical mechanisms of cell motility, intercellular adhesion, and microrheology. He is Editor-in-Chief of Cell Health and the Cytoskeleton and serves on the editorial boards of Biophysical Journal, Physical Biology, and Cell Adhesion and Migration. He is the founder and Associate Director of the Johns Hopkins Institute for NanoBioTechnology (INBT).

Appendix B. Site Visit Reports – Europe

Site visit reports are arranged in alphabetical order by organization name.

P. Janmey et al. (eds.), *Physical Sciences and Engineering Advances in Life Sciences and Oncology*, Science Policy Reports,
DOI 10.1007/978-3-319-17930-8

École Polytechnique Fédérale de Lausanne (EPFL)

Site Address:	School of Life Sciences and School of Engineering Station 15, CH-1015 Lausanne, SWITZERLAND Tel.: +41 21 693 96 68 Fax: +41 21 693 96 65 ibi@epfl.ch
Date Visited:	8 May 2012
WTEC Attendees:	Sharon Gerecht, Parag Mallick, Owen McCarty, Lance Munn (report author), Hassan Ali
Host(s):	Prof. Jeffrey Hubbell School of Life Sciences/IBI/LMRP, Station 15 Tel.: +41 21 693 9681 Fax: +41 21 693 9685 jeffrey.hubbell@epfl.ch http://epfl.ch/
	Dr. Jennifer Munson Post-doctoral fellow (Georgia Tech) Prof. Melody Swartz's lab; EPFL SV IBI-SV LLCB AI 3137 (Bâtiment AI), Station 15 Tel.: +41 21 693 31693 jennifer.munson@epfl.ch http://swartz-lab.epfl.ch/
	Asst. Prof. Matthias Lutolf Faculty of Life Sciences and Institute of Bioengineering, EPFL Office: AI1109 Tel.: +41 21 69 31876, 37281 matthias.lutolf@epfl.ch http://lscb.epfl.ch
	Prof. Henry Markram Neural Microcircuitry Laboratory EPFL, SV BMI LNMC, AAB 110 (Bâtiment AAB), Station 15 Tel.: +41 21 69-39537, 39569 henry.markram@epfl.ch

Overview

Prof. Jeffrey Hubbell opened the meeting with an overview of the EPFL Institute of Bioengineering (IBI). The life sciences program is relatively young, initiated in 2001. Hubbell was recruited to develop the IBI in 2003. The institute has a translational focus and is dedicated to interdisciplinary training: all biologist undergraduates get bioengineering training. There are 490 personnel, 33 of whom are faculty. Few of the faculty members are from Switzerland.

The WTEC panel heard presentations from four groups with projects related to cancer: Hubbell, Melody Swartz (presented by Jennifer Munson), Matthias Lutolf, and Henry Markram.

Research and Development Activities

Prof. Jeffrey Hubbell

Hubbell is a pioneer in biomaterials and protein engineering. Although some of his projects directly apply to cancer, most are geared toward more general issues of tissue engineering, drug delivery, and immunotherapeutics. By designing and producing novel hydrogel and nanoparticle biomaterials and novel protein therapeutics, he is trying to improve regenerative medicine, immunotherapeutics, and delivery of small molecule and gene drugs.

His strategies include: (1) exploring molecular variants of growth factors and adhesion protein morphogens; (2) developing new release vehicles for hydrophobic immunosuppressant, anticancer, and anti-proliferative small molecule drugs; and (3) investigating new product forms of nitric oxide. They also develop novel nonviral vectors for delivering siRNA and plasmid DNA.

An ongoing productive area of his research centers on immobilized growth factors incorporated into hydrogels or engineered matrices to convey bioactivity. For example, he has shown that hydrogels elicit better blood vessels if they are decorated with vascular endothelial growth factor (VEGF) or platelet derived growth factor (PDGF).

His group designed a fragment of fibronectin (FN) which includes a factor XIIIa substrate fibrin-binding sequence, the 9th to 10th type III FN repeat (FN III9-10) containing the major integrin-binding domain, and the 12th to 14th type III FN repeat (FN III12-14), which binds many different growth factors, including VEGF-A165 and PDGF-BB. Using this matrix component, they show synergistic signaling between $\alpha5\beta1$ integrin and the growth factor receptors only when FN III9-10 and FN III12-14 are arranged in close proximity in the FN molecule (Fig. B.1). This approach has many applications, including healing of skin and bone. He is currently implementing a similar approach in a clinical trial, treating diabetic ulcers.

Dr. Jennifer Munson

Munson, a post-doctoral fellow in Prof. Melody Swartz's lab, presented a summary of work from their group. The Swartz lab has a long-standing interest in the regulation of lymphatic transport and cancer metastasis through the lymphatic system. Also related to cancer are issues of immune cell trafficking and adaptive immunity, vaccines, and immunotherapy. They approach these studies with an impressive toolbox that includes *in vivo*, *in vitro*, and *in silico* approaches.

Munson reviewed past research highlights, including the discovery of autologous chemotaxis (cancer cells produce and follow their own chemogradients toward lymphatic vessels) and the activation of leukocyte adhesion molecules in lymphatic vessels via shear stress and cytokines. Recent work uses microfluidic devices to create well-defined gradients of CCL21. This chemokine induces migra-

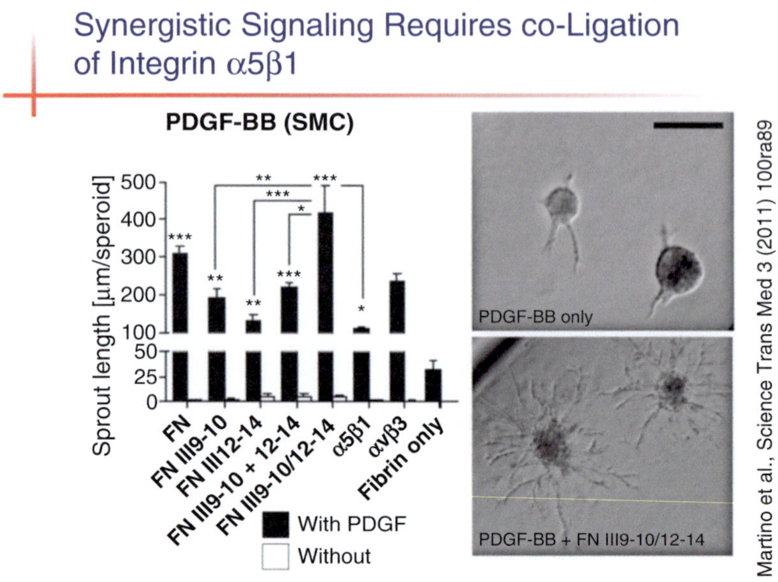

Fig. B.1 Fibronectin fragment including a factor XIIIa substrate fibrin-binding sequence (Courtesy of Jeffrey Hubbell, École Polytechnique Fédérale de Lausanne)

tion of dendritic cells, and is likely important in immunomodulation in the tumor microenvironment.

The group is also interested in immune tolerance contributed by the lymphatic system. In mice immunized with OVA, a foreign protein, tumors that express this protein have a growth delay. However, when the tumors also overexpress VEGF-C, this growth delay disappears, indicating that VEGF-C is inducing immunotolerance, possibly by affecting lymphatic function and altering the T cell population (Fig. B.2).

Asst. Prof. Matthias Lutolf

Lutolf described his work on stem cell biology and bioengineering. He is interested in how protein components of tissue-specific niches control the behavior of stem cells. Thus, his research is more directly related to stem cell biology than cancer, but nonetheless has great potential for extension to tumor biology.

His approach is to reproduce multiple microenvironments in 2D microwells molded in hydrogel (Fig. B.3). A major focus is on the neural stem cell niche, in which he has shown that notch, jagged, and dll4 are involved in self-renewal of stem cells in his devices.

An important aspect of Lutolf's methodology is the ability to adjust the stiffness of the gels in addition to the biochemical composition. Doing this, he has verified that intermediate stiffness enhances stem cell renewal. Thus, the devices allow

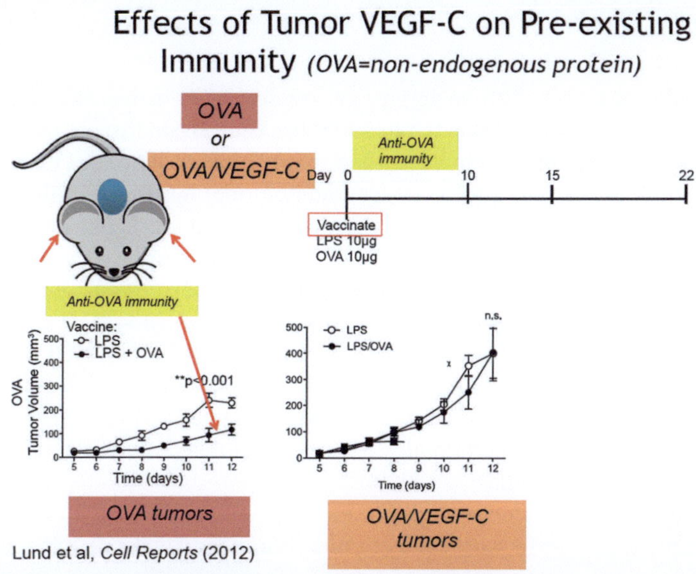

Fig. B.2 VEGF-C expression in tumors interferes with the normal immune response to OVA (Courtesy of Jennifer Munson, École Polytechnique Fédérale de Lausanne)

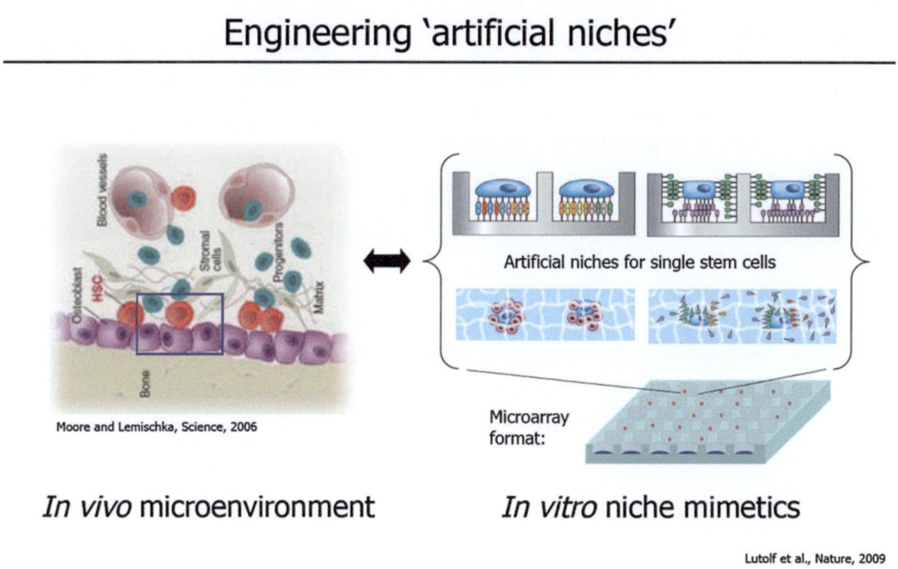

Fig. B.3 Creating artificial niches for stem cell culture using micropatterned assays (From Lutolf et al. 2009)

deliberate searches for the correct microenvironment to produce cells that will, for example, help regenerate damaged muscle. Ongoing work focuses on adapting the system to provide appropriate 3D environments for the cells.

Prof. Henry Markram

Markram ended the visit with an impressive presentation of his efforts to model brain structure and function. Founded in 2005, the Blue Brain Project uses the concept of "liquid computing" to allow easy and reliable incorporation of multiple data formats into the model. The goal is to accurately model human cognition and disease states—rather than starting with a simpler organism—because data are more readily available, and the basic structural components are similar across species.

The overall strategy is to simplify the problem. With current computational power, one dedicated processor is needed to simulate the activity of every single neuron. Thus, the billions of neurons in the brain cannot be modeled neuron-by-neuron. To overcome this limitation, the project uses multi-level simulations in which only highly active groups of neurons are simulated in detail. The resulting "virtual brain," living in supercomputers, will incorporate "all the data that neuroscience has generated to date."

The Human Brain Project is an extension of the Blue Brain Project, bringing together a consortium of 13 partners from nine European Union member states. In preparation for a full-scale flagship project, the consortium partners are each developing one specific pillar of activity that will be integrated in the final project.

Translation

Within the IBI, Hubbell's work is the most translational. He is actively working with clinicians to apply his engineered materials to patients.

Sources of Support/Funding

Each faculty member receives $1.25 million Euros per year, and is expected to match this with grants. Sources of external funding include the Swiss National Science Foundation, Oncosuisse, and the National Centre for Competence in Research in Molecular Oncology.

Summary and Conclusions

EPFL has a world-class bioengineering institute, with some emphasis on cancer. Of the sites visited, EPFL is most similar to U.S. institutions in terms of the various research directions and approaches. This is likely due to the fact that all of the PIs represented at our visit had extensive training in the United States. There is an impressive range of projects, spanning from genes to organisms, and inter-group collaborations are extensive. All of this is actively supported by the EPFL, which is obviously dedicated to nurturing these bioengineering efforts.

References

Bott, K., Upton, Z., Schrobback, K., Ehrbar, M., Hubbell, J.A., Lutolf, M.P., Rizzi, S.C.: The effect of matrix characteristics on fibroblast proliferation in 3D gels. Biomaterials. **31**, 8454–8464 (2006)

Di Crescenzo, A., Velluto, D., Hubbell, J.A., Fontana, A.: Biocompatible dispersions of carbon nanotubes: a potential tool for intracellular transport of anticancer drugs. Nanoscale. **3**, 925–928 (2011)

Eming, S.A., Hubbell, J.A.: Extracellular matrix in angiogenesis: dynamic structures with translational potential. Exp. Dermatol. **20**, 605–613 (2011)

Hubbell, J.A.: Drug development: longer-lived proteins. Nature. **467**, 1051–1052 (2010)

Velluto, D., Thomas, S.N., Simeoni, E., Swartz, M.A., Hubbell, J.A.: PEG-b-PPS-b-PEI micelles and PEG-b-PPS/PEG-b-PPS-b-PEI mixed micelles as non-viral vectors for plasmid DNA: tumor immunotoxicity in B16F10 melanoma. Biomaterials. **32**, 9839–9847 (2011)

Swartz

Haessler, U., Teo, J.C., Foretay, D., Renaud, P., Swartz, M.A.: Migration dynamics of breast cancer cells in a tunable 3D interstitial flow chamber. Integr. Biol. (Camb.). **4**, 401–409 (2012)

Miteva, D.O., Rutkowski, J.M., Dixon, J.B., Kilarski, W., Shields, J.D., Swartz, M.A. Transmural flow modulates cell and fluid transport functions of lymphatic endothelium. Circ. Res. **106**, 920–931 (2010)

Shieh, A.C., Swartz, M.A.: Regulation of tumor invasion by interstitial fluid flow. Phys. Biol. **8**, 015012 (2011)

Swartz, M.A., Iida, N., Roberts, E.W., Sangaletti, S., Wong, M.H., Yull, F.E., Coussens, L.M., DeClerck, Y.A. Tumor microenvironment complexity: emerging roles in cancer therapy. Cancer Res. **72**, 2473–2480 (2012)

Swartz, M.A., Lund, A.W.: Lymphatic and interstitial flow in the tumour microenvironment: linking mechanobiology with immunity. Nat. Rev. Cancer. **12**, 210–219 (2012)

Lutolf

Bichsel, C.A., Gobaa, S., Kobel, S., Secondini, C., Thalmann, G.N., Cecchini, M.G., Lutolf, M.P.: Diagnostic microchip to assay 3D colony-growth potential of captured circulating tumor cells. Lab Chip. **12**, 2313–2316 (2009)

Lutolf, M.P., Gilbert, P.M., Blau, H.M.: Designing materials to direct stem-cell fate. Nature. **462**, 433–441 (2009)

Roccio, M., Gobaa, S., Lutolf, M.P.: High-throughput clonal analysis of neural stem cells in microarrayed artificial niches. Integr. Biol. (Camb.). **4**, 391–400 (2012)

Sala, A., Hanseler, P., Ranga, A., Lutolf, M.P., Voros, J., Ehrbar, M., Weber, F.E.: Engineering 3D cell instructive microenvironments by rational assembly of artificial extracellular matrices and cell patterning. Integr. Biol. (Camb.). **3**, 1102–1111 (2011)

Vannini, N., Roch, A., Naveiras, O., Griffa, A., Kobel, S., Lutolf, M.P.: Identification of *in vitro* HSC fate regulators by differential lipid raft clustering. Cell Cycle. **11**, 1535–1543 (2012)

Markram

Berger, T.K., Silberberg, G., Perin, R., Markram, H.: Brief bursts self-inhibit and correlate the pyramidal network. PLoS Biol. **8**(9), e1000473 (2010). doi:10.1371/journal.pbio.1000473

Khazen, G., Hill, S.L., Schurmann, F., Markram, H.: Combinatorial expression rules of ion channel genes in juvenile rat (*Rattus norvegicus*) neocortical neurons. PLoS One. 7, e34786 (2012)

Lasserre, S., Hernando, J., Hill, S., Schumann, F., Anasagasti Pde, M., Jaoude, G.A., Markram, H.: A neuron membrane mesh representation for visualization of electrophysiological simulations. IEEE Trans Vis Comput Graph. **18**, 214–227 (2012)

Markram, H.: Newsmaker interview: Henry Markram. Blue Brain founder responds to critics, clarifies his goals. Interview by Greg Miller. Science. **334**, 748–749 (2011)

European Institute of Oncology

Site Address:	via Ripamonti 435, 20141 Milan, ITALY (The meeting took place at the Venetian Institute of Molecular Medicine, University of Padua)
Date Visited:	11 May 2012
WTEC Attendees:	Sharon Gerecht, Parag Mallick (report author), Owen McCarty, Hassan Ali
Host(s):	Alberto d'Onofrio Institute of Molecular Oncology Foundation – European Institute of Oncology Italian Foundation for Cancer Research Tel.: +39 02 574 891 Fax: +39 02 5748 9208 alberto.donofrio@ifom-ieo-campus.it www.ifom-ieo-campus.it/research/donofrio.php

Overview

The European Institute of Oncology (IEO) project was launched in 1987 as a comprehensive cancer center with research laboratories and clinical services. It is the largest oncology research institute in Italy. Research started in 1991, and clinical work in the current site began in 1994. It has been a private non-profit Scientific Institute for Research, Hospitalization and Health Care (IRCCS) since 1996. IEO has three core activity areas: clinical work, research, and training. IEO is one of Italy's 44 research hospitals and treatment centers that deal with specific disease sectors. It also has an official agreement with Italy's National Health Service (NHS), making the IEO equivalent, from the patients' point of view, to an NHS structure.

Research and Development Activities

Dr. Alberto d'Onofrio

The research of the d'Onofrio lab is focused on the application of a wide spectrum of computational and analytical tools of physics and mathematics in the basic science of cancer and clinical oncology. In particular, their research focuses on systems biomedicine. Unlike systems biology, systems biomedicine focuses on: (1) cells interplay and organ physiopathology (as "emergent properties"); (2) single- and multi-scale modeling of therapies; and (3) stressing similarities between bio-processes that are typically considered significantly distinct. Dr. d'Onofrio's use of metaphor is an important component of his research and overlaps significantly with the goals of the physical sciences in oncology research initiative. For example, it was described at the outset that the goals of the initiative might be to identify new metaphors that might be broadly applicable, in the same way that "signaling," which originated in communication theory, has now become pervasive.

Dr. d'Onofrio described a number of exciting research directions and metaphors including a metaphor between tumor-immune interactions, and the ecologic analogy of predator-prey relationships. He also highlighted the importance of noise and of bi-stability in biological systems. We briefly discussed the potential impacts of anti-angiogenic therapies and approaches for drug scheduling.

Sources of Support/Funding

Sources of support and funding come from the local area and—mainly—from the European Union.

Summary and Conclusions

Despite significant funding barriers and a climate that in Italy is not extremely conducive to interdisciplinary studies, the research being conducted at IEO is charting exciting new horizons at the interface between oncology and the physical sciences.

References

http://www.ifom-ieo-campus.it/research/donofriopub.php

Hubrecht Institute

Site Address:	Uppsalalaan 8 3584 CT Utrecht, NETHERLANDS www.hubrecht.eu/
Date Visited:	7 May 2012
WTEC Attendees:	Paul Janmey, Cindy Reinhart-King (report author), Dan Fletcher, Nastaran Kuhn, Nicole Moore, Hemant Sarin
Host(s):	Johan de Rooij Group Leader of Dynamics of Cell Adhesion j.derooij@hubrecht.eu +31 (0)30 212 19 60)
	Jacco Van Rheenen Group Leader of Cancer Biophysics j.vanrheenen@hubrecht.eu +31 (0)30 212 19 05)

Overview

The Hubrecht Institute is a research institute of the Royal Netherlands Academy of Arts and Sciences (KNAW), located on the Utrecht University campus. KNAW's research focus is on developmental and stem cell biology. Most of the university faculty as well as the University Medical Center Utrecht are located at the same site. The Hubrecht Institute also houses an imaging center, the Hubrecht Imaging Center (HIC), which was founded in 2009. Since its founding, the center has acquired a number of advanced microscopes that can be used for simple phase-contrast imaging of cells to high-resolution imaging of living tissue. HIC is headed by the microscopy manager Anko de Graaff as part of the groups of Johan de Rooij and Jacco van Rheenen.

Research and Development Activities

We heard presentations from Drs. Johan de Rooij and Jacco van Rheenen, whose major works are described below.

Dr. Johan de Rooij

The De Rooij lab is interested in the mechanics of tissue remodeling with specific emphasis on mechanotransduction mechanisms at cell-cell junctions. His lab is addressing several main questions:

- Is mechanical force a signal? How is it transduced into biochemical signals? (cadherin-based mechanotransduction)
- Is mechanical force involved in HGF-induced epithelial cell plasticity *in vivo*?
- Can we identify signaling that is intermediated directly by cell-cell junctions (not through cell migration or cytoskeletal rearrangements)?
- Which cell-cell adhesion complexes are targeted by HGF?

The de Rooij lab has focused on both e-cadherin and VE-cadherin (Fig. B.4) as force sensors. They have used magnetic tweezers to identify cadherins as mechano-sensitive and live imaging of cell-cell junctions to examine junction dynamics and

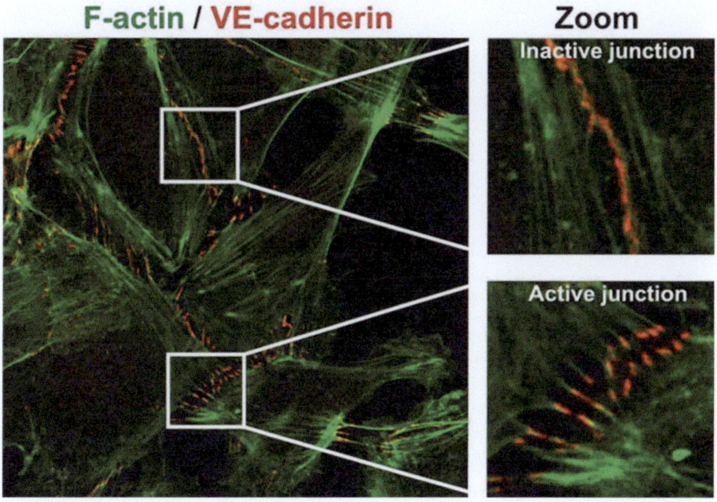

Fig. B.4 VE-cadherin regulation of cell-cell junctions (From www.hubrecht.eu/research/derooij/research.html)
Immunofluorescent image of HUVECs stained for F-actin (*green*, fibers making up most of cell body) and VE-cadherin (*red*, junctions and lines at the cell medians). The right panels show that the organization of the actin cytoskeleton influences the organization of VE-cadherin at cell-cell junctions

the relative role of focal adherens proteins in junction-sensing ability, with specific focus on vinculin. Their major goals have been to examine the role of mechanotransduction at junctions utilizing simplified 2D systems, and more complicated tumor organoid and *in vivo* models for the study of tissue morphogenesis in zebrafish. Future questions involve asking how matrix stiffness affects HGF transformation and whether focal adherens dynamics can affect metastatic capabilities.

Dr. Jacco van Rheenen

The van Rheenen group uses state-of-the-art imaging and animal models with imaging windows to study how tumor heterogeneity is formed and maintained (Fig. B.5). They are addressing four major questions:

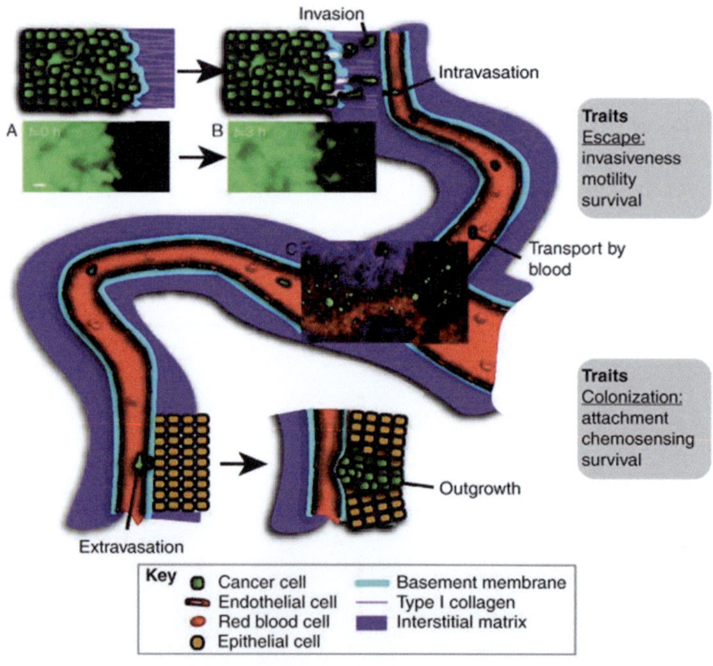

Fig. B.5 Intravital microscopy (IVM) of individual steps of metastasis (From Beerling et al. 2011)
A schematic and corresponding microscopy image of the metastatic process is shown. In the first stages, tumor cells (*green*) escape from the primary tumor and then ultimately move to and proliferate in a distant site. Cells invade the interstitial matrix (*purple*), move through the BM (*blue*), and into the blood (*red*). Cells are transported to a distant site where they form metastatic foci. IVM can be used to image metastatic processes, as illustrated by the IVM images of tumor cells (*green*), type I collagen (*purple*) and blood (*red*). IVM images A and B represent different time points of invasion of a polyoma middle T (PyMT) mammary tumor. IVM image C shows the tumor cells present in a vessel that collects blood from a C26 colorectal tumor. Scale bar = 10 μm

- How is healthy tissue formed and maintained by stem cells (focused on intestinal and mammary tissue)?
- How is heterogeneity of tumors formed and maintained (e.g., imaging of cancer stem cells)?
- How and why do tumor cells escape from primary tumors?
- How and why can tumor cells form metastasis at a distant organ?

The lab utilizes multiphoton microscopy, fluorescence lifetime imaging, and optical parameter oscillator (OPO) techniques to examine the development of tumors created by orthotopically injected tumor models and genetic models of breast and colorectal tumors. The group is addressing the question of heterogeneity by observing tumor formation from single cells *in vivo* and lineage tracing tumors by intravital imaging of individual cells.

Sources of Support

The de Rooij lab receives funding from Netherlands Cancer Society, Netherlands Scientific Organization (biophysical), Netherlands Center for Systems Biology (Imaging), and the NWO (innovational research schemes). The van Rheenen Lab receives support from Netherlands Cancer Society and NWO.

Collaborations and Possibilities

The de Rooij lab has collaborations with both Deborah Leckband and Ning Wang in the United States The van Rheenen lab collaborates with the Universitair Medisch Centrum, Utrecht, Netherlands (Onno Kraanenburg, Inne Borel Rinkes, Patrick Derksen, Rene Mederma); Albert Einstein Medical College, United States (Condeelis and Segall); The David H. Koch institute for Integrative Cancer Research at MIT, United States (Gertler); Nederlands Kanker Instituut-Antoni van Leeuwenhoek, Netherlands (Jos Jonkers, Karin de Visser, Carmen Gerlacj, Ton Schumacher); and the Hubrecht Institute (De Rooij, De Koning, Clevers, Snippert).

Summary and Conclusions

The Hubrecht Institute's emphasis on multi-scale research, state-of-the-art imaging in animal models, and tumor heterogeneity work is uncovering novel insights into mechanotransduction and tumor formations.

References

Barker, N., Huch, M., Kujala, P., van de Wetering, M., Snippert, H.J., van Es, J.H., Sato, T., Stange, D.E., Begthel, H., van den Born, M., Danenberg, E., van den Brink, S., Korving, J., Abo, A., Peters, P.J., Wright, N., Poulsom, R., Clevers, H.: Lgr5(+ve) stem cells drive self-renewal in the stomach and build long-lived gastric units *in vitro*. Cell Stem Cell. **6**, 25–36 (2010)

Beerling, E., Ritsma, L., Vrisekoop, N., Derksen, P.W., van Rheenen, J.: Intravital microscopy: new insights into metastasis of tumors. J. Cell Sci. **124**, 299–310 (2011)

Itzkovitz, S., Blat, I.C., Jacks, T., Clevers, H., van Oudenaarden, A.: Optimality in the development of intestinal crypts. Cell. **148**, 608–619 (2012)

le Duc, Q., Shi, Q., Blonk, I., Sonnenberg, A., Wang, N., Leckband, D., de Rooij, J.: Vinculin potentiates E-cadherin mechanosensing and is recruited to actin-anchored sites within adherens junctions in a myosin II-dependent manner. J. Cell Biol. **189**, 1107–1115 (2010)

Sato, T., Vries, R.G., Snippert, H.J., van de Wetering, M., Barker, N., Stange, D.E., van Es, J.H., Abo, A., Kujala, P., Peters, P.J., Clevers, H.: Single Lgr5 stem cells build crypt-villus structures *in vitro* without a mesenchymal niche. Nature. **459**, 262–265 (2009)

Snippert, H.J., Haegebarth, A., Kasper, M., Jaks, V., van Es, J.H., Barker, N., van de Wetering, M., van den Born, M., Begthel, H., Vries, R.G., Strange, D.E., Toftgard, R., Clevers, H.: Lgr6 marks stem cells in the hair follicle that generate all cell lineages of the skin. Science. **327**, 1385–1389 (2010)

Zomer, A., Beerling, E., Vlug, E.J., van Rheenen, J.: Real-time intravital imaging of cancer models. Clin. Transl. Oncol. **13**, 848–854 (2011)

www.hubrecht.eu/research/clevers/research.html

www.hubrecht.eu/research/index.html

www.hubrecht.eu/research/clevers/research.html

www.hubrecht.eu/research/derooij/research.html

www.hubrecht.eu/research/vanrheenen/research.html

www.hubrecht.eu/research/vanoudenaarden/index.html

Institut Curie

Site Address:	26 rue d'Ulm 75248 Paris Cedex 05, FRANCE http://curie.fr/en
Date Visited:	9 May 2012
WTEC Attendees:	Paul Janmey, Cindy Reinhart-King, Dan Fletcher (report author), Jerry Lee, Nastaran Kuhn, Hemant Sarin

(Continued)

Host(s):	Daniel Louvard Group Leader of Morphogenesis and Signaling Cell Director of Research at Curie Institute daniel.louvard@curie.fr
	Jean-François Joanny Physical Approach of Biological Problems Professor, Exceptional Class University jean-francois.joanny@curie.fr +33 1 56 24 67 55
	Jacques Prost Physical Approach of Biological Problems Research Director, Exceptional Class CNRS jacques.prost@curie.fr Tel.: +33 1 56 24 64 72
	Bruno Goud Group Leader of the Molecular Mechanisms of Intracellular Transport bruno.goud@curie.fr +33 1 56 24 63 98
	Jean-Louis Viovy Group Leader of Macromolecules and Microsystems in Biology and Medicine jean-louis.viovy@curie.fr Tel.: +33 1 56 24 67 52

Overview

The Curie Institute, founded in 1909, is a recognized public utility foundation. It is solely devoted to multidisciplinary cancer research and treatment, and brings together over 3,000 researchers, physicians, and caregivers for this purpose. Being the birthplace of radiotherapy, it continues to be an innovator for techniques in high-precision radiation therapy, proton therapy, brachytherapy, imaging, and oncogenetics. Its highly-recognized Proton Therapy Centre is located in Orsay and is perfectly adapted for radiotherapy of childhood tumors. Being one of the largest European research centers in cancer, it is composed of 84 teams and 15 units with platforms available for advanced cell imaging, bioinformatics, genomics, and proteomics. Since 1993, Prof. Daniel Louvard has been the director of research of the Curie Institute, where he has also functions as head of the morphogenesis and cell signaling team in the 144 UMR CNRS – Institut Curie.

Research and Development Activities

Louvard gave an overview of the history and organization of research at the Curie Institute. His priorities for strengthening the Institute include: (1) creating incentives for collaboration; (2) training young researchers; and (3) establishing core facilities. The success of these efforts in establishing the Curie Institute as one of the premier research institution at the intersection of physical science and biology was evident in both the research we heard about and the facilities we toured.

Jean-François Joanny is head of physical chemistry unit (UMR 168) at the Curie Institute and a theoretician working on physical approaches to biological problems. Bruno Goud is head of the cell biology unit (UMR 144) and is working on molecular mechanisms of intracellular transport. Joanny and Goud presented an overview of activities in the physical chemistry and cell biology units, both of which are characterized by a highly interdisciplinary group of researchers with expertise in soft matter physics, biochemistry, cell biology, and biophysics (Fig. B.6). Major initiatives include CelTisPhyBio and LabEx.

The research in cell biology and physical chemistry is supported by an integrated effort to advance interactions between cell and tissue biology and physical sciences known as "CelTisPhyBio." Researchers at the Curie have pioneered the idea that homeostatic pressure in tissues and tumors regulates their growth. They have developed a theory to describe this homeostatic pressure. Researchers are also pursuing vesicle-based reconstitution of membrane remodeling, curvature sensing, and artificial cell contraction, which is helping to identify fundamental mechanisms that animate cells. A relatively new research direction involves studying the collective behavior of cell aggregates, which show spreading and wetting behaviors that are dependent on cell-cell and cell-matrix adhesion properties.

Jean-Louis Viovy is the leader of the group studying macromolecules and microsystems in biology and medicine. He described the development of microfluidic technologies for using basic biological research and clinical applications. Viovy has developed a pressure-based flow control system with switching times under 100 ms that is being commercialized by Fluigent. Viovy also described the development of manufacturing technologies, including nanoparticle assembly, magnetic particle arrays, and template self-assembly. To support further development of microfluidic technologies, the de Genne Institute for Microfluidics will be opening in January 2014 and will provide new clean room space for microfluidics research projects. Clinical applications of microfluidics that are underway at the Curie Institute include genetic testing using "FISH in chips" and identification of circulating tumor cells with microfluidic devices.

Our tour of the Curie Institute included: (1) Joanny's theoretical group, where Edouard Hannezo described mechanical models of morphogenesis and tumorigenesis; (2) the laboratory of Pierre Nassoy, where he demonstrated the encapsulation of cells in alginate; (3) the laboratory of Francoise Brochard, where she described the behavior of multi-cellular aggregates; and (4) the core Nikon Imaging facility, with an extensive set of microscopes available for Curie Institute researchers.

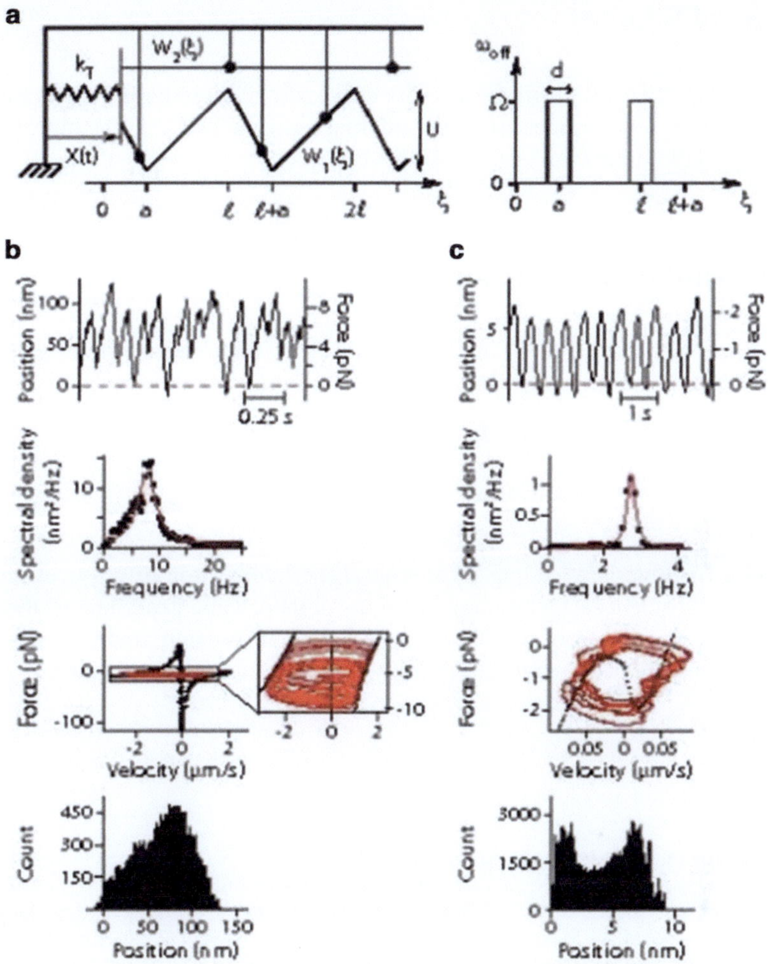

Fig. B.6 Motor oscillations. Oscillations of molecular motor assemblies obtained from stochastic simulations (Courtesy of Jean-François Joanny and Jacques Prost, Institute Curie)
The model is sketched in (*a*) which shows the potentials seen by the motors. The oscillations in the position of the filament interacting with the motors, the power spectrum and the histogram of the positions of the filament are then shown for two sets of parameters

Translational Efforts

Close interactions exist between the clinical units at the Curie Institute and the basic research units. These interactions are fostered by internal funding.

Sources of Support

Researchers at the Curie Institute are very successful at winning European funding, as well as French government funding. Internal funds at the Curie Institute are used to foster collaborations.

Collaborations and Possibilities

Many topics of mutual interest are being pursued by the Curie Institute researchers and would be of great interest for U.S. collaborations.

Summary and Conclusions

The Curie Institute is a leading institution at the intersection of physical sciences and biological sciences. Combining strengths in theory and experiments, the Curie Institute is at the forefront of both basic and clinical research, and they serve as a model of how productive multidisciplinary collaborations can be formed.

References

Basan, M., Risler, T., Joanny, J.F., Sastre-Garau, X., Prost, J.: Homeostatic competition drives tumor growth and metastasis nucleation. HFSP J. **3**, 265–272 (2009)

Dendale, R., Lumbroso-Le Rouic, L., Noel, G., Feuvret, L., Levy, C., Delacroix, S., Meyer, A., Nauraye, C., Mazal, A., Mammar, H., Garcia, P., D'Hermies, F., Frau, E., Plancher, C., Asselain, B., Schlienger, P., Mazeron, J.J., Desjardins, L.: Proton beam radiotherapy for uveal melanoma: results of Curie Institut-Orsay proton therapy center (ICPO). Int. J. Radiat. Oncol. Biol. Phys. **65**, 780–787 (2006)

Fre, S., Pallavi, S.K., Huyghe, M., Lae, M., Janssen, K.P., Robine, S., Artavanis-Tsakonas, S., Louvard, D.: Notch and Wnt signals cooperatively control cell proliferation and tumorigenesis in the intestine. Proc. Natl. Acad. Sci. USA. **106**, 6309–6314 (2009)

Lizarraga, F., Poincloux, R., Romao, M., Montagnac, G., Le Dez, G., Bonne, I., Rigaill, G., Raposo, G., Chavrier, P.: Diaphanous-related formins are required for invadopodia formation and invasion of breast tumor cells. Cancer Res. **69**, 2792–2800 (2009)

Schauer, K., Duong, T., Bleakley, K., Bardin, S., Bornens, M., Goud, B.: Probabilistic density maps to study global endomembrane organization. Nat. Meth. **7**, 560–566 (2010)

http://curie.fr/en

http://protontherapie.curie.fr/fr/institut-curie

http://umr144.curie.fr/fr/les-equipes/mecanismes-moleculaires-du-transport-intracellulaire-bruno-goud/mecanismes-moleculaires-

http://umr144.curie.fr/fr/equipes-de-recherche/morphogenese-et-signalisation-cellulaires-daniel-louvard/morphogenese-et-signal

Instituut-Lorentz

Site Address:	(meeting at the Hubrecht Institute) Uppsalalaan 8 3584 CT Utrecht, NETHERLANDS www.leidenuniv.nl/
Date Visited:	7 May 2012
Host(s):	Helmut Schiessel, Group Leader Statistical Physics of Biological Matter Instituut-Lorentz for Theoretical Physics Niels Bohrweg 2 2333 CA Leiden, NETHERLANDS lorentz@lorentz.leidenuniv.nl Tel.: +31 71 5275505

Overview

There is a long, rich tradition of physics in Leiden. The Leiden Institute of Physics (LION) was established in the fall of 1993 in order to promote and stimulate research of the highest level in a number of priority fields in experimental and theoretical physics; and to create and sustain the necessary infrastructure for an outstanding graduate education in experimental and theoretical physics. In addition, LION is responsible for teaching physics at the undergraduate level. The institute consists of the Kamerlingh Onnes Laboratorium, the Huygens Laboratorium, and the Instituut-Lorentz founded in 1921. The research groups of LION participate in three research schools that include the Casimir Research School with Delft University of Technology, Dutch Research School for Theoretical Physics, and the Graduate School for the Structure and Function of Bio-macromolecules between research groups at the Leiden Institute for Chemistry and LION.

Research and Development Activities

While at the Hubrecht Institute, we met with Helmut Schiessel from the Instituut-Lorentz. He works in the area of chromatin mechanics, including the mechanics of single base-pair steps and the structure of chromosomes and chromatin fibers. He has developed a theory that can explain data from the Seidel Lab of the extension of DNA when twisted using magnetic tweezers. In addition, he has found that there is an energy barrier to nucleosome unwrapping from DNA and force-induced strengthening that prevents unwinding (Fig. B.7). A change in conformation is necessary for nucleosome unwrapping, and just simply pulling is not sufficient to induce unwrapping.

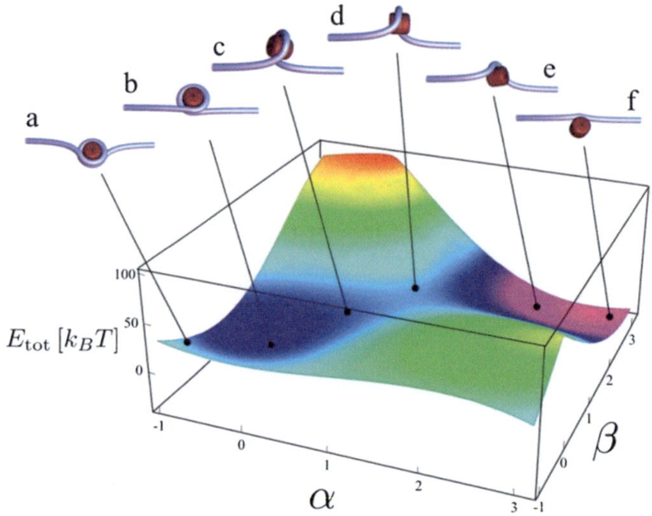

Fig. B.7 Nucleosome dynamics. (From http://www.lorentz.leidenuniv.nl/~schiessel/ResearchPages/nucleosomedynamics.htm)
Thermal fluctuations can lead to spontaneous DNA unwrapping from one of the ends of the wrapped portion. Schiessel studies how one can theoretically learn about nucleosome energetics by measuring the accessibility of DNA-binding proteins to their target sequence inside a nucleosome. The findings can be interpreted in the light of new experiments on force-induced unwrapping

Schiessel has authored a book entitled *Biophysics for Beginners: A Journey Through the Cell Nucleus* through Pan Stanford Publishing.

Collaborations and Possibilities

The Schiessel group collaborated with the late Jonathan Widom at Northwestern University, United States, in addition to collaborations in France, Iran, Japan, and Leiden.

Summary and Conclusions

Dr. Schiessel's successful collaborations with and incorporation of data from experimentalists will continue to make significant strides in our understanding of nucleosome dynamics and could have important implications for cancer research.

References

Lanzani, F.G., Schiessel, H.: Out of register: How DNA determines the chromatin fiber geometry. Europhys. Lett. **97**, 1–6 (2012)
Prinsen, P., Schiessel, H.: Nucleosome stability and accessibility of its DNA to proteins. Biochimie **92**, 1722–1728 (2010)
www.leidenuniv.nl/
www.lorentz.leidenuniv.nl/~schiessel/index.htm
www.lorentz.leidenuniv.nl/~schiessel/ResearchPages/nucleosomedynamics.htm
www.physics.leidenuniv.nl/institute/organization/organization.asp

Max Planck Institute for Dynamics and Self-Organization, Göttingen

Site Address:	(Meeting at University of Leipzig) Linnéstrasse 5 04103 Leipzig Germany www.ds.mpg.de/english/research/index.php
Date Visited:	11 May 2012
WTEC Attendees:	Paul Janmey, Cindy Reinhart-King (report author), Dan Fletcher, Lance Munn, Hemant Sarin
Host(s):	Oskar Hallatschek Biological Physics and Evolutionary Dynamics Max Planck Research Group Max Planck Institute for Dynamics and Self-Organization Bunsenstrasse 10 D-37073 Göttingen, Germany +49 0 551 5176 670 oskar.hallatschek@ds.mpg.de

Overview

The Max Planck Institute for Dynamics and Self-Organization, Göttingen, houses three specialized departments and several researcher groups studying the physical principles underlying biological interactions. The departments include the theory department of Theo Geisel (non-linear dynamics), and experimental departments of Eberhard Bodenschatz (fluid dynamics, pattern formation, and biocomplexity) and Stephan Herminghaus (dynamics of complex fluids).

The theory department focuses on theoretical and computational neuroscience, nonlinear dynamics, and transport phenomena in complex systems. Bodenschatz's experimental department investigates turbulence and other pattern formation phenomena in fluids, the physics of clouds, and self-organization in biological systems. In Herminghaus' experimental department, dry and wet granular systems are studied as paradigm systems far from thermal equilibrium, as well as self-organization in emulsions, soft autonomous microsystems, and geophysical systems. The research program of these three departments is complemented by independent groups, with research on dynamical networks (Marc Timme); turbulence in shear flows (Björn Hof); evolution on the cellular scale (Oskar Hallatschek); transport phenomena in emulsion systems (Jean-Christophe Baret); self-organized collective behavior of heart muscle cells (Stefan Luther), theoretical neurophysics (Fred Wolf); and polymers, complex fluids, and disordered systems (Annette Zippelius).

Research and Development Activities

While in Leipzig, we met with Dr. Oskar Hallatschek, the Group Leader of Biological Physics and Evolutionary Dynamics at the Max Planck Institute for Dynamics and Self-Organization. His presentation topic was "From colony growth to tumor growth." He is interested in range expansions—the movement of populations to different areas where they evolve separately. These patterns allow investigation of reproductive noise. This was first done in bacteria with Dr. David Nelson, but has been adapted to colon cancer and clonal expansion in neoplastic tissues. In these cases, the work allows for mutations to come in that confer a certain growth rate advantage. This model may be good for understanding the growth of intestinal epithelial cells out of the crypt (Figs. B.6 and B.8).

Summary and Conclusions

The group presented their research of interesting transitions from bacteria work to tumor growth using the ideas of selective pressures and advantage to understand pattern formation.

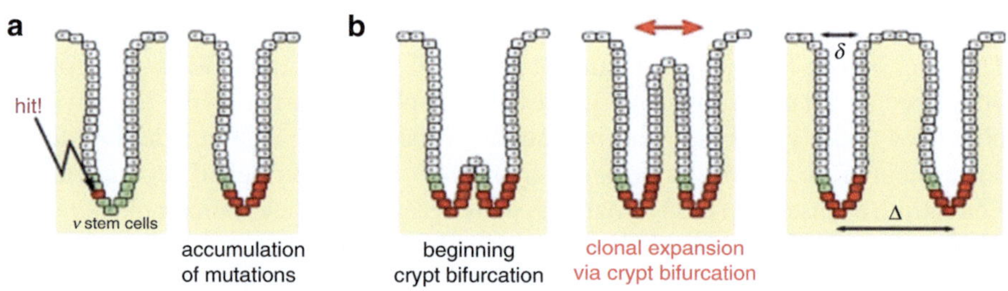

Fig. B.8 Spatial structure increases the waiting time for cancer (From Martens et al. 2011, www.evo.ds.mpg.de/)

References

Hallatschek, O., Hersen, P., Ramanathan, S., Nelson, D.R.: Genetic drift at expanding frontiers promotes gene segregation. Proc. Natl. Acad. Sci. USA. **104**, 19926–19930 (2007)

Lehe, R., Hallatschek, O., Peliti, L.: The rate of beneficial mutations surfing on the wave of a range expansion. PLoS Comput. Biol. **8**, e1002447 (2012)

Martens, E.A., Kostadinov, R., Maley, C.C., Hallatschek, O.: Spatial structure increases the waiting time for cancer. New. J. Phys. **13** (2011)

www.ds.mpg.de/english/research/index.php

www.ds.mpg.de/pdf/MPIDS_Research_Report_2011.pdf

www.evo.ds.mpg.de/

Max Planck Society, Dresden

Site Address:	Max Planck Institute for the Physics of Complex Systems Dresden (meeting at the University of Leipzig) Linnéstrasse 5 04103 Leipzig, GERMANY www.mpg.de/155526/physik_komplexer_systeme
Date Visited:	11 May 2012
WTEC Attendees:	Paul Janmey, Cindy Reinhart-King (report author), Dan Fletcher, Lance Munn, Hemant Sarin
Host(s):	Dr. Guillaume Salbreux, Group Leader Physics of the Cytoskeleton Max Planck Institute for the Physics of Complex Systems Nöthnitzer Strasse 38 01187 Dresden, GERMANY salbreux@pks.mpg.de Tel.: +49 351 871 1214

Overview

The Max Planck Society (MPG) operates a number of research institutions in Germany and internationally. These institutes are independent and autonomous in the selection and conduct of their research pursuits as long as they meet MPG's excellence criteria. The Max Planck Institutes (MPI) carry out basic research in the life sciences, natural sciences, and the social and human sciences. In 1999, MPG launched the Inter-Institutional Research Initiatives program to promote the interdisciplinary basic research amongst its institutes, including the sharing of laboratory infrastructures.

One of the projects under the Inter-Institutional Research Initiatives program is the "identification of clinical predictive markers and drug development by large-scale translational genomic analysis of lung adenocarcinoma" (2009-2015), in which the Lung Cancer Genome Project (CLCGP), the MPI for Neurological Research (Cologne), and the MPI of Biochemistry (Martinsried) collaborate. It will undertake an in-depth analytical characterization of 600 adenocarcinomas genomes to identify new predicative and prognostic markers and therapeutic targets.

Research and Development Activities

We met with Guillaume Salbreux who works in the MPI for Physics of Complex Systems directed by Frank Jülicher. They have a joint research program with MPI Molecular Cell Biology and Genetics, which is also located in Dresden. The Center for Systems Biology is expanding and getting a new building. They want to add informatics and image analysis to the already strong experimental and theoretical activities. The biological physics department of the MPI for Physics of Complex Systems is interested in studying the molecular, cellular, and collective behaviors of molecules, cells, and tissues with four main research thrusts:

- Active molecular processes (force generation)
- Collective behaviors of motors and filaments (movements and flows)
- Spatiotemporal processes in cells (cell shape control)
- Dynamic organization of tissues (growth and patterns)

Salbreux's research is focused on cytoskeleton physics and how cells and tissues change shape (Fig. B.9). His group is interested in actin-myosin contractility, creating models of blebbing, shape changes during cytokinesis, and deformations of tissues during development.

Collaborations and Possibilities

Salbreux collaborates with E. Paluch group in the MPI-CBG in Dresden to provide the theory behind blebbing and cytokinesis (Sedzinsky et al. 2011). They also collaborate with S. Grill in the MPI-CBG in Dresden and C.P. Heisenberg in IST Austria in Vienna, Austria, modeling zebrafish epiboly, and with J. Solon at the CRG in Barcelona, modeling dorsal closure during *Drosophila* development.

Summary and Conclusions

Overall, it appears that the interaction between the two MPIs creates a good environment for interactions between theorists and experimentalists.

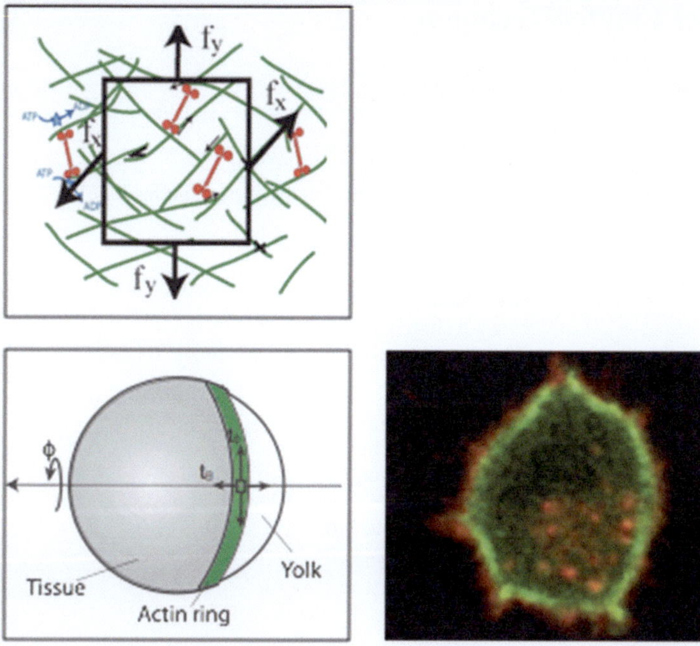

Fig. B.9 How are cell and tissue shapes dictated by the generation of forces inside the cell at the molecular level? How are the original properties of active matter, driven out of equilibrium by ATP hydrolysis, related to biological processes?
The Physics of the Cytoskeleton group in MPI for the Physics of Complex Systems addresses such questions with the help of physical analysis, numerical simulations, and through close collaborations with biologists. (Bottom right image courtesy of J.Y. Tinevez and E. Paluch, MPI Cell Biology and Genetics, Dresden, Germany)

References

Sedzinski, J., Biro, M., Oswald, A., Tinevez, J.Y., Salbreux, G., Paluch, E.: Polar actomyosin contractility destabilizes the position of the cytokinetic furrow. Nature. **476**, 462–466 (2011)
www.mpg.de/155526/physik_komplexer_systeme
www.mpipks-dresden.mpg.de
www.mpipks-dresden.mpg.de/index_en.html
www.mpipks-dresden.mpg.de/pages/forschung/frames_forsch_en.html
www.mpg.de/770854/Cooperation_with_Fraunhofer
www.mpg.de/804961/Research_Initiatives
www.mpg.de/institutes
www.pks.mpg.de/~salbreux/Physics_of_the_cytoskeleton/Home.html

Novocure Limited/Technion

Site Address:	Novocure Limited Advanced Technology Centre, Topaz Bldg. 31905, Haifa, ISRAEL www.novocure.com/ (The meeting was held at the Weizmann Institute) Technion Israel Institute of Technology Technion City, 32000 Haifa, ISRAEL www1.technion.ac.il/en/about (The meeting was held at the Weizmann Institute)
Date Visited:	14 May 2012
WTEC Attendees:	Paul Janmey and Sharon Gerecht (report author)
Host(s):	Yoram Palti (Technion-NovoCure Ltd.) Emeritus professor, Faculty of Medicine Technion Israel Institute of Technology Tel.: +972 04 8295270 palti@tx.technion.ac.il

Overview

Novocure is a commercial-stage oncology company dedicated to the advancement of tumor treating fields (TTF) therapy for patients with solid tumors. The company pioneered the concept that the electric properties of cells can be used as effective targets for anti-neoplastic therapy. Founded by Dr. Yoram Palti in 2000, NovoCure has grown to become a global organization with employees in six countries. NovoCure is headquartered in the Jersey Isle, U.K. NovoCure's U.S. operations are based in Portsmouth, NH, and the company maintains a research facility in Haifa, Israel.

Research and Development Activities

Dr. Yoram Palti

Palti is a veteran scientist and entrepreneur in the fields of biophysics, biosensors, and electrophysiology; and founder of NovoCure and Carmel Biosensors Ltd. During our visit, Palti described the challenges he faced as an entrepreneur during his academic career. Driven by a translational approach, Palti utilized a novel concept that a cell's physical properties can serve as targets for an anti-cancer therapy. Alternating electric fields have been shown to disrupt mitotic spindle microtubule assembly, resulting in dielectrophoretic dislocation of intracellular macromolecules and organelles during cytokinesis (Fig. B.10; Kirson et al. 2007). These processes

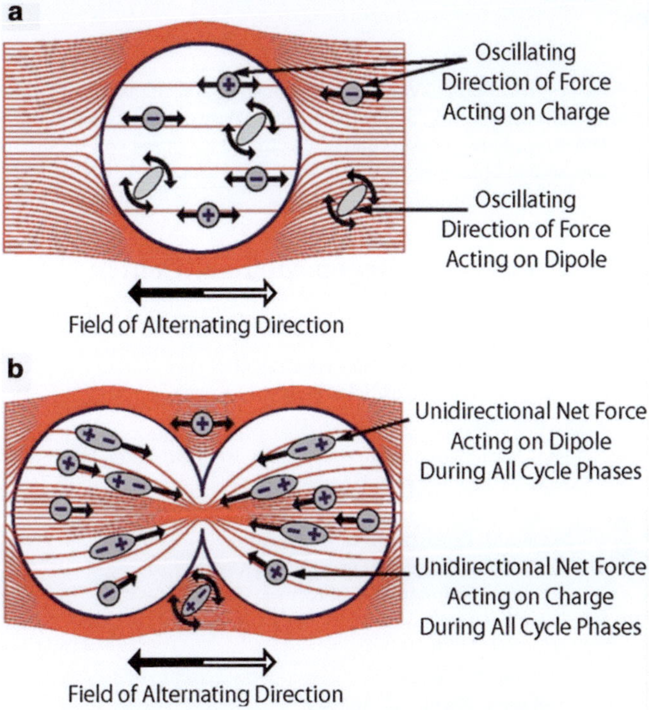

Fig. B.10 AC field distribution in and around quiescent (*A*) and dividing (*B*) cells (From Kirson et al. 2007)
Inside quiescent cells, the field is uniform, and the oscillating electric forces result only in "vibration" of ions and dipoles (the forces associated with each half cycle are indicated with white and gray arrows). In contrast, the non-uniform field within dividing cells (*B*) induces forces pushing all dipoles toward the furrow. At frequencies of 0.1–1.0 MHz, the cell membrane impedance is relatively high, so only a small fraction of the currents penetrate the cells, as seen from the density of lines

lead to physical disruption of the cell membrane and programmed cell death (apoptosis). A TTF therapy has been developed where the frequency used for a particular treatment is specific to the cell type being treated without affecting healthy cells. TTF therapy is delivered using non-invasive, insulated transducer arrays that are placed directly on the skin in the region surrounding the tumor. TTF therapy does not deliver any electric current to the tissue nor does it stimulate nerves or muscles or heat tissue. TTF therapy creates an alternating electric field within the tumor that exerts electric forces on the charged components of the proliferating cells.

Another technology developed by Palti is the transthoracic parametric Doppler—a non-invasive pulsed Doppler ultrasound technology incorporating three new modes of Doppler action. It is designed to parametrically analyze movement and flow in vital body systems. Through another company, EchoSense (http://echosense.co.il/Index.aspx), this technology is being utilized to diagnose abnormal changes pulmonary blood flow and cardiac contractility.

References

Kirson, E.D., Dbaly, V., Tovarys, F., Vymazal, J., Soustiel, J.F., Itzhaki, A., Mordechovich, D., Steinberg-Shapira, S., Gurvich, Z., Schneiderman, R., Wasserman, Y., Salzberg. M., Ryffel, B., Goldsher, D., Dekel, E., Palti, Y.: Alternating electric fields arrest cell proliferation in animal tumor models and human brain tumors. Proc. Natl. Acad. Sci. USA. **104**, 10152–10157 (2007)

St Radboud University Nijmegen Medical Center

Site Address:	Geert Grooteplein 26–28 6525 GA Nijmegen, NETHERLANDS www.umcn.nl/Pages/default.aspx
Date Visited:	7 May 2012
WTEC Attendees:	Paul Janmey (report author), Cindy Reinhart-King, Dan Fletcher, Nastaran Kuhn, Nicole Moore, Hemant Sarin
Host(s):	Peter Friedl Professor, Department of Cell Biology Chair, Microscopic Imaging of the Cell Microscopic Imaging Centre P.Friedl@ncmls.ru.nl Tel.: +31 24 36 10 907
	Wilhelm T.S. Huck Professor, Faculty of Science Group Leader of Physical Organic Chemistry w.huck@science.ru.nl Tel.: +31 24 36 52 676
	Katarina Wolf Scientific Researcher K.Wolf@ncmls.ru.nl Tel.: +31 24 36 14 296

Overview

The mission of the Radboud University Nijmegen Medical Centre (RUNMC) is to advance human knowledge by conducting biomedical, translational, and clinical research. The center's major strengths are in medical life sciences and clinical practice. RUNMC is made up of departments as well as several transdisciplinary institutes, which include the Research Institute for Oncology; Nijmegen Institute for Infection, Inflammation, and Immunity; Institute for Genetic and Metabolic Disease; Donders Centre for Neuroscience; Nijmegen Centre of Molecular Life Sciences; and Nijmegen Centre for Evidence-Based Practice. The major areas of research in the center's department of cell biology include cellular mobility, metabolism and immunity, cancer cell metastasis and invasion, and mechanisms of neurodegenerative disease. Its infrastructure is comprised of state-of-the-art technology platforms and translational research facilities. These include the Microscopic Imaging Centre (MIC) for fluorescence and electron microscopy of basic cellular processes. It also works in close collaboration with the Preclinical Imaging Centre (PRIME), which is a partnership between the clinical departments of radiology, nuclear medicine, cell biology and rheumatology of the Medical Center (Bakker et al. 2012; Gritsenko et al. 2012).

Research and Development Activities

Drs. Peter Friedl and Katarina Wolf

Friedl is chair for Microscopical Imaging of the Cell at the Nijmegen Centre of Molecular Life Sciences (NCMLS) which includes the core facility for microscopy at RUNMC. He also heads the cell dynamics lab and has a joint-faculty position as head of the imaging section at the David H. Koch Center, Department of Genitourinary Medical Oncology, M.D. Anderson Cancer Center, United States. Wolf is a scientific researcher in the department of cell biology at RUNMC.

The Friedl laboratory specializes in developing and applying new technologies to enable imaging deep into tumors to evaluate cell motility. The centerpiece of this technology at the RUNMC is a multiphoton excitation microscope for live animal imaging at subcellular resolution up to one millimeter deep within the tissue. This technology—and equally important—the animal care and cell biologic facilities within this center allow (for example) imaging of cancer cell migration out of tumors and along collagen fibers or nerve fibers (Fig. B.11). The instrumentation

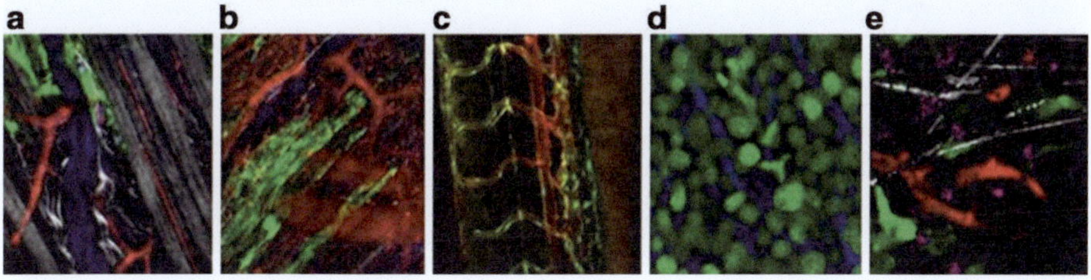

Fig. B.11 Examples of 3D tissue reconstructions obtained by intravital multiphoton microscopy and FLIM. (From www.umcn.nl/Research/Departments/cdl/PRIME/Pages/MultiPhotonMicroscopy.aspx) (*A*) B16 melanoma cells (green, top left) invading along and between muscle fibers (SHG, grey, top right and bottom left) and a nerve (THG, blue—thick central line) of the mouse dermis. (*B*) Collective invasion of B16 melanoma xenograft (green, center) along blood vessels (red). (*C*) Developing blood (red, center; yellow, right of red) and lymph vessels (green, right of yellow) in a zebrafish embryo. (*D*) Mitotic activity of B16 melanoma tumors growing in the mouse dermis by monitoring the dynamics of nuclear Histone-2B-EGFP (green, dots). Blue (lines): collagen fibers (SHG). (*E*) Tumor cells (green, lower left and upper right), blood vessels (red, central horizontal lines), macrophages (purple, dots), and collagen (gray, horizontal slashes) in the mouse dermis recorded by FLIM in one channel and discriminated by real-time phasor analysis has been developed in close collaboration with industrial partners including LaVision Biotec, Germany, and Coherent-APE. The facility allows simultaneous excitation by multiple wavelengths and imaging by fluorescence, second and third harmonic generation, and fluorescence lifetime (FRET/FLIM) techniques.

Wolf investigates the mechanism by which cancer cells migrate through 3D matrices, using *in vitro* systems of reconstituted collagen networks as a simplified extracellular matrix (ECM) mimic. The limitations of studying cell biology on smooth, flat, rigid surfaces are increasingly being overcome by studying cells embedded in soft, 3D fibrous networks that are ubiquitous in soft tissue and provide many of the micro-environmental cues for normal and pathological cell function. Recent work from Friedl, Wolf, and colleagues has examined the interplay between proteolysis of the ECM by enzymes secreted or attached to the cell surface, and the deformation of both cell and ECM by cell-generated forces that are required to push or pull the cytoplasm and nucleus though the mesh of the network. Imaging localized proteolysis using fluorescent probes and cell migration along fibers with the imaging facilities at RUNMC is shown in Fig. B.12 for a HT1080 fibrosarcoma cell migrating within a collagen gel.

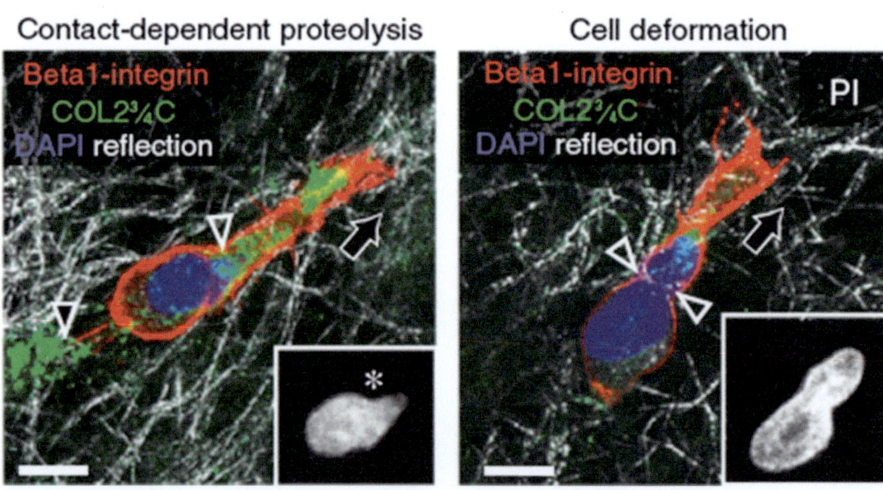

Fig. B.12 Protease-dependent and non-proteolytic cancer cell migration. (From Wolf and Friedl 2011)
Contact-dependent proteolysis of migrating HT1080/MT1-MMP cell within a 3D collagen lattice (*left*), compared with a cell in the presence of a broad-spectrum protease inhibitor cocktail (PI, *right*). With contact-dependent proteolysis intact (empty arrowhead, left image), the nucleus retains an ellipsoid, poorly deformed shape (*, inset). Proteolytic path (black arrowhead). With proteases inhibited, the inability to degrade the ECM leads to deformation of the nucleus (empty arrowheads). Thick black arrows *(a, b, d)* indicate the direction of movement. Green ECM denotes proteolytic degradation zones

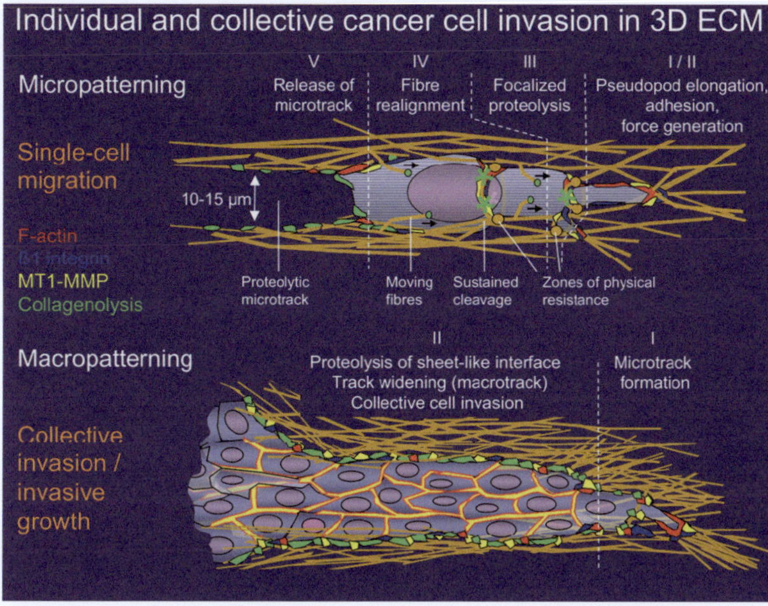

Fig. B.13 Cells can migrate individually or collectively as multicellular groups. (From Wolf et al. 2007, Friedl and Wolf 2008)
(*Top*) Single-cell migration involves five processes that change the cell shape, its position, and the tissue structure through which it migrates. (*Bottom*) Collectively migrating cells form two major zones: 1) a "leader cell" generates a proteolytic microtrack at the front of the migrating group; and 2) in which the subsequent cells then widen this microtrack to form a larger macrotrack

Migration of cells within 3D matrices is also strongly altered by cell-cell adhesions, resulting in different modes of movement for single cells and multicellular aggregates (Fig. B.13).

Wilhelm T.S. Huck

Huck is professor of chemistry at the Institute for Molecules and Materials (IMM). His lab employs microprinting and microfluidic methods to create substrates of controlled topology, chemistry, and mechanics for studies of cells on surfaces and to create picoliter cell environments for single cell and multiple cell studies in 3D. An example of the power of patterning cell substrates—a method now in use in many labs—is shown in Fig. B.14. Here, the size of a collagen-coated island strongly regulates cell proliferation, monitored by KI-67 and keratinocyte differentiation, as

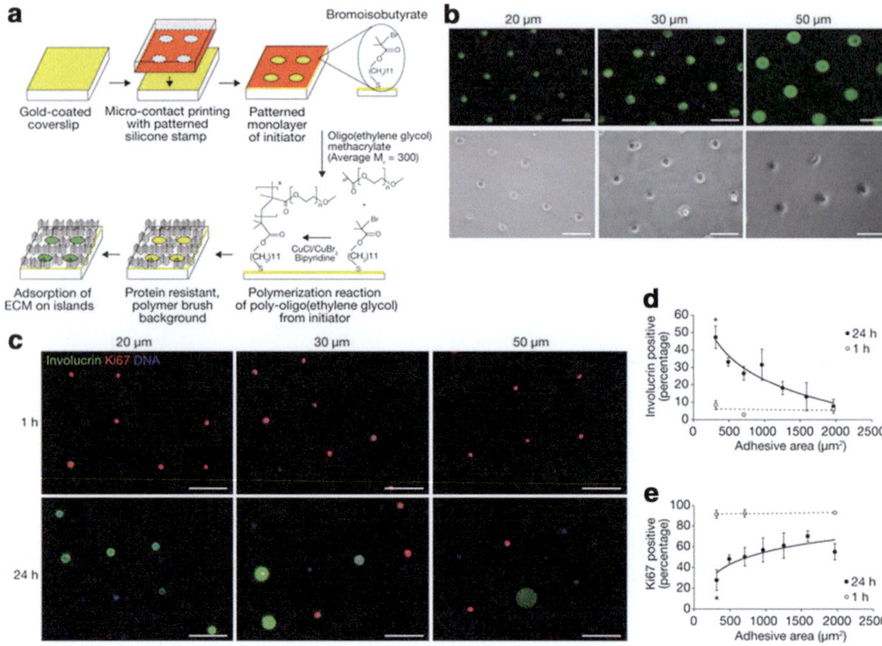

Fig. B.14 Regulation of keratinocyte shape and differentiation on micropatterned substrates (*a*) Overview of the micropatterning strategy. (*b*) Immunofluorescence microscopy images of type I collagen (*top*) and phase-contrast microscopy images of primary human keratinocytes (bottom) on 20, 30, and 50 μm diameter islands. (*c*) Representative immunofluorescence microscopy images of involucrin (green) and Ki67 (red) expression on substrates with 20, 30, or 50 μm diameters at 1 h and 24 h after seeding. Scale bars, 100 μm. (*d*) Quantification of positive cells at 1 h and 24 h on substrates with adhesive areas ranging from 314 μm² to 1963 μm² (20-50 μm diameters). Data represent means ± s.e.m. (n = 4 experiments, asterisk indicates P = 0.0001, compared with the cells on substrates consisting of 50 μm diameter islands). (*e*) Quantification of Ki67-positive cells at 1 h and 24 h. Data represent means ± s.e.m. (n = experiments; asterisk indicates P = 0.0472, compared with the cells on substrates consisting of 50 μm diameter islands). Involucrin-positive cells at 1 h and 24 h on substrates with adhesive areas ranging from 314 μm² to 1963 μm² (20-50 μm diameters). Data represent means ± s.e.m. (n = 4 experiments, asterisk indicates P = 0.0001, compared with the cells on substrates consisting of 50 μm diameter islands). (From Connelly et al. 2010)

assessed by involucrin staining. Studies of precisely patterned substrates help elucidate how physical cures such as area confinement are transduced into transcriptional and translational changes in the cells.

Other recent projects at the physics/biology interface include comparisons of different substrates with tunable stiffness that show interesting differences between two commonly used materials: polyacrylamide and polydimethylsiloxane (PDMS) as supports for integrin ligands, with a loss of stiffness responses on collagen fiber-coated PDMS that is attributed to the differences in the manner by which the collagen is linked to the surfaces. (Trappmann et al. 2012). Initiatives include the New Life Sciences and the Cornell Genomics.

Sources of Support

RUN is funded by the Netherlands Organization for Scientific Research's (NWO) Gravity Program (30 million for 10 years), as are fellow Dutch Universities and Institutes. Investigators can apply for NWO grants like the Vernieuwingsimpuls. RUN also receives award support from the Royal Netherlands Academy of Arts and Sciences (KNAW), an advisory body to the Dutch Government that manages and reviews grants and funding programs on behalf of the Dutch Ministry of Education, Culture, and Science. KNAW awards a large number of scientific and scholarly honors, for example, the prestigious biennial Heineken Prizes. Investigators at RUNMC are also eligible to receive grants by submitting proposals to the European Union's Seventh Framework Programme, the complementary Competitiveness and Innovation Framework Programme, and from the European Science Foundation—an independent, non-governmental organization dedicated to international collaboration. International sources of funding include the U.S. National Institutes of Health, the Human Frontier Science Program, and the International Agency for Research on Cancer.

Additional funding information is available at www.umcn.nl/Research/ResearchInstitutes/PDP/Pages/Grants.aspx.

Collaborations and Possibilities

RUN has collaborations within the Nijmegen Center for Molecular Life Sciences (NCMLS), which is the largest one of the university's 18 research institutes shared by the Medical Faculty and the Faculty of Natural Sciences.

Summary and Conclusions

The facilities and staff at RUN are very strongly positioned for fundamental progress in applying physical sciences to cancer biology and, more generally, biomedicine. The infrastructural funding appeared very strong, with both a significant commitment from central institutional funds and a spirit of cooperation among different divisions, for example, to pool funds from various sources in order to develop an integrated facility that combines multiple imaging modes including light, atomic force, and electron microscopy, as well as magnetic resonance imaging. There also appeared to be a strong attitude of optimism for the future of this work, with researchers seeing activity at the physics/biology interface being in a rapidly growing stage both in the Netherlands and within Europe.

References

Alexander, S., Friedl, P.: Cancer invasion and resistance: interconnected processes of disease progression and therapy failure. Trends Mol. Med. **18**, 13–26 (2012)

Bakker, G.J., Andresen, V., Hoffman, R.M., Friedl, P.: Fluorescence lifetime microscopy of tumor cell invasion, drug delivery, and cytotoxicity. Meth. Enzymol. **504**, 109–125 (2012)

Connelly, J.T., Gautrot, J.E., Trappmann, B., Tan, D.W., Donati, G., Huck, W.T., Watt, F.M.: Actin and serum response factor transduce physical cues from the microenvironment to regulate epidermal stem cell fate decisions. Nat. Cell Biol. **12**, 711–718 (2010). doi:10.1038/ncb2074

Friedl, P.: To adhere or not to adhere? Nat. Rev. Mol. Cell Biol. **11**, 3 (2010)

Friedl, P., Alexander, S.: Cancer invasion and the microenvironment: plasticity and reciprocity. Cell. **147**, 992–1009 (2011)

Friedl, P., Wolf, K.: Tube travel: the role of proteases in individual and collective cancer cell invasion. Cancer Res. **68**, 7247–7249 (2008)

Gautrot, J.E., Wang, C., Liu, X., Goldie, S.J., Trappmann, B., Huck, W.T., Watt, F.M.: Mimicking normal tissue architecture and perturbation in cancer with engineered micro-epidermis. Biomaterials. **33**, 5221–5229 (2012)

Gritsenko, P.G., Ilina, O., Friedl, P.: Interstitial guidance of cancer invasion. J. Pathol. **226**, 185–199 (2012)

Trappmann, B., Gautrot, J.E., Connelly, J.T., Strange, D.G., Li, Y., Oyen, M.L., Cohen Stuart, M.A., Boehm, H., Li, B., Vogel, V., Spatz, J.P., Watt, F.M., Huck, W.T.S.: Extracellular-matrix tethering regulates stem-cell fate. Nat. Mater. **11**, 642–649 (2012)

Wolf, K., Friedl, P.: Extracellular matrix determinants of proteolytic and non-proteolytic cell migration. Trends Cell Biol. **21**, 736–744 (2011)

Wolf, K., Wu, Y.I., Liu, Y., Geiger, J., Tam, E., Overall, C., Stack, M.S., Friedl, P.: Multi-step pericellular proteolysis controls the transition from individual to collective cancer cell invasion. Nat. Cell Biol. **9**, 893–904 (2007)

http://www.umcn.nl/Pages/default.aspx

http://www.ncmls.eu/principal-investigators/friedl/

http://www.umcn.nl/Research/Departments/cdl/PRIME/Pages/MultiPhotonMicroscopy.aspx

http://www.umcn.nl/Research/Departments/cellbiology/Pages/default.aspx

Technical University of Munich

Site Address:	James-Franck-Str. 1 85748 Garching, GERMANY http://portal.mytum.de/welcome/
Date Visited:	9 May 2012
WTEC Attendees:	Sharon Gerecht, Parag Mallick (report author), Owen McCarty, Hassan Ali
Host(s):	Andreas Bausch Department of Cellular and Molecular Biophysics; Physik-Department E27 Tel.: +49 89 289 124 80 Fax: +49 89 289 144 69 abausch@ph.tum.de http://bio.ph.tum.de/home/e27-prof-dr-bausch/research-areas.html
	Franz Pfeiffer Phone: +49 89 289 12552 Fax: +49 89 289 12548 franz.pfeiffer@tum.de www.e17.ph.tum.de/
	Martin Zacharias Physik-Department (T38) Tel.: +49 89 289 12335 martin.zacharias@ph.tum.de www.t38.physik.tu-muenchen.de/
	Hendrik Dietz Laboratory for Biomolecular Nanotechnology Am Coulombwall 4a, Walter Schottky Institut/ZNN Tel.: +49-89 289 11615, Fax: +49-89 289 11612, dietz@tum.de http://bionano.physik.tu-muenchen.de/coordinates.html
	Friedrich C. Simmel Physics Department and ZNN/WSI Am Coulombwall 4a, Room 2.016. Tel.: +49 89 289 11610 (secretary), 11611 (office) Fax: +49 89 289 11612. simmel@tum.de www.e14.ph.tum.de/index.php/group-members/friedrich-c-simmel
	Thorsten Hugel Department Biophysics and Institute for Medical Engineering Tel.: +49 89 289 16781 Fax: +49 89 289 10805 hugel@imetum.tum.de
	Matthias Rief Chair of Biophysics E22 Tel.: +49 89 289 12472 Fax: +49 89 289 12523 mrief@ph.tum.de www.bio.ph.tum.de/index.php?id=44&tx_wtdirectory_pi1

Overview

The Technical University of Munich (TUM) is one of the most research-focused universities in Germany and Europe. This claim is supported by relevant rankings, such as the DFG-Förderranking (DFG Funding Rankings) or the research rankings of the Centrum für Hochschulentwicklung (CHE-Center for Higher Education Development). TUM was one of three universities which were successful in obtaining funding in all three funding lines from the Excellence Initiative in 2006. Along with the IGSSE Graduate School and TUM's participation in five Clusters of Excellence, of which TUM is a leading institution, the strategic plan "TUM. The Entrepreneurial University" is also being developed. In addition, the university takes part in 23 collaborative research centers, of which TUM is the leading institution in nine. In the seventh European Union Research Framework Program, TUM coordinates thus far nine projects and also received six Starting Independent Researcher Grants and five Advanced Investigator Grants.

The scientists we visited commented on the importance of physics in the German curriculum. Whereas many U.S. institutions typically graduate tens of students, TUM graduates hundreds per year. There is significant demand in Germany for students with physics undergraduate degrees in all sectors of industry.

In addition, TUM has been fortunate to be an active participant in Germany's Excellence Initiative, which provides a significant funding base for diverse research as diagrammed below.

Our visit focused on researchers in the physics department. There are likely additional researchers taking a physical sciences approach to biology whom we were unable to visit with due to time constraints.

Research and Development Activities

Dr. Andreas Bausch

Bausch's group highlighted exciting emergent properties in fibrillar self-assembly. In one example, he explored the emergence of collective motion in a high-density motility assay that consists of highly concentrated actin filaments propelled by immobilized molecular motors in a planar geometry. Above a critical density, the filaments self-organize to form coherently moving structures with persistent density modulations, such as clusters, swirls, and interconnected bands (Fig. B.15). These polar nematic structures are long-lived and can span length scales orders of magnitudes larger than their constituents. This property of context-dependent state transitions is likely a general property that applies ubiquitously throughout biology. In another example, he highlighted the existence of absorbant states in natural systems. Specifically, they consider a 2D system of actin filaments. A combination of active directed motion and steric repulsion causes the system to produce dynamic patterns

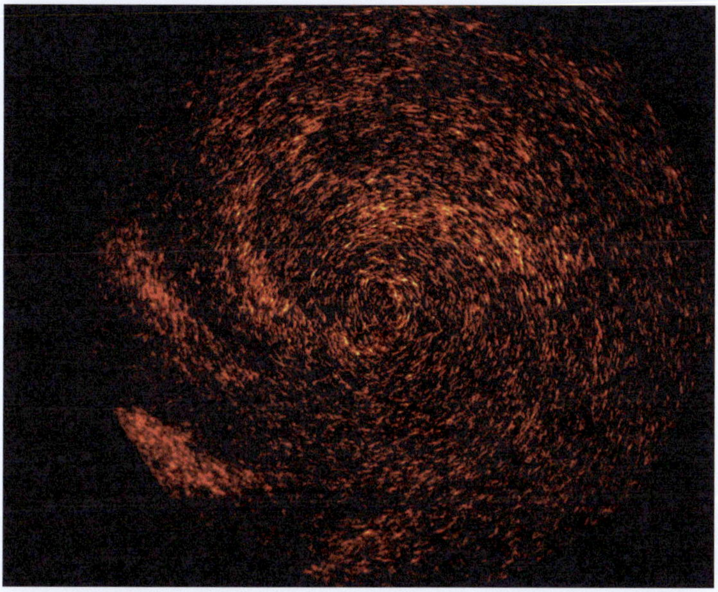

Fig. B.15 Self-organizing actin filaments (From Schaller et al. 2010)

in the form of density fluctuations and waves; however, these are short-lived structures that appear and disappear, so the system has similarities to other fluctuating nonequilibrium liquid states. When a critical amount of protein is reached, cross-linking between actin filaments leads the system to self-organize into a distinct moving state characterized by all the hallmarks of an absorbed or dynamically frozen state. Bausch showed a third example of such a phase transition governed by pH as well.

Dr. Franz Pfeiffer

The work of the Pfeiffer group is focused on the translation of modern X-ray physics concepts to biomedical sciences and clinical applications. They are particularly interested in advancing conceptually new approaches for biomedical X-ray imaging and therapy, and working on new kinds of X-ray sources, contrast modalities, and images processing algorithms. Their activities range from fundamental research using state-of-the-art, large-scale X-ray synchrotron and laser facilities to applied research and technology transfer projects aiming at the creation of improved biomedical device technology for clinical use. From a medical perspective, their work currently targets early cancer and osteoporosis diagnostics (Fig. B.16). In one study, they highlighted the role of basic physical principles to improve phase contrast imaging. This technique may ultimately allow differentiation of pathologic from non-pathologic tissues as well as better molecular annotation (Fig. B.17). A

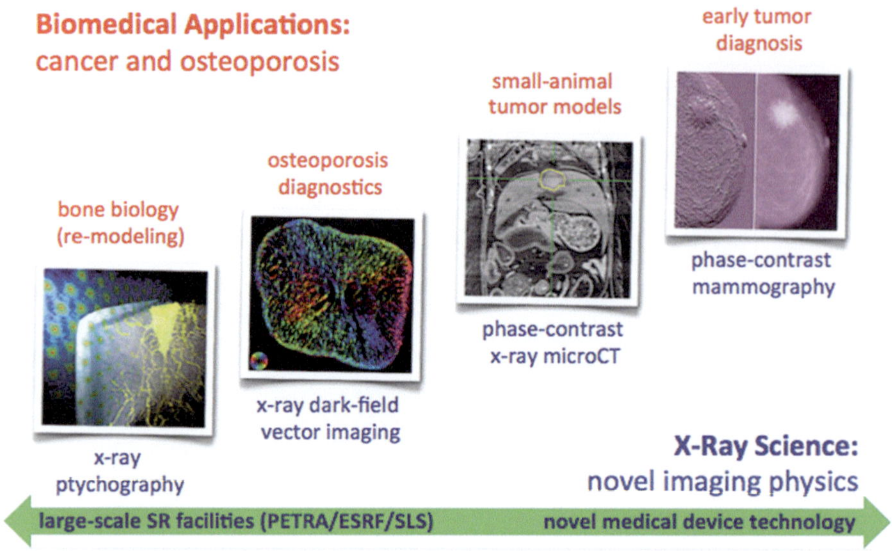

Fig. B.16 Targeting early cancer and osteoporosis diagnostics (Courtesy of Franz Pfeiffer, Technical University of Munich)

Fig. B.17 A conventional X-ray (*left*) and DPC image (*right*) of a fish (From Pfeiffer et al. 2006)

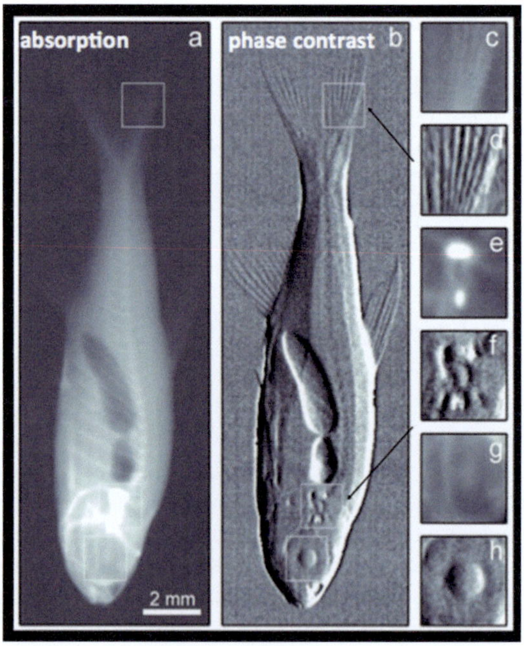

conventional X-ray transmission image reveals the skeleton of the fish and other highly absorbing structures, such as the calcified ear stones (otoliths). However, small differences in the density of the soft tissue (e.g., the different constituents of the eye) are hardly visible in the conventional absorption image, but clearly evident in the corresponding differential phase contrast (DPC) image. They have additionally developed approaches to apply these approaches in 3D.

Dr. Martin Zacharias

The function of proteins and nucleic acids in living systems is strongly coupled to the molecular motion and dynamics of these biomolecules. The Zacharias group uses computer simulation methods to study the structure, function, and dynamics of biomolecules. Their primary tool in these inquiries is classical molecular dynamics. This approach allows them to extract thermodynamic and kinetic properties of biomolecular systems to look at the impact of mutation on binding of diverse drugs and nucleic acids. For example, they look at the dynamics of peptide binding in MHC class 1 molecules (Fig. B.18), an allele dependent binding mechanism. In particular, some alleles appear more likely to require helper proteins to load their binding clefts and local flexibility appears to be a critical mediator of function.

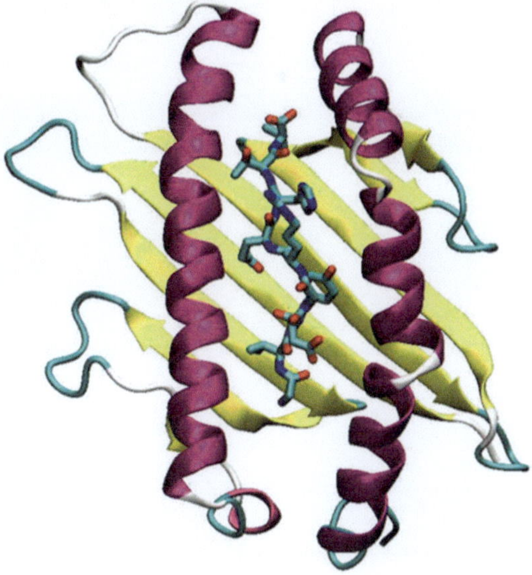

Fig. B.18 MHC class 1 molecule (Courtesy of Martin Zacharias, Technical University of Munich)

Dr. Hendrik Dietz

The Dietz lab is interested in using DNA and protein as building blocks for constructing diverse structures. They refer to these efforts as DNA origami. They are able to assemble a wide array of complex shapes by using scaffold and staple molecules (Figs. B.19 and B.20).

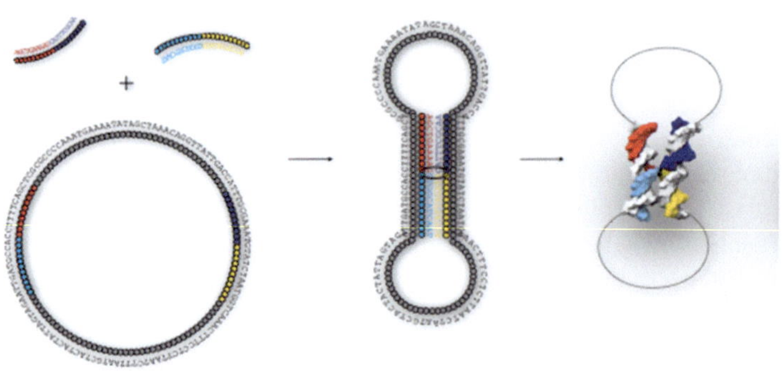

Fig. B.19 Scaffold and staple molecules (Courtesy of Hendrik Dietz, Technical University of Munich)

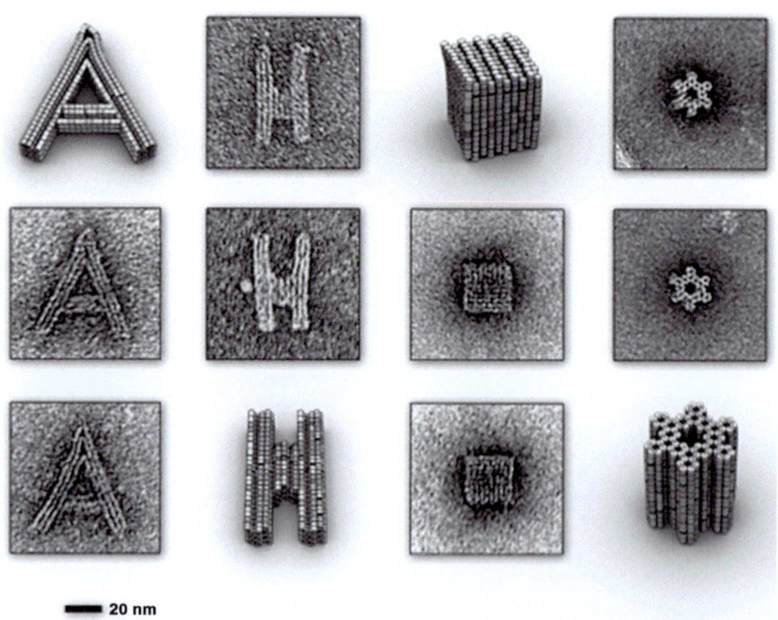

Fig. B.20 Complex shapes obtained by using scaffold and staple molecules (From Franco et al. 2011)

They are also able to introduce bends and curvature to further expand the space of structures they can develop. This work may ultimately lead to novel delivery strategies, such as suggested by the Church lab. In addition, such precise control may allow creation of novel nanostructures that would otherwise be unattainable.

An extensive resource on biomolecular nanotechnology can be found here: http://bionano.physik.tu-muenchen.de/biomolecular_nanotechnology.html.

Dr. Friedrich Simmel

The Simmel Lab works closely with the Dietz lab to develop self-organizing molecular systems that are able to respond to their environment, compute, move, and take action. Their goal is to develop reconfigurable, autonomous systems that can learn, evolve, or develop. Simmel described a number of research areas in his group including super-resolution imaging of DNA origami structures for investigating binding and unbinding kinetics. As DNA origami structures allow the organization of small molecules, proteins, aptamers, or nanoparticles into specified geometries, they represent promising scaffolds for molecular computation, artificial molecular machines, molecular assembly lines, nanorobots, and factories. Such applications imply dynamic processes and require dynamic functional imaging in real time with high spatial resolution. They introduce a single-molecule assay for dynamic binding and dissociation of short fluorescently labeled DNA oligonucleotides to single-stranded docking strands protruding, e.g., from DNA nanostructures. This allowed them to determine kinetic rates and how those rates impact concentrations, temperature, and binding site location on the nanostructures.

Another exciting area of research is the development of a synthetic transcriptional circuit that can be used as a molecular clock for timing biochemical processes *in vitro*. They used this approach to drive a set of DNA tweezers. Figure B.21 shows the regulatory circuit. When switch SW21 is turned on, RNA polymerase transcribes regulatory RNA (rI2) from the genelet template T21. RNA strand rI2 inhibits transcription from switch SW12 by removal of DNA strand A2 from template T12, resulting in an incomplete promoter region. On the other hand, RNA species rA1, which is transcribed from SW12, activates transcription from SW21 by releasing A1 from the A1•dI1 complex. RNA levels in the system are controlled by RNase H-mediated RNA degradation. By fluorescently labeling strand T21 with Texas Red or TYE665 (red dot), strand T12 with TAMRA or TYE563 (green dot), and activation strands A1 and A2 with Iowa Black RQ quenchers (black dots), the genelet states can be monitored by fluorescence measurements—high signals correspond to low transcription activity.

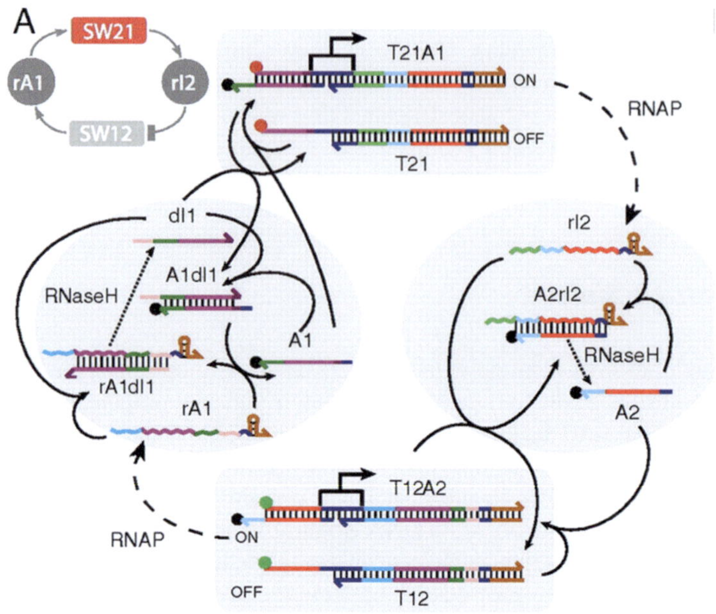

Fig. B.21 Synthetic transcriptional circuit (From Franco et al. 2011)

Dr. Thorsten Hugel

The Hugel lab uses single molecule methods to gain a thorough understanding of complex biological processes. These methods allow real-time observation of molecular machines at work and their specific manipulation. Results of such experiments yield new insights into problems from fundamental physics at the nanoscale to the development of new drugs.

Recently, the Hugel group in cooperation with the Buchner group (Biotechnology, TUM), discovered a rocking motion of the heat shock protein and molecular chaperone Hsp90 (Fig. B.22). Hsp90 is eminently important because it plays a decisive role in many basic cellular processes—in humans as well as in bacteria or yeasts. For example, it is decisive in folding polypeptide chains into functioning proteins with very precisely defined spatial structures. Especially when cells are exposed to stress through heat or poisonous substances, Hsp90 production increases to keep the damage in check.

Particularly interesting is that the double scissor movements at the N and C terminals are closely coupled: The Hsp90 dimer obviously opens and closes in alternation at each end, like a rocker. That explains the great stability of the dimmer—otherwise

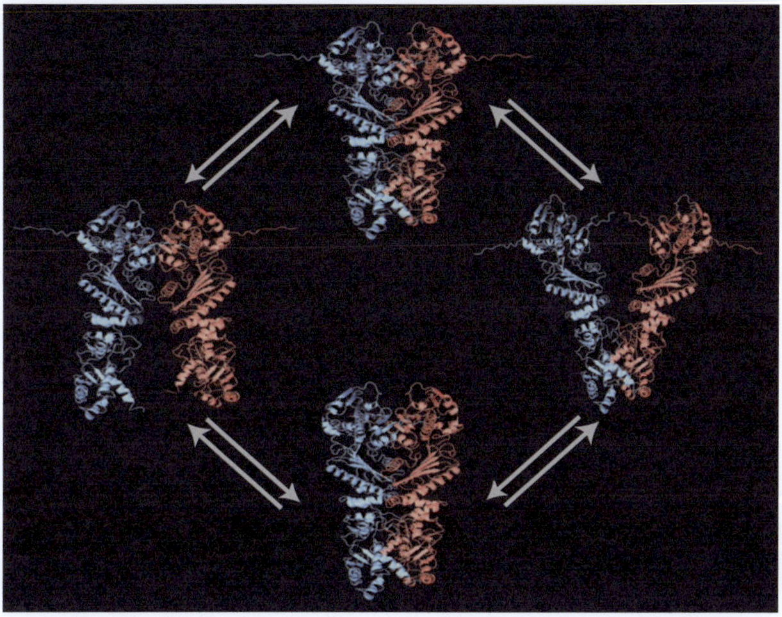

Fig. B.22 Diagram of the rocking motion of yeast Hsp90 (From Ratzke et al. 2010)

Hsp90 would fall apart much faster. The observed movement and communication patterns are interesting not only for basic research, but also for medical research since Hsp90 is a new drug target in cancer therapy. The most promising drug candidates to date block the binding of ATP at the N terminal domains of the anti-stress protein. However, these compounds may have undesirable side effects. Thanks to their new insights, the TUM researchers can now concentrate on the C terminal dimerization of Hsp90, where there are unique docking points for drugs that should function without side effects.

Dr. Matthias Rief

Proteins are fascinating examples of self-organized molecular machines. Without any help, a polypeptide strand can fold into functional 3D structures. Reif's lab is interested in studying the function and folding process of proteins on the single molecule level. Examples are single molecule folding/unfolding studies or the motility of molecular motors in optical traps.

Translation

The Pfeiffer group is actively working towards translation of their novel imaging modality. However, the majority of groups we met with are focused more towards basic sciences.

Sources of Support/Funding

Support is from the European Research Council (ERC), the collaborative research center Forces in Biomolecular Systems (SFB 863), Institute for Advanced Study (IAS), and the Center for Integrated Protein Science (CIPS).

Summary and Conclusions

The groups at TUM are focused heavily on exploring fundamental molecular biophysics. Their unique approach focuses on identifying specific questions to ask, developing novel experimental systems to ask and answer those questions and then answering them. Through their studies, they have uncovered numerous exciting properties of biomolecules and have a particular strength in studying group behaviors.

References

Franco, E., Friedrichs, E., Kim, J., Jungmann, R., Murray, R., Winfree, E., Simmel, F.C.: Timing molecular motion and production with a synthetic transcriptional clock. Proc. Natl. Acad. Sci. USA. **108**, E784–793 (2011)

Ratzke, C., Mickler, M., Hellenkamp, B., Buchner, J., Hugel, T.: Dynamics of heat shock protein 90 C-terminal dimerization is an important part of its conformational cycle. Proc. Natl. Acad. Sci. USA. **107**(37), 16101–16106 (2010). doi:10.1073/pnas.10000916107

Schaller, V., Weber, C., Semmrich, C., Frey, E., Bausch, A.R.: Polar patterns of driven filaments. Nature. **467**, 73–77 (2010)

Pfeiffer, F., Weitkamp, T., Bunk, O., David, C.: Phase retrieval and differential phase-contrast imaging with low-brilliance X-ray sources. Nat. Phys. **2**, 258–261 (2006)

Smith, D.M., Schuller, V., Forthmann, C., Schreiber, R., Tinnefeld, P., Liedl, T.: A structurally variable hinged tetrahedron framework from DNA origami. J. Nucleic Acids. **2011**, 360954 (2011)

The Royal Institute of Technology (KTH)

Site Address:	Life Science Technology Platform (meeting at Uppsala University) Dag Hammarskjölds väg 20 752 37 Uppsala, SWEDEN www.kth.se/en
Date Visited:	8 May 2012
WTEC Attendees:	Paul Janmey, Cindy Reinhart-King, Dan Fletcher (report author), Nastaran Kuhn, Nicole Moore, Hemant Sarin
Host(s):	Wouter van der Wijngaart Professor, KTH School of Electrical Engineering Micro- and Nanosystems Department Osquldas väg 10, 100 44 Stockholm, SWEDEN wouter@kth.se Tel.: +46 9 790 6613
	Jerker Widengren Professor, Experimental Biomolecular Physics, Department of Applied Physics jerker@biomolphysics.kth.se Tel.: +46 8 55 37 8030
	Ozan Öktem Associate Professor, Department of Mathematics ozan@kth.se Tel.: +46 9 790 6606)
	Christian Gasser Department of Solid Mechanics School of Engineering Sciences tg@hallf.kth.se Tel.: +46 8 790 7793
	Jochen Schwenk Associate Professor, Platform Manager at Science for Life Laboratory Human Protein Atlas Project jochen.schwenk@scilifelab.se Tel.: +46 8 524 814 82

Overview

KTH is Sweden's premier technical university. One-third of Sweden's technical research and engineering education takes place here (13,000 undergraduates, 4,500 post-graduates and employees, and 64 research teams). KTH's core research strengths are categorized into five platforms: (1) energy; (2) materials; (3) life science technologies; (4) information and communication technology; and (5) transport. These platforms serve as the basis for multidisciplinary research initiatives of KTH units as well as with its external partners—which include industry, healthcare, academia, and the general public. Within Life Sciences, KTH has six focus areas: (1) bioimaging; (2) biomolecular tools and biomaterials; (3) fundamental research in life science; (4) medical devices; (5) mathematical and computational sciences; and (6) infrastructure in health.

Research and Development Activities

Dr. Wouter van der Wijngaart, the Life Science Technology Platform Director, provided an overview of biomedical research in Sweden and of his research in microsystems with a focus on cancer treatment. The Stockholm-Uppsala region is the third largest medical technology research concentration in Europe, counting over 700 business and academic organizations active within the field. KTH is serving as the central academic technology partner. About 15 to 20 percent of research at KTH is focused on life science related technologies, with six focus areas (described above). These strengths at KTH support a broad range of large efforts, including the Stockholm Brain Project, the Science for Life Lab, the Centre for Technology in Medicine and Health, and the Human Protein Atlas Project.

Dr. van der Wijngaart is Professor at the KTH Micro- and Nanosystems Department, one of the internationally leading academic MEMS players (Fig. B.23). The Department develops, amongst others, micro- and nanotechnologies to advance and accelerate biomedical research. Examples of cancer treatment related devices under development include a lab-on-a-chip cell encapsulation system, an internal radiation therapy device, a microwave sensor for skin cancer diagnosis, and circulating tumor cell technology.

Dr. Jerker Widengren is the experimental biomolecular physics group leader and presented his recent research on ultrasensitive spectroscopy and imaging. Widengren's research includes development and use of fluorescence correlation spectroscopy (FCS) and fluorescence cross-correlation spectroscopy (FCCS) to study molecular interactions and protein densities. A new and potentially powerful imaging method pioneered by Widengren is the use of the triplet state as a source of contrast. This method, known as TRAST, has been used to image metabolism, including O_2 concentrations in live cells. Widengren is also applying his broad

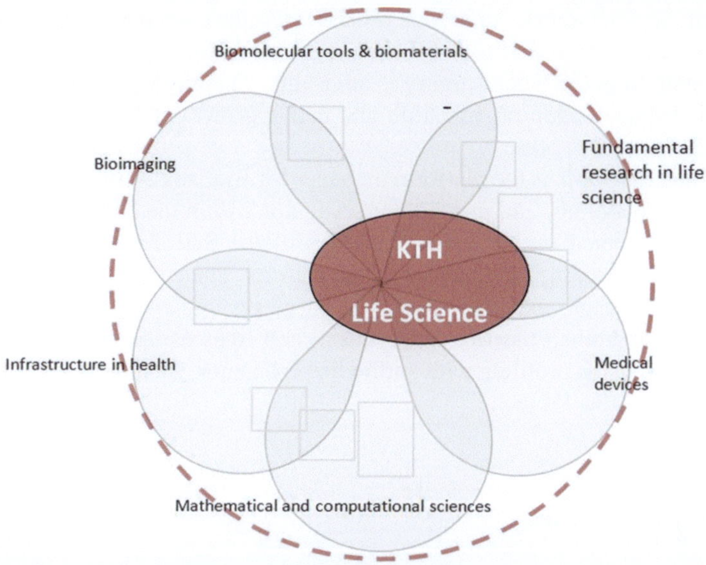

Fig. B.23 Overview of the KTH Life Science Technologies Platform (From Wouter van der Wijngaart, courtesy of The Royal Institute of Technology)

range of fluorescence imaging modalities to cancer diagnostics, where individual cells from fine needle aspiration samples are being analyzed with advanced spectroscopic and microscopy techniques.

Dr. Ozan Öktem from the department of mathematics gave an overview of bioimaging technologies in the life science technology platform. Regarding instrumentation, KTH has developed several medical imaging technologies that have now been commercialized, including X-ray phase contrast imaging (eXcillum) and mobile 3D SPECT (Adolesco). KTH is also developing advanced photon-counting and energy-resolving detectors. Regarding algorithms for image reconstruction and signal processing, new methods of image processing are needed to extract the most useful clinical information from these and other imaging modalities. Öktem is, together with others at the department of mathematics at KTH, developing image reconstruction and signal processing based on sparse signal processing. The department of mathematics also pursues applied research on methods for optimization of radiation therapy. This is done jointly with an industrial partner, RaySearch Laboratories.

Dr. Christian Gasser in the department of solid mechanics presented his research on soft biological tissue modeling, which is aimed at addressing cardiovascular diseases, specifically abdominal aortic aneurysm. A large fraction of the population has this condition, and it is not dangerous until is ruptures, so the clinical challenge is to determine when repair is necessary. Gasser has developed an integrated rupture risk assessment method that uses CT data together with peak wall stress computations that help to guide when surgery is advisable. The mechanical modeling takes into account the collagen organization and active growth of the vessels to predict time and location of rupture.

Dr. Jochen Schwenk is the platform manager for Biobank Profiling at the science for life lab (SciLifeLab) and principle investigator within the Human Protein Atlas project. He presented on the major activities of the SciLifeLab, which included genomics, RNA profiling, and bioinformatics efforts. The ambitious Human Protein Atlas provides protein expression profiles based on immunohistochemistry for a large number of human tissues, cancers, and cell lines using antibodies raised in white rabbits. Plasma profiling with antibodies is a major, yet not public effort in the Human Protein Atlas Project.

Translational Efforts

Strong connections with industry are enabling translation of new technology. This is true for most of the medical and biomedical focused research. Furthermore, KTH is one of the driving forces in unifying the Stockholm region with respect to medical and biomedical innovations.

Sources of Support

Major sources of funding include:

- The Swedish Research Council
- The European Commission through the Framework Programs (FP6, FP7, Horizon 2020)
- Through industrial collaborations
- The Knut and Alice Wallenberg Foundation
- The Swedish Human Protein Atlas Project (funded by the Knut and Alice Wallenberg Foundation).

Within the coming years, the Science for Life Lab will be supported by industry and the Swedish national government with up to 100 M€ annually.

Collaborations and Possibilities

The extensive data collected as part of the Human Protein Atlas Project are publically available. The large European biobanking efforts would provide an exciting opportunity for collaboration with U.S. researchers.

Summary and Conclusions

KTH is a leader in biomedical research and has a particular strength in collaborative biomedical projects. By levering new technologies and biobanking resources in Sweden, KTH is advancing the state-of the art in biomedical research.

References

Uhlen, M., Oksvold, P., Fagerberg, L., Lundberg, E., Jonasson, K., Forsberg, M., Zwahlen, M., Kampf, C., Wester, K., Hober, S., Wernerus, H., Björling, L., Ponten, F.: Towards a knowledge-based Human Protein Atlas. Nat. Biotechnol. **28**, 1248–1250 (2010)
www.kth.se/en
www.kth.se/lifescience
www.proteinatlas.org/
www.scilifelab.se
www.ctmh.se

University Medical Center Utrecht

Site Address:	(Meeting at the Hubrecht Institute) Uppsalalaan 8 3584 CT Utrecht, NETHERLANDS www.uu.nl/
Date Visited:	7 May 2012
WTEC Attendees:	Paul Janmey, Cindy Reinhart-King (report author), Dan Fletcher, Nastaran Kuhn, Nicole Moore, Hemant Sarin
Host(s):	Philip de Groot Prof. of Thrombosis and Hemostasis, Department of Hematology University Medical Center Utrecht Heidelberglaan 100 3584 CX Utrecht, NETHERLANDS ph.g.degroot@umcutrecht.nl Tel.: +31 88 7557769

Overview

The University Medical Center, Utrecht (UMC Utrecht), is comprised of the Faculty of Medicine of Utrecht University, the former Academic Hospital, and the Wilhelmina Kinderziekenhuis Children's Hospital. Faculty members conduct interdisciplinary research in biomedical genetics, biomedical image sciences, and clinical epidemiology. The medical center's research areas include brain, infection and immunity, circulatory health, personalized cancer care, regenerative medicine, and child health. UMC Utrecht and the University closely liaise in large-scale research programs such as TI Pharma, the Centre for Translational and Molecular Medicine, and Biomedical Materials. The development of Science Park Utrecht at De Uithof has also increased collaboration between both institutions.

Research and Development Activities

While we were at the Hubrecht Institute, our group met with Dr. Philip de Groot from UMC Utrecht. Even though he is a biochemist by training, he spoke about work at the intersection of hemostasis and engineering. The department includes both a diagnostic lab that employs 300 people. The research labs cover four major topics and extensively collaborate with engineers in several major areas:

- Flow models for hemostasis
- Point-of-care diagnostic assays
- Improved assays to detect microparticles
- The study of the intersection of primary and secondary hemostasis

Questions that continue to be a focus for the research labs include:

- Why do some patients bleed?
- Can we create a reliable model for thrombosis that incorporates elements of flow and adhesion, which are essential for drug testing?
- Can we create better point-of-care tests for bleeding?

The use of novel antibodies made from llamas was discussed. Llama antibodies are unique because they only have the heavy chain, so the antigen interaction site is smaller, making it more specific. As such, they have the ability to make antibodies against specific conformations of antigens (Fig. B.24).

Summary and Conclusions

The combination of diagnostic labs, biochemists, and engineers is a strength of UMC Utrecht that will continue to allow them to make significant advances in blood research.

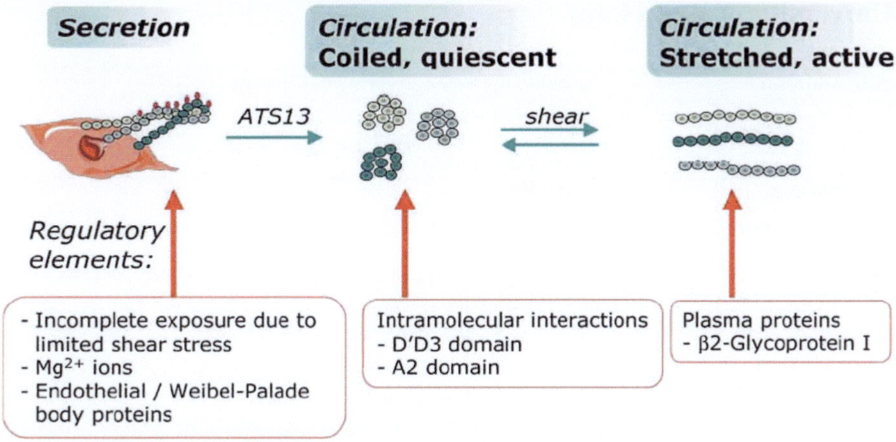

Fig. B.24 Pathways that inhibit formation of VWF-platelet aggregates. (From Lenting et al. 2010) "The formation of VWF-platelet aggregates may be inhibited at the level of VWF by several pathways. First, limited exposure to shear stress, the presence of Mg2+ ions and/or the inhibition of VWF-platelet interaction by proteins that are co-localized with VWF in the Weibel-Palade bodies (or eventually proteins located in the endothelial cytoplasm) may result in a reduced capacity of VWF to bind to platelets. Second, proteolysis of VWF at the endothelial surface by ADAMTS13 relieves VWF from wall shear stress, and allows the transition from an elongated, platelet-binding configuration into a globular quiescent form. In this globular form, intra-molecular interactions between the A1 domain and its adjacent regions (i.e., the amino -terminal D′-D3 domains and the carboxyterminal A2 domain) are in place to reduce platelet accessibility. Finally, under conditions where circulating globular VWF adopts an active platelet-binding conformation, β2-GPI may act as a 'first line of defense' to prevent undesired platelet aggregation"

References

de Groot, P.G., Rolf, T., Urbanus, R.T.: Screening: Guidelines for antiphospholipid antibody detection. Nat. Rev. Rheumatol. **8**, 125–126 (2012)

de Laat, B., de Groot, P.G.: Autoantibodies directed against domain I of beta2-glycoprotein I. Curr. Rheumatol. Rep. **13**, 70–76 (2011)

Lenting, P.J., Pegon, J.N., Groot, E., de Groot, P.G.: Regulation of von Willebrand factor-platelet interactions. Thromb. Haemost. **104**, 449-455 (2010)

www.uu.nl/

University of Barcelona

Site Address:	Institute for Bioengineering of Catalonia Baldiri Reixac, 10-12, 08028 Barcelona, Spain www.ibecbarcelona.eu/
Date Visited:	7 May 2012
WTEC Attendees:	Sharon Gerecht, Parag Mallick, Owen McCarty (report author), Lance Munn, Hassan Ali
Host(s):	Prof. Josep A. Planell, Director Tel.: +34 93 403 97 06 Fax: +34 93 403 97 02 japlanell@ibecbarcelona.eu
	Dr. Jordi Alcaraz, Principal Investigator Translational Lung Biomechanics Laboratory Unitat de Biofísica i Bioenginyeria Facultat de Medicina Casanova 143, 08036 Barcelona Tel.: (office) +34 93 402 4515, (lab): +34 93 403 9764 Fax: +34 93 403 5278 jalcaraz@ub.edu www.ub.edu/biofisica/jalcaraz.htm
	Prof. Xavier Trepat ICREA Research Professor and Group Leader Edifici Hèlix Baldiri Reixac 15-21, 08028 Barcelona Tel.: +34 93 402 0265 xtrepat ibecbarcelona.eu www.ibecbarcelona.eu/xtrepat
	Dr. Jerome Solon Centre for Genomic Regulation C/ Dr. Aiguader 88, 08003 Barcelona Tel. +34 93 316 01 00 Fax +34 93 316 00 99 Jerome.Solon@crg.eu http://pasteur.crg.es/portal/page/portal/Internet/02_Research
	Prof. Pere Roca-Cusachs, Principal Investigator C/ Baldiri Reixac 10-12, 08028 Barcelona Tel.: +34 934 020 863 rocacusachs@ub.edu www.ub.edu/biofisica/procacusachs.htm
	Dr. Marc Marti-Renom Genome Biology Group Leader, Structural Genomics Team National Center for Genomic Analysis Parc Cientific de Barcelona - Torre I Baldiri Reixac, 4 - 2a.p – 08028, Barcelona Tel.: +34 934 033 743 Fax: +34 934 037 279 mmarti@pcb.ub.cat http://marciuslab.org or http://cnag.cat

Overview

The Institute for Bioengineering of Catalonia (IBEC) is an interdisciplinary research center in Barcelona, Spain, focused on bioengineering and nanomedicine. IBEC was established in 2005 by the Ministries of Innovation, Universities and Enterprises and Health of the Generalitat de Catalunya (Autonomous Government of Catalonia), the University of Barcelona (UB), and the Technical University of Catalonia (UPC). Today, IBEC's relationship with the UB and UPC researchers continues to operate under a framework agreement signed in 2008. IBEC's mission is to conduct high-quality research that creates knowledge while contributing to a better quality of life, improving health, and creating wealth. The institute establishes close links with international research centers, universities, hospitals, and industry to exchange talent and develop and execute projects.

The institute currently has 15 research groups and 250 researchers and staff from 20 different countries. IBEC's groups and their activities are organized into six research programs:

- Cellular biotechnology
- Biomechanics and cellular biophysics
- Nanobiotechnology
- Biomaterials, implants, and tissue engineering
- Medical signals and instrumentation
- Robotics and biomedical engineering

The location of IBEC in the Parc Científic de Barcelona offers a highly stimulating biomedical environment in which the institute can work closely with organizations from the public and private sector interested in the biomedical application of nanotechnology. In addition, IBEC has access to powerful technological facilities, including the nanotechnology platform that offers services in nanofabrication, nanomanipulation and nanocharacterization.

Research and Development Activities

Dr. Jordi Alcaraz

The Alcaraz group is focused on determining the role that tissue and cell mechanics play in regulating cellular functions in health and disease. Dr. Alcaraz has shown how epithelial cell phenotype is regulated by the biochemical and mechanical extracellular matrix (ECM) microenvironment, and how changes in these parameters drive breast cancer development (Alcaraz et al. 2008). They have recently shown the role that increased collagen deposition and decreased ECM degrading activity by matrix metalloproteinases (MMPs) plays in driving abnormal extracellular matrix mechanical properties (Alcaraz et al. 2011). They have developed a novel cylindrical flat-ended atomic force microscopy (AFM) tip in order to define how cells respond to force bidirectionality (Acerbi et al. 2012). As shown in Fig. B.25,

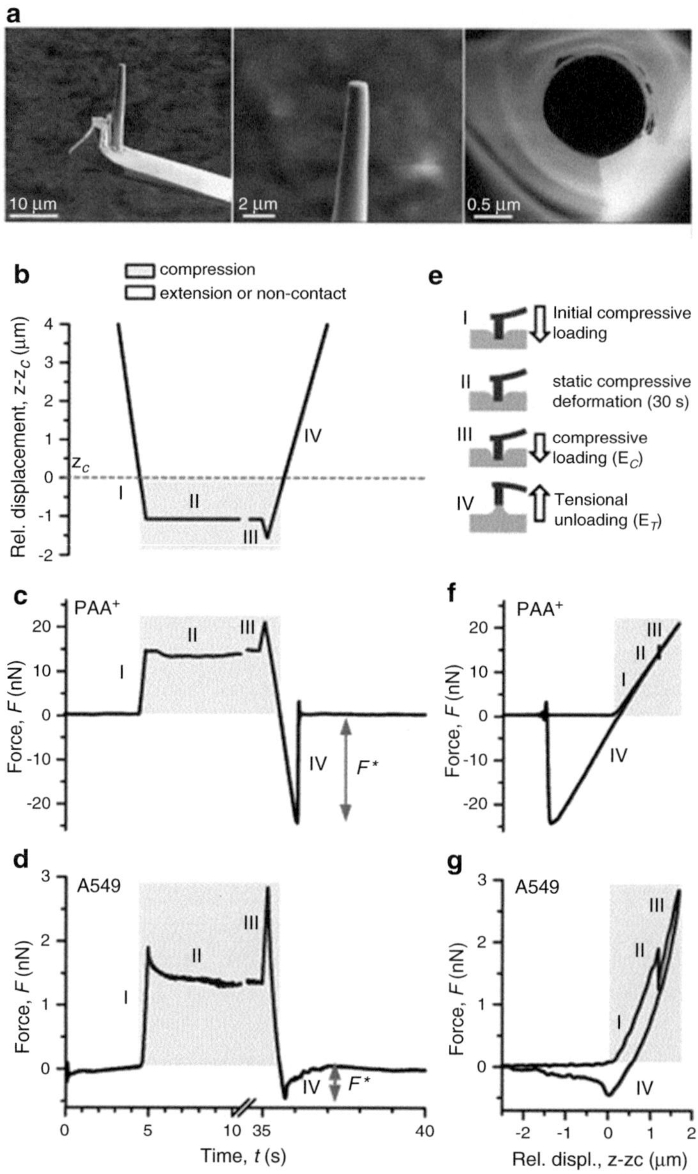

Fig. B.25 Illustration of the 4-step protocol based on FE-AFM tips used to probe cell mechanoresponses to compression and extension (Adapted from Acerbi et al. 2012)
(*A*) Representative SEM images of a nanofabricated cylindrical FE-AFM tip. A whole FIB-milled cylindrical tip is shown in the left panel, and detailed lateral and top view images of the tip are shown in the middle and right panels, respectively. (*B*) Driving signal of the piezotranslator in z as a function of time (t) used to probe the sample mechanoresponse to compression and extension. Corresponding F recordings as a function of t on a PAA+ gel and a single A549 cell are shown in (*C*) and (*D*), respectively. A common t axis was used in (*B–D*). F* was obtained from step IV as illustrated in (*C, D*). (*E*) Cartoon describing the tip-sample mechanical interactions corresponding to the 4-steps of the experimental protocol. EC and ET were calculated using signals from step III and IV. F signals from (*C*) and (*D*) were plotted against z in (*F*) and (*G*), respectively. The parts of the z and F signals obtained in compression were highlighted in gray. All F data were scaled relative to the corresponding zero force ($k \cdot d_0$)

they have fabricated flat-ended cylindrical AFM tips that have a cross-sectional area on the order of 1 μm^2. Tips are coated with either the integrin-specific (RGD) or non-specific (RGE/BSA) peptides in order to characterize the integrin-specific mechanoresponses to compression and extension in lung cells. Their results show that lung cells exhibit an asymmetric resistance to force directionality. The team is focusing future research on determining if asymmetric mechanoresponses play a role in driving collective cell migration during lung development and repair.

Dr. Xavier Trepat

The Trepat group is focused on defining how cell and tissue dynamics are integrated to drive function. In particular, his group is one of the leaders in the emerging field of plithotaxis—the emergent mechanism of innately collective cell guidance (Trepat and Fredberg 2011). To study this process, they have created a novel technique, monolayer stress microscopy, to characterize the local state of stress within a monolayer (Tambe et al. 2011). This technology allows the measurement of stresses within and between cells comprising a monolayer for the first time (Fig. B.26). The team's results show the key role that local orientation of maximal principal stress plays in regulating local cellular migrations. The correlation between the orientation of the maximal principal stress and that of cellular velocity is greatest in regions were stress anisotropy is strongest. Based on their observation that migrations of both endothelial and epithelial monolayers conform to this behavior, as do breast cancer cell lines before but not after the epithelial-mesenchymal transition, plithotaxis is perhaps not a particular property of any constituent cell but rather an emergent phenomenon and unifying physiological principle of a collective system.

Dr. Jerome Solon

The Solon group is focused on characterizing the biomechanics of morphogenesis. Their group uses an integrative approach of high-resolution imaging, automated image processing, and physical modeling to characterize phenomena such as cell pulsing during embryogenesis (Fig. B.27). They utilize the dorsal closure (DC) as a model system—a morphogenetic movement occurring at a late stage of *Drosophila* gastrulation. DC comprises the closure of a gap in the epidermis at the dorsal side of the embryo. The process begins with the dorsal convergence of two lateral, epidermal cell layers and terminates with the dorsal zippering of the leading cells from both layers. DC combines many cellular behaviors including cooperative cell movement, tissue force generation, and cell shape changes that are fundamental to the development and functioning of multicellular organisms. By coupling quantitative analysis, combined with laser cutting experiments and simulations, they were able to show that tension-based dynamics and cell coupling control the force pulses that

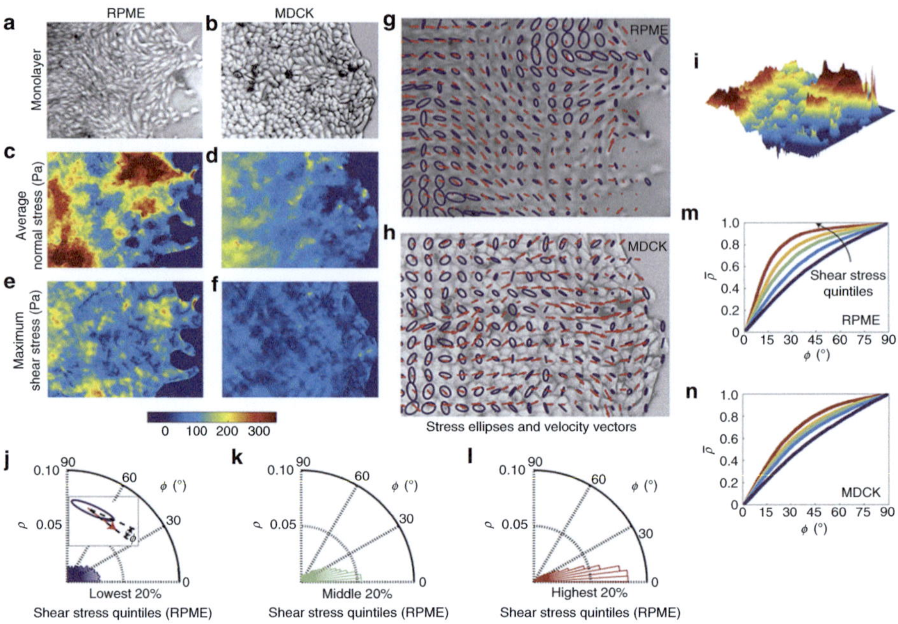

Fig. B.26 Transmitted light image of the RPME cell monolayer (*a*) and the MDCK cell monolayer (*b*). (Adapted from Tambe et al. 2011)
Corresponding to these images are the maps of average normal stress (*c, d*), maximum shear stress (*e, f*) and principal stress ellipses (blue) and cell velocity vectors (red) (*g, h*). Note that for the MDCK cell monolayer, the average tensile stress (*d*) increased systematically with increasing distance from the advancing front, thus contributing to the state of global tug-of-war. The map of average normal stress for the RPME cell monolayer is predominately tensile, but forms a rugged stress landscape (*i*). The alignment angle, φ, between the major axis of the principal stress ellipse and the direction of the cellular motion (*j*, inset) shows that the greater the local maximum shear stress the narrower is the distribution of φ (*j–l*). The cumulative probability distribution varied strongly and systematically with stress anisotropy (*m*); curves, from blue (bottom) to red (top), are in the order of higher quintiles. The cumulative probability distribution for the MDCK cell monolayer is also shown (*n*). Vertical size of the images of cell monolayers: RPME-545 μm, MDCK-410 μm. Each curve in *m* and *n*, and distributions in *j, k*, and *l* have >8,000 observations

drive dorsal closure in the developing *Drosophila* embryo (Solon et al. 2009). Their current aim is to unravel how the forces driving such collective cell movements are generated and coordinated.

Dr. Pere Roca-Cusachs

The Roca-Cusachs group is focused on defining the mechanisms by which molecules detect and respond to forces, triggering downstream cellular response. In particular, they utilize biophysical techniques to characterize the mechanical link between integrins and the actin cytoskeleton. The group has focused defining the

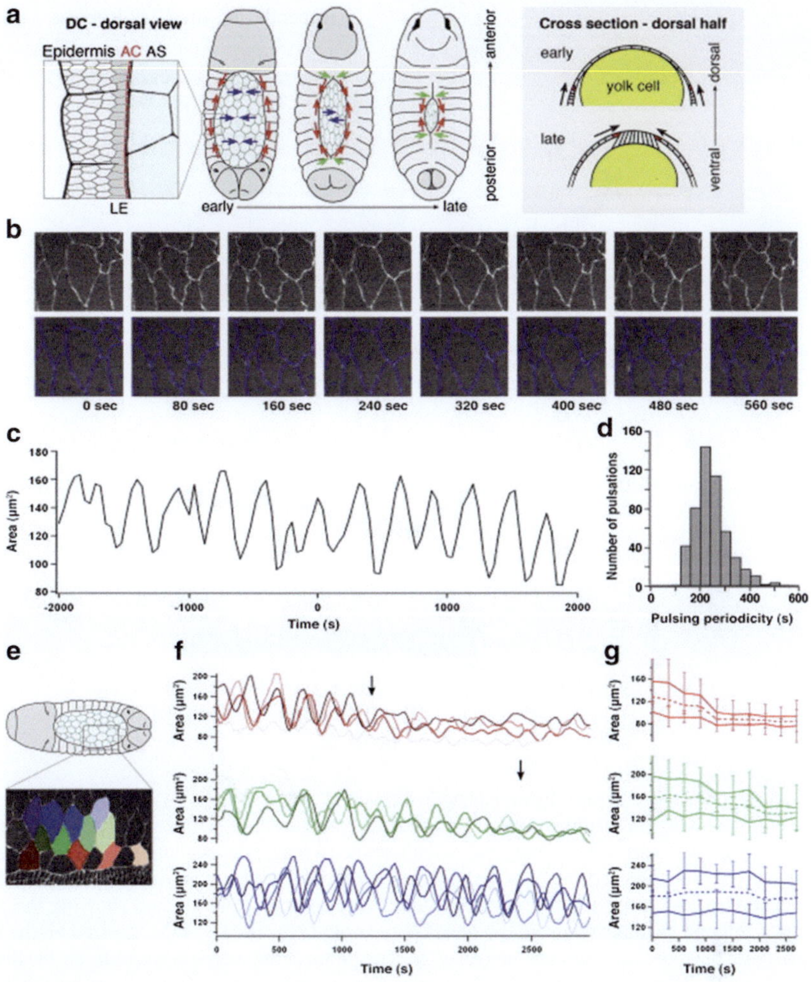

Fig. B.27 AS cell dynamics. (From Solon et al. 2009)
(A) Cartoon of DC embryos. Colored arrows depict forces produced by AS cells (blue, two center images—central arrows), AC (red, arrows following the curve of the ellipse), and zippering (green, arrows pointing to the median). Black arrows show the direction of LE movement. *(B)* Typical apical surface area pulsations of an AS cell in a GFP-Arm expressing embryo. The upper panel shows raw data, the lower panel shows the superimposed segmented image. *(C)* Apical cell surface fluctuations of an AS cell. Time point zero depicts the approximate onset of dorsal closure. *(D)* The period distribution of 505 pulses measured in 35 AS cells in two embryos is shown. The distribution is narrow and centered at 230 ± 76 s. *(E)* Image of a GFP-Arm-expressing embryo showing the epidermis (small cells at the bottom), the LE, and part of the AS tissue. AS cells are color-coded depending on their distance from the LE; the ventral-most cell row in red, the second row in green, and the third row in blue. *(F)* Analysis of the apical surface fluctuations of the AS cells highlighted in *(E)*. Cells sequentially cease pulsing (indicated by arrows): cells contacting the LE (red) are first (top), followed by the cells in the second row (green, middle). Cells in the third row (blue) continue pulsing throughout the analyzed time period (bottom). *(G)* The mean of the apical surface maxima and minima is shown for the different rows of AS cells in *(F)*. The minima remain virtually constant over time while the maxima decrease sequentially. Consequently, the average surface (dashed lines) decreases mainly as function of the maxima reduction

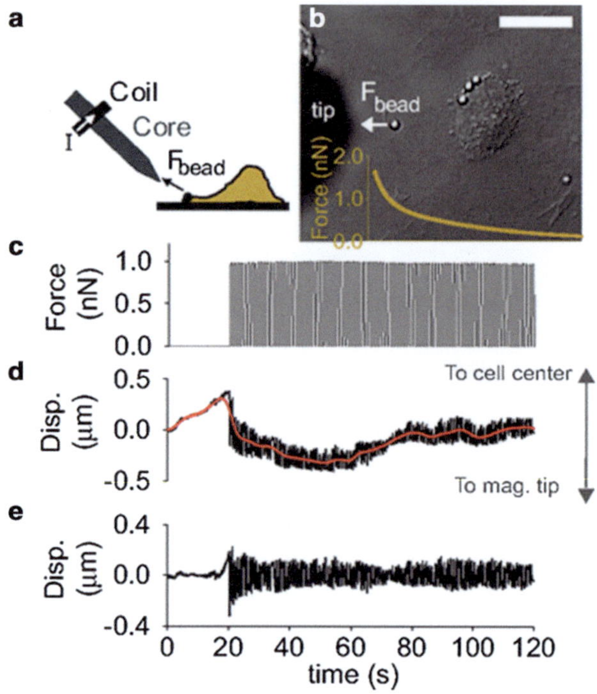

Fig. B.28 Experimental setup. (From Roca-Cusachs et al. 2009)
(*A*) Diagram showing the magnetic tweezers apparatus. A current (I, white arrow) goes through a set of coils placed around a magnetic core, which creates a magnetic gradient around the core tip. The force exerted on the bead (F_{bead}), which increases with this gradient, is stronger as the bead and the magnetic tip get closer. (*B*) Differential interference contrast (DIC) image showing the magnetic tip, a cell and an attached magnetic bead coated with FN7–10. The force exerted on the bead by the magnet pulls the bead toward the cell edge. The graph at the image bottom shows the dependency of applied force on distance to the tip. (Scale bar = 20 µm.) (*C*) Force sequence applied to the measured bead. No force is applied during the first 20 s of recording, and then subsequent pulses of 0.5 s of force/0.5 s without force are applied. Force is calculated from tip-bead distance. (*D*) Corresponding bead displacement in the direction toward the cell center. Before force is applied, beads move toward the cell center and away from the magnetic tip. When force is applied, beads that do not detach start pulsating accordingly and temporarily revert their movement toward the tip. Actual bead movement shown in black (jagged), red (straight) line is filtered to account for bead pulsation. (*E*) Subtraction between black and red lines from (*D*), showing only the pulsatory bead response. As time progresses the adhesion around the bead stiffens, decreasing the amplitude of movement

role that integrin clustering and integrin-talin linkages play in regulating adhesion strength and mechanotransduction (Fig. B.28). This study was enabled by the development of a magnetic tweezers apparatus able to exert forces of 1 nN on 2.8-µm diameter magnetic beads coated with a four-domain segment of fibronectin responsible for cell binding and containing the RGD and PHSRN motifs (Roca-Cusachs et al. 2009). Their results suggest a model for the mechanics of fibronectin-cell contacts in which fibronectin clustering leads to integrin binding, clustering, and recruitment.

Dr. Marc Marti-Renom

Dr. Marti-Renom is the Genome Biology Group Leader for the National Center for Genomic Analysis (CNAG). He and his structural genomics team are interested in the molecular mechanisms that regulate cell fate. To study such mechanisms, they employ the laws of physics and the rules of evolution to develop and apply computational methods for predicting the 3D structures of macromolecules and their complexes. Their three areas of research include:

- *Protein-ligand interactions.* They have developed methods for comparative docking of small chemical compounds and their target proteins. Such methods have already been applied to identify drug targets in 10 genomes that cause tropical diseases.
- *Comparative RNA structure prediction.* The recent interest in RNA, specifically non-coding RNA molecules, has prompted the team to develop a series of tools for the alignment of RNA structures and the prediction of their functions.
- *Structure determination of genomes.* More recently, they have engaged collaboration with experimentalists to study the 3D organization of the chromatin. Such work is resulting in the first ever structures of genomic domains and entire genomes.

Sources of Support/Funding

This work is supported in part by the European Union European Research Council (ERC) and the Spanish Ministerio de Ciencia e Innovación. In addition, the work at the CNAG has received funding from the Tropical Disease Initiative, the Marie Curie Actions grant program, and a Generalitat Valenciana research grant.

Summary and Conclusions

IBEC has established a multidisciplinary research of excellence in biomedical engineering and has become the technological counterpart to hospitals, biomedical research centers, and universities in Catalonia.

References

Acerbi, I., Luque, T., Gimenez, A., Puig, M., Reguart, N., Farre, R., Navajas, D., Alcaraz, J.: Integrin-specific mechanoresponses to compression and extension probed by cylindrical flat-ended AFM tips in lung cells. PLoS One. **7**, e32261 (2012)

Alcaraz, J., Mori, H., Ghajar, C.M., Brownfield, D., Galgoczy, R., Bissell, M.J.: Collective epithelial cell invasion overcomes mechanical barriers of collagenous extracellular matrix by a narrow tube-like geometry and MMP14-dependent local softening. Integr. Biol. (Camb.). **3**, 1153–1166 (2011)

Alcaraz, J., Xu, R., Mori, H., Nelson, C.M., Mroue, R., Spencer, V.A., Brownfield, D., Radisky, D.C., Bustamante, C., Bissell, M.J.: Laminin and biomimetic extracellular elasticity enhance functional differentiation in mammary epithelia. EMBO J. **27**, 2829–2838 (2008)

Roca-Cusachs, P., Gauthier, N.C., Del Rio, A., Sheetz, M.P.: Clustering of alpha(5)beta(1) integrins determines adhesion strength whereas alpha(v)beta(3) and talin enable mechanotransduction. Proc. Natl. Acad. Sci. USA. **106**, 16245–16250 (2009)

Solon, J., Kaya-Copur, A., Colombelli, J., Brunner, D.: Pulsed forces timed by a ratchet-like mechanism drive directed tissue movement during dorsal closure. Cell. **137**, 1331–1342 (2009)

Tambe, D.T., Hardin, C.C., Angelini, T.E., Rajendran, K., Park, C.Y., Serra-Picamal, X., Zhou, E.H., Zaman, M.H., Butler, J.P., Weitz, D.A., Fredberg, J.J., Trepat, X.: Collective cell guidance by cooperative intercellular forces. Nat. Mater. **10**, 469–475 (2011)

Trepat, X., Fredberg, J.J.: Plithotaxis and emergent dynamics in collective cellular migration. Trends Cell. Biol. **21**, 638–646 (2011)

University of Basel

Site Address:	Biozentrum, University of Basel Klingelbergstrasse 50/70 CH - 4056 Basel, SWITZERLAND (meeting took place at EPFL, Lausanne)
Date Visited:	8 May 2012
WTEC Attendees:	Sharon Gerecht, Parag Mallick, Owen McCarty, Lance Munn (report author), Hassan Ali
Host(s):	Cora-Ann Schoenenberger Tel.: +41 61 267 22 60 cora-ann.schoenenberger@unibas.ch www.biozentrum.unibas.ch/research/groups-platforms/overview/unit/schoenenberger/

Overview

The Biozentrum at the Universitat Basel, Center for Molecular Life Sciences, consists of 33 research groups with scientific focus on growth and development, infection biology, neurobiology, structural biology and biophysics, and computational and systems biology. A few of these groups are interested in the molecular underpinnings of cancer, and the relationship between cancer and stem cell biology. Two groups at this site (Prof. R. Lim, a nanobiologist and Dr. C-A. Schoenenberger) are overtly adapting concepts from nanomechanics to study oncology. The WTEC panel heard a presentation by Schoenenberger.

Research and Development Activities

Schoenenberger, originally trained as a cell biologist, now studies the structural and mechanical plasticity of healthy cells and tumors, with the premise that the cytoskeleton is an essential mediator of structural and nanomechanical plasticity of cells in health and disease. She believes that pathologically altered nanomechanical properties may serve as diagnostic markers for cancer.

Schoenenberger presented a summary of her early work which used atomic force microscopy (AFM) to image cytoskeletal dynamics in cultured cells. She then described studies pioneered by Plodinec et al., where breakthrough AFM-based nanotechnology known as ARTIDIS ("Automated and Reliable Tissue Diagnostics") is used to measure the nanomechanical properties of breast cancer biopsies. The feasibility of these studies was first demonstrated on tissue specimens of breast cancer in the mouse model, showing that different types of tumors and stages of tumor development can be distinguished on the basis of their stiffness signature. Schoenenberger is a close collaborator of a team headed by Prof. Lim investigating the potential of AFM measurements for the diagnosis and prognosis of breast cancer in humans. An interesting aspect of the data is the spatially-resolved measurements, which result in histograms of stiffness over the surface of the cylindrical biopsies. Distinct peaks in the histogram can be attributed to cancer cells, normal cells, and matrix components and can be used in diagnosis. In the final few slides, Schoenenberger outlined a more recent project aimed at understanding the relationship between tissue hypoxia, mechanical elasticity, and progression of cancer.

Besides her collaborations with Prof. Roderick Lim within the Biozentrum, Schoenenberger has outside collaborations with the Friedrich Miescher Institute, Basel (Dr. Mohamed Bentires-Alj) regarding the MMTV-PyMT mouse model. Collaboration with the Department of Gynecology and Gynecological Oncology, University Hospital, Basel, is the source of the clinical biopsies for elasticity analyzes. Histopathological aspects are covered by the Department of Pathology, University Hospital, Basel.

Translation

On the topic of translation, Schoenenberger pointed out that the inventors of ARTIDIS (Dr. med. Marko Loparic, Dr. Marija Plodinec and Prof. Lim), in partnership with the Swiss company Nanosurf AG, are actively working with the University Hospital, Basel to further develop it for the diagnosis of breast cancer. Patents have been filed on the technology and intellectual property related to breast cancer diagnostics by the University of Basel.

Sources of Support/Funding

Funding comes from the Swiss National Science Foundation; Swiss Nanoscience Institute, Basel; and Swiss Commission for Technology and Innovation.

Summary and Conclusions

Schoenenberger's research has at its foundation the concept that cell mechanical properties are a result of—and contribute to—cancer progression. Her interdisciplinary collaborations provide a good example of how physics and oncology can be integrated so as to promote the translation of these concepts to the bedside.

References

Plodinec, M., Loparic, M., Monnier, C.A., Obermann, E.C., Zanetti-Dallenbach, R., Oertle, P., Hyotyla, J.T., Aebi, U., Bentires-Alj, M., Lim, R.Y.H., Schoenenberger, C-A.: The nanomechanical signature of breast cancer. Nat. Nanotechnol. **7**, 757–765 (2012)

Plodinec, M., Loparic, M., Suetterlin, R., Herrmann, H., Aebi, U., Schoenenberger, C.A.: The nanomechanical properties of rat fibroblasts are modulated by interfering with the vimentin intermediate filament system. J. Struct. Biol. **174**, 476–484 (2011)

Plodinec, M., Schoenenberger, C.A.: Spatial organization acts on cell signaling: how physical force contributes to the development of cancer. Breast Cancer Res. **12**, 308 (2010)

Schoenenberger, C.A., Mannherz, H.G., Jockusch, B.M.: Actin: from structural plasticity to functional diversity. Eur. J. Cell Biol. **90**, 797–804 (2011)

Silvan, U., Boiteux, C., Sutterlin, R., Schroeder, U., Mannherz, H.G., Jockusch, B.M., Berneche, S., Aebi, U., Schoenenberger, C.A.: An antiparallel actin dimer is associated with the endocytic pathway in mammalian cells. J. Struct. Biol. **177**, 70–80 (2012)

University of Freiburg

Site Address:	Hermann-Herder Straβe 3, 79104 Freiburg im Breisgau Germany www.uni-freiburg.de/forschung-en (meeting held at the University of Heidelberg)
Date Visited:	10 May 2012
WTEC Attendees:	Sharon Gerecht, Parag Mallick (report author), Owen McCarty, Hassan Ali

(Continued)

Host(s):	Jens Timmer Institute of Physics Tel.: +49 761 203 5829 Fax: +49 761 203 8541 jeti@fdm.uni-freiburg.de http://webber.physik.uni-freiburg.de/~jeti/
	Hauke Busch Freiburg Institute for Advanced Studies School of Life Sciences Tel.: +49 761 203 97154 Fax: +49 761 203 97334 hauke.busch@frias.uni-freiburg.de www.frias.uni-freiburg.de/lifenet/fellows/hauke-busch

Overview

The Freiburg Institute for Advanced Studies (FRIAS) is a key component of the University of Freiburg. FRIAS is organized into four schools: (1) School of History; (2) School of Language & Literature; (3) School of Life Sciences – LifeNet; and (4) School of Soft Matter Research. There are also several interdisciplinary research groups in annual rotation. These schools and research groups work closely with the university's 11 faculties.

FRIAS's work pivots around four key concepts that also lie at the heart of the university's institutional strategy:

- "Windows for Research" – FRIAS provides leading researchers with privileged conditions to conduct their work.
- "New Universitas" – FRIAS opens up new opportunities for interdisciplinary contact and collaboration.
- "Internationalisation" – FRIAS reinforces the university's international networking and visibility.
- "Promotion of Early-Stage Researchers" – FRIAS offers outstanding conditions for young academics.

In 2009, an important component was added to the four-school architecture: the interdisciplinary research group competition. This program provides University of Freiburg professors with the opportunity to apply for a 10-month FRIAS fellowship in order to carry out an innovative, interdisciplinary research project at the institute. This can also include international partners, if applicable. Two or three research groups are supported each year.

The major FRIAS interdisciplinary symposia also offer new options for projects. They aim to promote dialogue between the humanities, social sciences, and natural sciences and to shed light on a topic of key academic and social relevance from different perspectives (2009: Evolution, 2011: Catastrophes). The same is true of many smaller, inter-school event formats such as monthly dinner speeches and after-hours conversations.

FRIAS School of Life Sciences-LifeNet focuses its research on the biology of complex systems. Its multidisciplinary, system-oriented approach includes a wide range of researchers whose expertise spans mathematics and physics to biology and medicine.

Research and Development Activities

Both the Timmer and Busch groups highlighted how mathematical modeling is used to create hypotheses, which, in turn, are validated using experiments in high-throughput biological systems (genomic, proteomic, and metabolomic methods) and imaging technologies. This system-oriented approach is designed to improve prediction of normal functions in plants, animals, and humans. Similarly, the approach can be used to disclose causes and progress of illnesses and to assess how successful different treatment options will be.

Dr. Jens Timmer

Dynamic processes are ubiquitous in the life sciences. They can be found from the regulation in cells up to oscillations in tremor. Malfunction of these dynamical processes can be a cause or a sign of diseases. In interdisciplinary projects, the Timmer group develops and applies mathematical methods to analyze and model these processes based on measured data. The final aim of these efforts is to help to turn the life sciences from a qualitative descriptive science into a quantitative predictive science. The group is exploring a number of exciting models of regulation ranging from testing the accuracy of sensor network interaction maps, to understanding the design principles underlying biological systems (particularly with regards to robustness), as well as understanding randomness in gene expression and pattern formation in development. One study that was highlighted investigates why biology has selected a particular level of complexity to chemotaxis networks. This study used a combined experimental and computational approach to look at the very simple and measurable system of Che proteins (Fig. B.29). Unlike traditional systems oriented approaches that are focused on tens of thousands of genes and on discovering the interactions between them, the Timmer group focused on a simple, compact system. Regulation of this system is driven by a handful of proteins (CheR, CheB, CheA, CheZ and CheY). After measuring the extent of noise using YFP and CFP reporter, the authors were able to look at both the protein covariation and noise. In order for

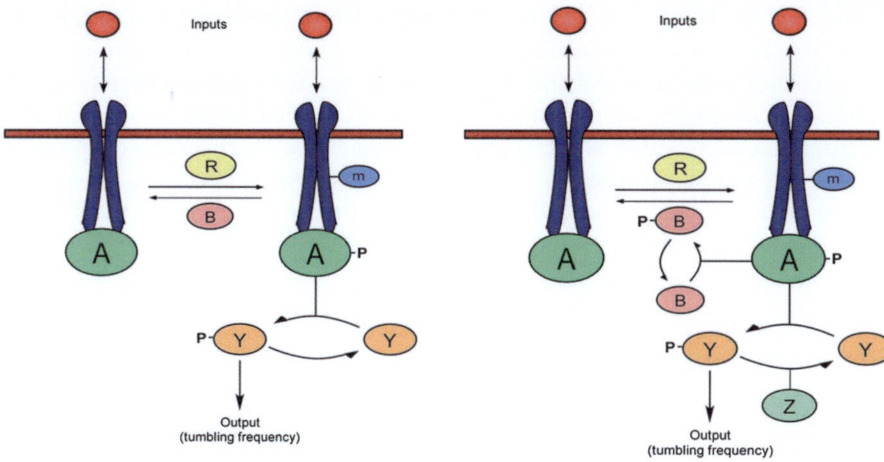

Fig. B.29 Experimental and computational approach to look at the very simple and measurable system of Che proteins (Adapted from Kollmann et al. 2005)

the system to function appropriately, it must be sensitive to signaling molecules (e.g., chemoattractants and chemorepellants), but insensitive to random noise fluctuations. After evaluating the common 'BL' model (Fig. B.29, left), the authors found that it was unable to accurately reproduce the measured data. However, more complex topologies (Fig. B.29, right) allow cells in the population to respond accurately to changes in ligand concentration. Furthermore, the topologies are sufficiently robust to compensate for co-variations in expression levels and some suppression of noise-based fluctuations. Though more complex structures may be envisioned, they appear not necessary for maintaining system responsiveness and stability. It is highly likely that this is a general principle of biological signaling networks.

Dr. Hauke Busch

The Busch group focuses on the development and verification of mathematical models for cellular behavior from an initial stimulus to the final phenotype. With a systems biology approach, he combines experimental research on cell-cell communication with the development of appropriate multi-scale dynamic models to investigate the necessary and sufficient control points that lead to cell proliferation, differentiation, migration or death. We adapt concepts from non-linear dynamics and complex systems to develop appropriate dynamic models unraveling self-organizing properties in cellular behavior.

One area that he is particularly interested in is the definition of cell states and the according state-transition functions. There is a recognized vast disparity between the number of genes and proteins (10^5), and the few cell fates (proliferation, differentiation, apoptosis and migration). In the formalism he described, cells inhabit a transcriptome/proteome phase space and homeostasis is governed by a set of attractors in that phase space. A result of this hypothesis is that there are a finite number of mutually exclusive cell states, each of which is driven by potentially vast and overlapping sets of genes and gene-products (Fig. B.30).

Sources of Support/Funding

Current

Federal Ministry of Education and Research Program "German Virtual Liver Network"

Federal Ministry of Education and Research Program "Medical Systems Biology:" LungSys, BreastSys, SARA

EU FP 7 STREP OpenTox

Systems Biology of Signalling in Cancer: SBCancer

EU IMI project MIP-DILI

Former

EU FP 7 STREP EPILEPSIAE

FRISYS in the frame of the Federal Ministry of Education and Research Program "FORSYS"

German Science Foundation: Graduate College 1305 "Signaling systems in plant model organisms"

EU FP 7 STREP CancerSys

EU FP 6 IP Sens-it-iv: Novel Testing Strategies for *In vitro* Assessment of Allergens

Bernstein Center for Computational Neuroscience Freiburg

Federal Ministry of Education and Research Program "Systems Biology of Hepatocytes" (HepatoSys)

German Science Foundation: Statistical Modelling in Neurology

Federal Ministry of Education and Research Program "QuantPro"

EU FP 6 STREP COSBICS: Computational Systems Biology in Cellular Signalling

Nationales Genomforschungsnetz NGFN Explorative Project: Microarray Validation of cardio-vascular Risk

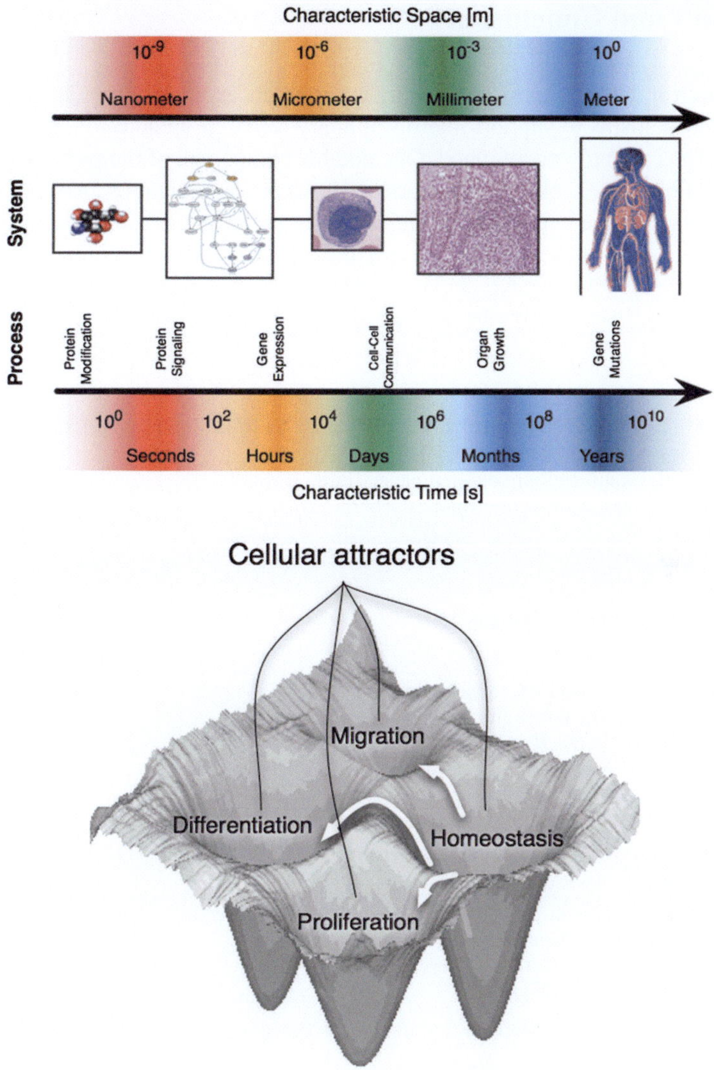

Fig. B.30 Definition of cell states (Courtesy of Hauke Busch, University of Freiburg)

German Research Foundation Priority Program 1114 "Mathematical methods for time series analysis and digital image processing"

State of Baden-Württemberg, Funding Initiative RNA/RNAi

Federal Ministry of Education and Research Program on Non-linear Dynamics

Graduate College "Nonlinear Differential Equations: Modelling, Theory, Numerics, Visualisation"

Summary and Conclusions

The research at the University of Freiburg is extremely exciting. There are a number of funding mechanisms to support interdisciplinary research. In addition, interdisciplinary research is strongly encouraged. One notable component of the proposed research is the significant focus on integrating experiment and computation. Among the most exciting aspects at the University of Freiburg was the focus on formalizing hypotheses mathematically and then identifying experiments to test those mathematical models.

References

Kollmann, M., Løvdok, L., Bartholomé, K., Timmer, J., Sourjik, V.: Design principles of a bacterial signaling network. Nature. **438**, 504–507 (2005)
http://webber.physik.uni-freiburg.de/~jeti/papers/Julie_JIM.pdf
www.frias.uni-freiburg.de/institute/frias-im-ueberblick-en/frias-overview

University of Heidelberg

Site Address:	BioQuant, In Neuheimer Feld 27 Ute Greenuier Saal D-69120, GERMANY www.bioquant.uni-heidelberg.de/
	The German Cancer Research Center (DKFZ) Im Neuenheimer Feld 280 69120 Heidelberg, Germany www.dkfz.de/en/index.html
	Center for Molecular Biology of the University of Heidelberg (ZMBH) Im Neuenheimer Feld 282 D-69120 Heidelberg, Germany
Date Visited:	10 May 2012
WTEC Attendees:	Sharon Gerecht (report author), Parag Mallick, Owen McCarty, Hassan Ali
Host(s):	Joachim Spatz Max Planck Institute (MPI) for Metals Research Heisenbergstr. 3 D - 70 569 Stuttgart, Germany Joachim.Spatz@mf.mpg.de www.mf.mpg.de/en/abteilungen/spatz/index.html University of Heidelberg INF 253 D - 69 120 Heidelberg, Germany Joachim.Spatz@urz.uni-heidelberg.de www.pci.uni-heidelberg.de/bpc/biophysik.html

(continued)

	Ulrich Schwartz Institute for Theoretical Physics Tel.: +49 6221 54 9399 Fax: +49 6221 54 9331 u.schwartz@phys.uni-heidelberg.de www.tphys.uni-heidelberg.de
	Evgeny Gladilin Group Leader, Cellular Mechanics, Bioquant DKFZ E.Gladilin@dkfz.de
	Ralf Kemkemer MPI for Intelligent Systems Stuttgart, Germany Tel.: +49 711 689 3516 Fax: +49 711 689 3612 ralf.kemkemer@is.mpg.de www.is.mpg.dle
	Harald Herrmann B065 Functional Architecture of the Cell Division of Molecular Genetics, Peter Lichter DKFZ
	Michael Knop Cell Morphogenesis and Signal Transduction Zentrum für Molekulare Biologie der Universität Heidelberg (ZMBH) ZMBH-DKFZ Alliance Im Neuenheimer Feld 282 69120 Heidelberg Tel.: +49 6221 54 4213 m.knop@zmbh.uni-heidelberg.de
	Marcel Schilling Systems Biology of Signal Transduction (A150) DKFZ Tel.: +49 6221 42 4485 m.schilling@dkfz.de www.dkfz.de/en/systembiologie/index.php

Overview

BioQuant (www.bioquant.uni-heidelberg.de/fileadmin/content/broschure/BioQuant_ Broschuere _080319.pdf) the Center for Quantitative Analysis of Molecular and Cellular Biosystems at Heidelberg University, was established in 2007 as an interdisciplinary research center dedicated to research and training in systems biology. The objective of BioQuant is to function as a platform for the development and constant refinement of mathematical models of complex biological systems as well as the swift validation of scientific hypotheses via experimental data. Currently, up to 40 university and non-university research groups (DKFZ, European Molecular Biology Lab, European Media Lab, and MPI for Medical Research) are affiliated with BioQuant. These research groups are instrumental in implementing numerous national and international systems biology funding initiatives.

In addition to advanced computational tools and methods for data analysis, image processing, and modeling, BioQuant's central technology platform provides cutting edge technologies for systematic functional imaging with an emphasis on high-throughput and high-content microscopy, high-resolution microscopy, and electron microscopy (conventional and cryo). The NIKON Imaging Center and the Hamamatsu Tissue Imaging and Analysis Center are both integral parts of BioQuant's technology platform.

Research and Development Activities

Dr. Joachim Spatz

Our visit at University of Heidelberg opened with a presentation by Spatz. Spatz is the Director of the MPI for Metals Research, Stuttgart, and a Professor of Biophysical Chemistry, University of Heidelberg. Spatz's research focuses on determining the important cues at the cell microenvironment and manipulating them using advanced materials to understand how they affect cellular behaviors. Of special interest are matrix geometry, spatial arrangement of membrane receptors, forces/mechanics, and dynamics. A recent technology developed in the lab utilizes patterned nanoparticles to enable the controlled-spacing presentation of trans-membrane receptors on surfaces (distance ranging from 1-100 nm), surrounded by non-adhesive regions. It was found that cell adhesion is receptor-space specific. An example was shown for integrin where spacing of <60 nm facilitated cell adhesion while no adhesion could be observed in 73 nm (Arnold et al. 2004). In the case of gradient spacing, cells migrate toward their optimal adhesion spacing (Hirschfeld-Warneken et al. 2008). Receptor over-expression did not seem to affect the optimal spacing. Current efforts are focused on synthetically tailoring the system to study the complexity of two signals (e.g., receptors and growth factors, Fig. B.31 (Shahal et al. 2012); and stiffness and spacing).

We also witnessed the group's recently developed "migration chip" made of microchannels of defined width next to reservoirs of soluble factors. They are fabricated atop glass slides allowing on-line (time-lapse) imaging of cell migration and velocity (Rolli et al. 2010). Using this system, the group found that the cancer cells' ability to migrate depends on space (tunnel width) and sphingosylphosphorylcholine, and that this response is cell-type dependent. Currently, the cellular responses to microenvironments confined by rough edges are being studied.

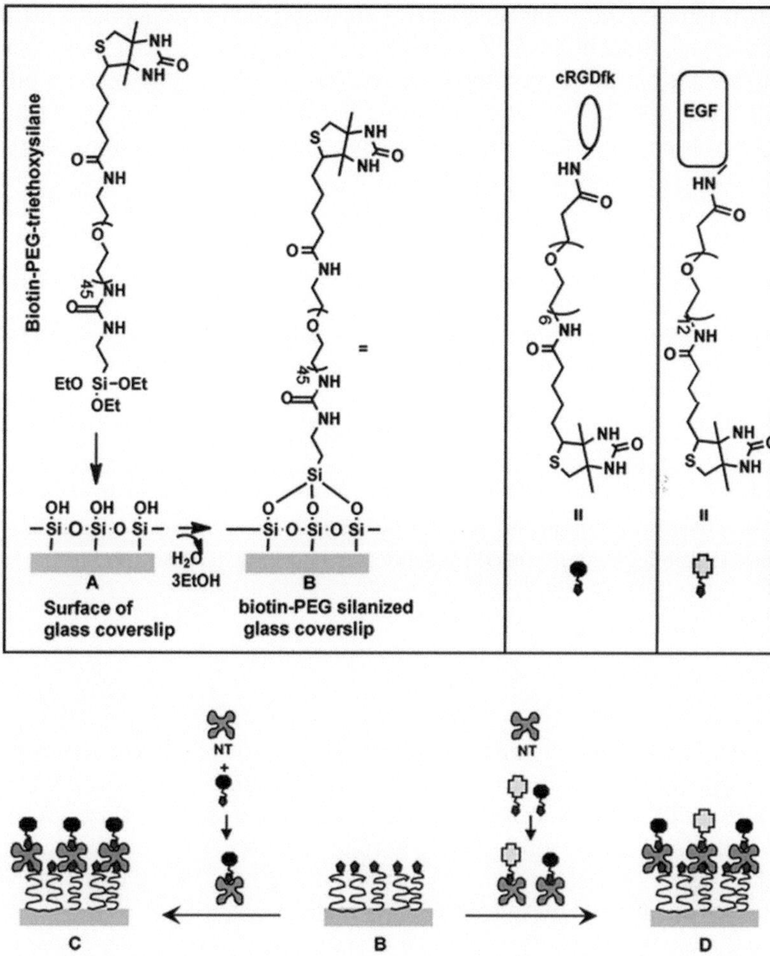

Fig. B.31 A schematic presentation of RGD and EGF surface functionalization. (From Shahal et al. 2012)
Plasma-etched glass slides (*a*) are incubated with triethoxysilane-PEG-biotin in toluene, resulting in the formation of biotinylated glass slides (b). The slides are further incubated with preformed NeutrAvidin (NT)-biotin-EGF/RGD complexes at various complex concentrations, resulting in the formation of glass cover slips biofunctionalized with RGD (*c*) or with a mixture of RGD and EGF at different densities and ratios (*d*)

Dr. Ulrich Schwarz

Schwarz presented information on the BioQuant location, its collaborative initiatives, and education program. Research in Schwarz's group aims at developing new theoretical concepts that enable the understanding of these processes in quantitative detail. Utilizing tools from material sciences, the physical cues of the cellular microenvironments are being mimicked—including adhesive geometry, stiffness, and topography. Modeling is being applied to predict cellular responses. An example of this is the use of traction-force microscopy to study cell movement on a soft substrate with embedded marker beads (Fig. B.32). Image processing and computational reconstruction of the cellular traction pattern revealed a strong correlation between adhesion structure and forces, and enabled the prediction of cell movement (Sabass et al. 2008; Munter et al. 2009). Another modeling example is the quantitative analysis of cell shape using micropatterned surfaces. It was found that for cells whose adhesion sites are restricted to small adhesive islands on a micropatterned substrate, their shape resembles a sequence of inward-curved circular arcs. This morphology is due to actively contracting cable networks (Bischofs et al. 2008) that do not have a reference shape (Guthardt Torres et al. 2012).

Overall, these studies demonstrate the ability to estimate forces without the need for traction-force microscopy. This work has led to new project: MEHTRICS: Micropattern – Enhanced High Throughput RNA Interference for Cell Screening. MEHTRICS combines RNAi high-throughput screens with micropatterns to achieve highly standardized conditions for cell culture. This project is funded by the European Union and is supported by academic and industrial collaboration.

Dr. Evgeny Gladilin

Mechanical factors play an important role in many basic biological phenomena on different spatial-temporal scales—from tissues and single cells to sub-cellular structures. The ability of cells to appropriately sense, process, and utilize mechanical energy and signals is essential for the normal function of the entire organism. A number of severe diseases, such as cancer and progeria, are known to be related to altered mechanical properties of the cellular matter.

To reveal the mechanisms behind the observed mechanobiological phenomena, Gladilin's lab develops novel approaches to quantitative analysis of cellular mechanics using a multimodal image- and numerical model-based framework. Image analysis is utilized to identify specific properties of the cells. The group applies this to different measurements including: (1) microplate cell stretching; (2) mechanics of embryonic stem cell division; (3) drug-induced nuclear deformation; (4) substrate cell stretching; and (5) optical cell stretching. Future directions include combining chemical perturbation (drugs, knockdowns) and mechanical cell phenotyping, 3D culture experiments with fluorescent microscopy, high-throughput methods based on a combination of microfluidics and different deformation induction techniques, development of algorithms for unbiased data assessment (automatization) processing and analysis of 3D microscopic images, and numerical solvers based on mesh-free methods.

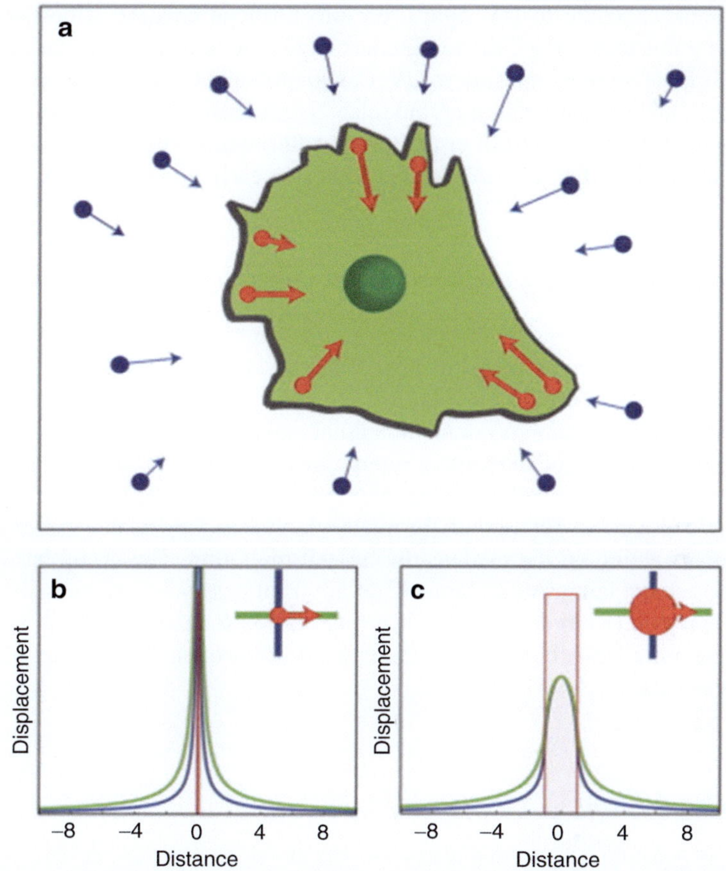

Fig. B.32 Traction force microscopy studies of cell movement. (From Sabass et al. 2008) (*A*) Schematic representation of traction force microscopy on flat elastic substrates. Marker beads in the substrate and the corresponding displacement vector field are shown in blue. Sites of adhesion and the corresponding force vector field are shown in red. (*B*) If force is assumed to be strongly localized, one can use the concept of point forces, which leads to a divergent displacement field at the site of force application. Here the magnitude of the displacement is plotted in two perpendicular directions. When relating force to displacement, the mathematical divergence can be avoided by using a simple cutoff rule. (*C*) If force is assumed to be spatially extended (here showing constant traction over a circular site of adhesion), then displacement is finite inside the adhesion area

Dr. Ralf Kemkemer

The research of Kemkemer focuses on the understanding of the mechanics of cells as it relates to their cytoskeleton and the study of cell migration. To address these topics, materials are being utilized to develop advanced tissue culture tools. For example, a photo-switchable wound healing assay where non-adhesive regions are cleaved upon exposure to UV light (365 nm) result in creating adhesive sites to cells. Using this technology, the researchers were able to measure expansion kinetics of cell clusters upon exposure to UV. Using epithelial cells, they have found that cluster size and boundary curvature modulate the expansion of the cell sheet and the formation of leader cells (Rolli et al. 2011). Additional studies utilize inert hydrogels to explore the effect of stiffness/mechanics on early tumor growth in 3D.

Dr. Harald Herrmann

Located in the Division of Molecular Genetics at DKFZ, Herrmann's group researches the structure and assembly of intermediate filament proteins. Intermediate filaments build two distinct systems in animal cells, one in the nucleus and one in the cytoplasm. The major function of intermediate filaments is speculated to be that of a mechanical stress absorber and an integrating device for the entire cytoskeleton. Herrmann's research is focused on the nuclear lamins as they relates to extracellular matrix forces acting on the nucleus during cell migration. The group has recently demonstrated that lamin A and lamin C do not form heterodimers, but almost completely segregate (Kolb et al. 2011). Current studies focus on studying how fragmented genomic DNA binds to the lamin scaffold through comparison of *in vivo* and *in vitro* data.

Dr. Michael Knop

Research in Knop's lab focuses on the processes that regulate cellular morphogenesis and cell signaling. They also study the cellular response to external stimuli that are processed via mitogen-activated protein kinase (MAPK) signaling pathways and lead to specific adaptations. The group utilizes high-throughput genetics, microscopy, and systems biology with yeast for functional imaging of protein behavior. An example of a systems approach to studying MAPK singling was shown. To quantify the abundance of complexes in the cytoplasm among different MAPKs in yeast pheromone signaling, the group used fluorescence cross-correlation spectroscopy (FCCS). They have found that specific MPAKs—Fus 3—forms a gradient of activity across the cell via a reaction-diffusion mechanism (Maeder et al. 2007). The group built on the FCCS technology to develop a microscope based on light-sheet illumination that allows massively parallel fluorescence correlation spectroscopy measurements (Fig. B.33). Based on knowledge about the optical properties of the setup, it is possible to calculate spatially resolved maps of protein concentrations and mobilities—especially maps of diffusion coefficients and interaction properties. Using this technology, the group reported the diffusion and interactions of proteins in mammalian cells and in isolated fly tissue (Capoulade et al. 2011).

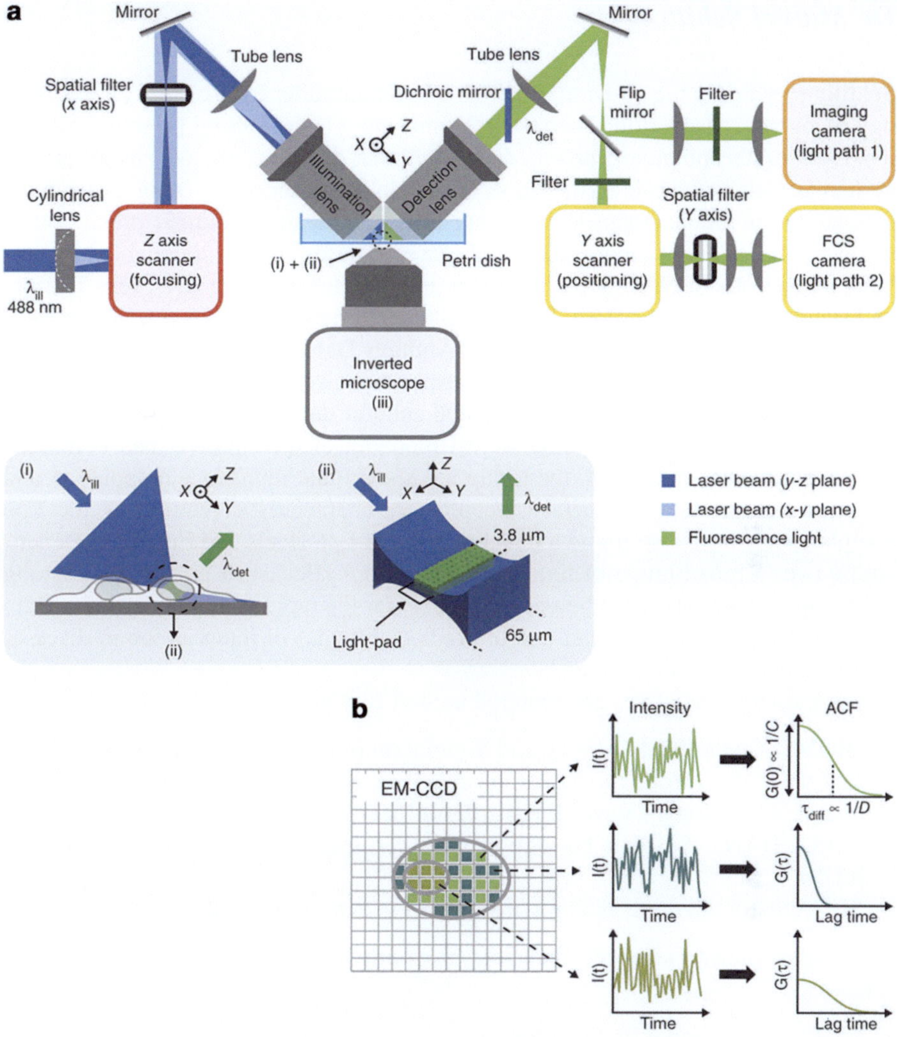

Fig. B.33 Microscope based on light-sheet illumination that allows massively parallel fluorescence correlation spectroscopy measurements. (From Capoulade et al. 2011)
(*a*) Front view of the main components of the light-pad microscope. The specimen is contained in a Petri dish. The first objective lens illuminates a thin slice in the specimen. Optical sectioning is performed under 45° with respect to the bottom of the Petri dish. The high aperture angle of the illumination light-sheet (74°) leads to weak illumination of cells lying outside the light-pad area [inset (*i*)]. Fluorescence is detected at a right angle to the illumination plane by the detection lens. Spatial filters in the illumination path and the detection path confine the observed area to a rectangular array of volume elements, the light-pad [inset (*ii*)]. An inverted microscope allows convenient positioning of the specimen. (b) Each individual pixel of the EM-CCD records a fluctuating fluorescence signal over time. These fluctuations are analyzed by temporal correlation analysis resulting in one ACF for each pixel. The ACF provides information about the diffusion coefficient D (dashed line) and the concentration C of diffusing fluorescently labeled molecules (amplitude of the curve)

Dr. Marcel Schilling

Schilling's work in the division of Dr. Ursula Klingmüller focuses on systems biology of signal transduction. The research aims are: (1) unraveling principal mechanisms of erythropoietin (Epo)-mediated cellular decisions in the hematopoietic system as well as the role of the EpoR system in lung cancer; (2) bridging from the cellular to the whole organ level during liver regeneration; and (3) prediction of strategies for efficient intervention in diseases. These projects are enabled by collaborative efforts of biology (Klingmüller and Schilling), theory (Drs. Jens Timmer and Hauke Busch), technology (Drs. Dirk-Peter Herten and Wolf-Dieter Lehmann), and medicine (Drs. Michael Thomas and Anthony D. Ho). This approach allows the conversion of time-resolved quantitative experimental data to mathematical models. As an example the study of Epo-mediated cellular decisions in the hematopoietic system was described. By mathematical modelng of quantitative data combined with experimental validation, the group has shown that rapid ligand depletion and replenishment of the cell surface receptor are characteristic features of the Epo receptor. They have also found a linear relation of Epo levels and Epo receptor activation over a broad range of ligand concentrations (Becker et al. 2010). Ongoing efforts apply data-based mathematical models for the rapid testing of hypotheses to uncover deregulation in cancer and to predict strategies of intervention in diseases towards personalized medicine.

Several large initiatives are centered around BioQuant:

- BIOMS: Center for Modeling and Simulation in Bioscience; www.bioms.de/
- CellNetworks: From molecular mechanisms to quantitative understanding of complex functions; www.cellnetworks.uni-hd.de/
- VIROQUANT: Systems biology of virus-cell interactions; www.viroquant.uni-hd.de/
- SBCANCER: Systems biology of signaling in cancer; www.dkfz.de/en/sbcancer/

Interdisciplinary science education programs are offered for Heidelberg life sciences:

- EMBL international Ph.D. program
- DKFZ Helmholtz International Graduate School for Cancer Research
- HBIGS molecular and cell biology
- HGS MathComp (computational sciences)
- HGS fundamental physics (plans to include biophysics in second funding period)

Summary and Conclusions

BioQuant is Europe's first quantitative biology center. It supports interdisciplinary collaboration between research groups from the biological and biomedical sciences and research from chemistry, physics, mathematics, and computer sciences. Emphasis on collaborative effort, through large initiatives and interdisciplinary education programs, enables the exploration and development of pioneering research approaches.

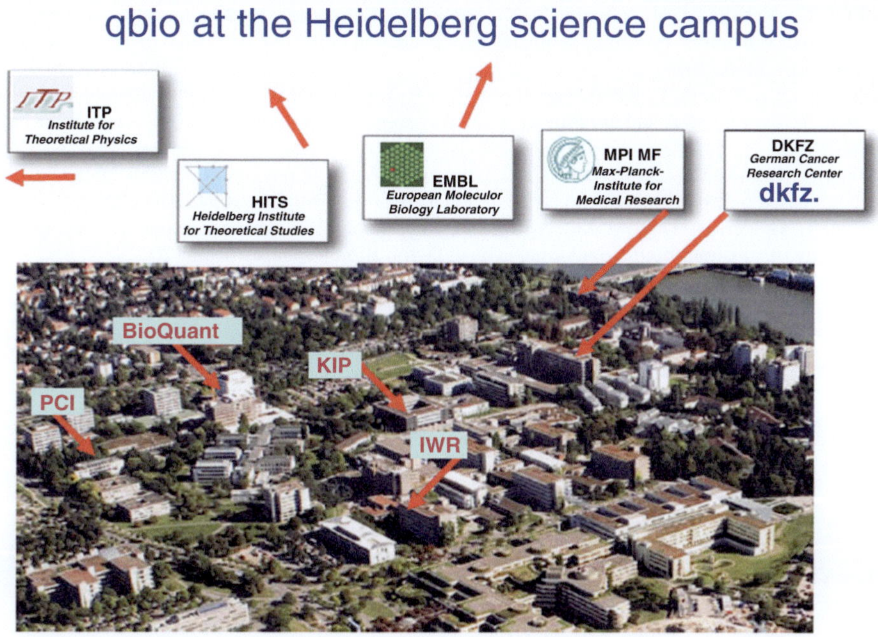

References

Arnold, M., Cavalcanti-Adam, E.A., Glass, R., Blummel, J., Eck, W., Kantlehner, M., Kessler, H., Spatz, J.P.: Activation of integrin function by nanopatterned adhesive interfaces. Chemphyschem. Eur. J. Chem. Phys. Phys. Chem. **5**, 383–388 (2004)

Becker, V., Schilling, M., Bachmann, J., Baumann, U., Raue, A., Maiwald, T., Timmer, J., Klingmuller, U.: Covering a broad dynamic range: information processing at the erythropoietin receptor. Science. **328**, 1404–1408 (2010)

Bischofs, I.B., Klein, F., Lehnert, D., Bastmeyer, M., Schwarz, U.S.: Filamentous network mechanics and active contractility determine cell and tissue shape. Biophys. J. **95**, 3488–3496 (2008)

Capoulade, J., Wachsmuth, M., Hufnagel, L., Knop, M.: Quantitative fluorescence imaging of protein diffusion and interaction in living cells. Nat. Biotechnol. **29**, 835–839 (2011)

Guthardt Torres, P., Bischofs, I.B., Schwarz, U.S.: Contractile network models for adherent cells. Phys. Rev. E., Stat. Nonlin. Soft. Matter Phys. **85**, 011913 (2012)

Hirschfeld-Warneken, V.C., Arnold, M., Cavalcanti-Adam, A., Lopez-Garcia, M., Kessler, H., Spatz, J.P.: Cell adhesion and polarisation on molecularly defined spacing gradient surfaces of cyclic RGDfK peptide patches. Eur. J. Cell Biol. **87**, 743–750 (2008)

Kolb, T., Maass, K., Hergt, M., Aebi, U., Herrmann, H.: Lamin A and lamin C form homodimers and coexist in higher complex forms both in the nucleoplasmic fraction and in the lamina of cultured human cells. Nucleus. **2**, 425–433 (2011)

Maeder, C.I., Hink, M.A., Kinkhabwala, A., Mayr, R., Bastiaens, P.I., Knop, M.: Spatial regulation of Fus3 MAP kinase activity through a reaction-diffusion mechanism in yeast pheromone signalling. Nat. Cell Biol. **9**, 1319–1326 (2007)

Munter, S., Sabass, B., Selhuber-Unkel, C., Kudryashev, M., Hegge, S., Engel, U., Spatz, J.P., Matuschewski, K., Schwarz, U.S., Frischknecht, F.: Plasmodium sporozoite motility is modulated by the turnover of discrete adhesion sites. Cell Host. Microbe. **6**, 551–562 (2009)

Rolli, C.G., Nakayama, H., Yamaguchi, K., Spatz, J.P., Kemkemer, R., Nakanishi, J.: Switchable adhesive substrates: revealing geometry dependence in collective cell behavior. Biomaterials. **33**, 2409–2418 (2011)

Rolli, C.G., Seufferlein, T., Kemkemer, R., Spatz, J.P.: Impact of tumor cell cytoskeleton organization on invasiveness and migration: a microchannel-based approach. PLoS One. **5**, e8726 (2010)

Sabass, B., Gardel, M.L., Waterman, C.M., Schwarz, U.S.: High resolution traction force microscopy based on experimental and computational advances. Biophys. J. **94**, 207–220

Shahal, T., Geiger, B., Dunlop, I.E., Spatz, J.P.: Regulation of integrin adhesions by varying the density of substrate-bound epidermal growth factor. Biointerphases. **7**, 23 (2012)

University of Leipzig

Site Address:	Soft Matter Physics Division Institute of Experimental Physics I Linnéstrasse 5 04103 Leipzig, GERMANY www.softmatterphysics.com
Date Visited:	11 May 2012
WTEC Attendees:	Paul Janmey, Cindy Reinhart-King (report author), Dan Fletcher, Lance Munn, Hemant Sarin
Host(s):	Josef A. Käs, Director of Experimental Physics I Institute of Experimental Physics I Faculty of Physics and Earth Sciences jkaes@physik.uni-leipzig.de Tel.: +49 341 97 32471

Overview

The University of Leipzig's research paradigm is based on an interdisciplinary bottom-up approach to modern materials, which was the basis for establishment of the Graduate School, Leipzig School of Natural Sciences – Building with Molecules and Nano-objects (BuildMoNa). The school's research focus is on the development of smart molecules and studying multifunctional scaffolds and complex nanostructures. This includes strong research in cell biophysics on the cytoskeleton. BuildMoNa is grouped into three focus areas that are quantum coherent structures, smart and active assemblies and physiochemical oncology. These focus areas have also helped to intensify relations with industrial partners and external research institutions. Moreover, a unique materials characterization facility is provided including micro- and nanostructures, nano-analysis, catalyst testing, biophotonics and magnetic resonance imaging.

Research and Development Activities

Dr. Josef A. Käs spent about half a day with the site team, giving us a tour of the facilities and demonstrating several experimental set-ups, including the optical stretcher setup developed and built in his lab, and atomic force microscopy. The optical stretching device can—in an automated fashion—measure the deformability of cells. The advantages of this system over systems like atomic force microscopy are its high throughput compared to many other systems (30 cell/min), which allows for the testing of many more cells. It can also operate with minimal user direction. Recent advances include the ability to image calcium dynamics during stretching (Gyger et al. 2011). Use of the optical stretching device has shown clear differences in the deformability of normal, malignant and metastatic cells, and future plans for the device include potential commercialization for use in the clinic for diagnosis. The group has a clear strength in understanding the challenges with translating technologies into the clinic and appears to be addressing them head-on.

Käs' presentation was titled "Are fundamental changes in a cell's material properties necessary for tumor progression?" The group has become very good at utilizing primary cells from patients and has demonstrated almost seamless collaborations with clinicians. His research has shown that with tumor progression the amount of cells that behave soft under small deformations increases (Fig. B.34). This is likely attributable (in part) to a reduction in the actin cortex. Notably, cell lines are much softer than primary cells, which may serve as a mechanical biomarker. Additionally, his research indicates that time in culture leads to an increase in deformability, indicating that culture creates artifacts in mechanical properties that others should be cautious of.

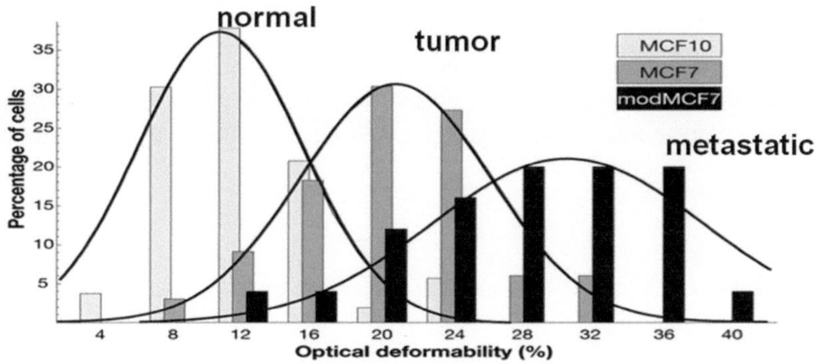

Fig. B.34 As the tumor progresses, the number of cells that are soft under small deformations increases (Courtesy of Josef Käs, University of Leipzig)

The group is currently addressing the question: How do soft cells grow against a rigid matrix? They have shown a role for intermediate filaments, in particular vimentin, through strain hardening. In separate work, their group is also exploring the role of the differential adhesion hypothesis for tumor progression using tumor spheroids with a mixture of normal and malignant cell populations, analogous to the experiments performed by Malcolm Steinberg in the 1960s.

Translational Efforts

There is a significant effort to translate the optical stretching device into clinical use. Käs has strong company ties and serves as a consultant for several companies.

Sources of Support

Support is provided by the German Research Foundation, German Federal Ministry of Education and Research, European Social Funds, and Era of Hope.

Collaborations and Possibilities

The Käs lab has several strong collaborations with surgeons and oncologists that are key for the group's successes.

Summary and Conclusions

Both the basic and translational science here are very strong. The group works well with clinicians, which is likely a key aspect in their success. The equipment and facilities are very good, and this group is poised to continue to make significant contributions to the cancer field.

References

Fritsch, A., Hockel, M., Kiessling, T., Nnetu, K.D., Wetzel, F., Zink, M., Kas, J.: Are biomechanical changes necessary for tumour progression? Nat. Phys. **6**, 730–732 (2010)

Gyger, M., Rose, D., Stange, R., Kiessling, T., Zink, M., Fabry, B., Kas, J.A.: Calcium imaging in the optical stretcher. Opt. Expr. **19**, 19212–19222 (2011)

Leal-Egana, A., Fritsch, A., Heidebrecht, F., Diaz-Cuenca, A., Nowicki, M., Bader, A., Kas, J.: Tuning liver stiffness against tumours: an *in vitro* study using entrapped cells in tumour-like microcapsules. J. Mech. Behav. Biomed. Mater. **9**, 113–121 (2012)
www.uni-leipzig.de/~pwm/web/
www.zv.uni-leipzig.de/en/
www.zv.uni-leipzig.de/en/university/profile-and-management/mission-profile-and-history/research-profile.html
www.dfg.de/en/dfg_profile/index.jsp

University of Milan

Site Address:	Department of Life Sciences Via Celoria 26 20133 Milan, ITALY National Research Council (CNR)-IENI Via R. Cozzi 53 20125 Milan, ITALY (meeting held at the Venetian Institute of Molecular Medicine, University of Padua)
Date Visited:	11 May 2012
WTEC Attendees:	Sharon Gerecht, Parag Mallick (report author), Owen McCarty, Hassan Ali
Host(s):	Stefano Zapperi Senior Researcher National Research Council-IENI Tel.: +39 02 66173 385 Fax: +39 02 66173 320 stefano.zapperi@cnr.it www.smmlab.it/research/quantitative-biology/

Overview

The University of Milan is a public teaching and research university. It has nine colleges and a teaching staff of 2,196, and is distinguished by its interdisciplinary focus. It is a leading institute in Italy and Europe for scientific productivity, and is the largest university in the region, with about 65,000 students.

The National Research Council (CNR) is the Italian coordinator for all public institutions devoted to science and research. It works as a general funding agency and as a research network across about one hundred institutes distributed throughout the Italian territory.

Research and Development Activities

Dr. Stefano Zapperi (CNR and Dr. Caterina La Porta, University of Milan)

La Porta and Zapperi have developed an interdisciplinary collaboration between oncology and theoretical physics focusing on the application of statistical physics approaches to investigate diverse biological challenges. In addition, they have been extremely active in outreach and community building developing both a quantitative biology seminar series and a number of exciting workshops in the area of physical sciences in oncology.

Zapperi presented results from a recent study on a novel approach to investigate tumor growth from a cancer stem cell perspective in melanoma. It is commonly believed that cell senescence—the loss of replicative capacity of cells—acts as a barrier for tumor growth. In their study, they followed the evolution of senescence markers in melanoma cells and found that while most cancer cells eventually turn senescent, this is irrelevant for the long-term growth rate of a tumor. To demonstrate this phenomenon they construct a mathematical population dynamics model (Fig. B.35, right) incorporating cancer stem cells, which is able to reproduce quantitatively the experimental data. Their results support the existence of cancer stem cells in melanoma and explain why it is difficult to fight cancer by inducing senescence in cancer cells. Only a fraction of the cells are susceptible to senescence, but those cells are irrelevant for tumor growth. A successful therapeutic strategy should instead target cancer stem cells, which are, however, likely to be strongly resistant to drug-induced senescence. This result is quite important and highlights the importance of evolutionary modeling of tumor growth as well as the possible insights that come from formal modeling approaches.

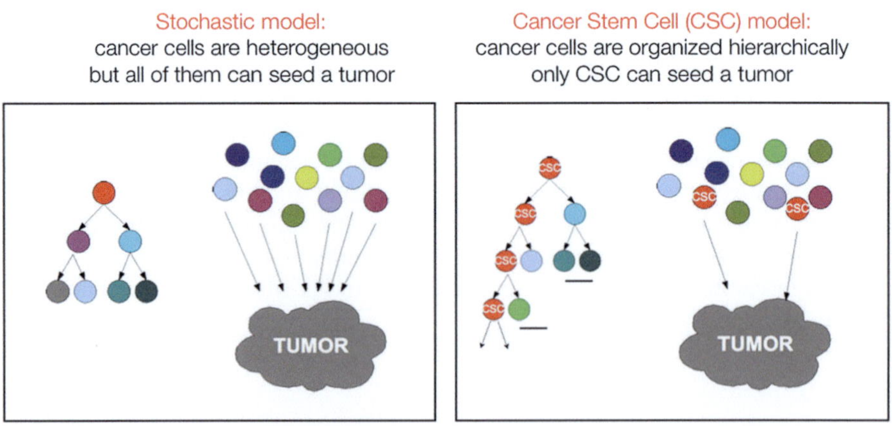

Fig. B.35 Stochastic and cancer stem cell model for tumors (From La Porta et al. 2012)

Sources of Support/Funding

European Science Foundation, Centre Européen de Calcul Atomique et Moléculaire

Summary and Conclusions

The interdisciplinary research underway in Milan is extremely exciting. They are looking at cancer through an evolutionary lens and developing rigorous approaches to describe evolutionary dynamics.

References

La Porta, C.A., Zapperi, S., Sethna, J.P.: Senescent cells in growing tumors: population dynamics and cancer stem cells. PLoS Comput. Biol. **8**, e1002316 (2012)
www.smmlab.it/
www.cecam.org/workshop-0-751.html
www.cancerphysics.unimi.it
www.ploscompbiol.org/article/info%3Adoi%2F10.1371%2Fjournal.pcbi.1002316
www.nature.com/srep/2012/120607/srep00441/full/srep00441.html

University of Mons

Site Address:	(meeting at Institute Curie, Paris) 26 rue d'Ulm 75248 Paris Cedex 05, FRANCE http://portail.umons.ac.be/en2/pages/default.aspx
Date Visited:	9 May 2012
WTEC Attendees:	Paul Janmey, Cindy Reinhart-King, Dan Fletcher (report author), Jerry Lee, Nastaran Kuhn, Hemant Sarin
Host(s):	Sylvain Gabriele Group Leader of Mechanobiology and Soft Matter University of Mons Campus of Sciences and Medicine, Mendeleïev Building, 1-059 Maistriau B-7000 Avenue Mons, BELGIUM sylvain.gabriele@umons.ac.be Tel.: +32 65 37 3824

Overview

The University of Mons is the founding partner of the Hainaut higher education consortium, which consists of the university and other higher education institutions. The University of Mons was represented by Dr. Sylvain Gabriele, Mechanobiology and Soft Matter group leader.

Research and Development Activities

Gabriele presented an overview of his basic and applied research projects. He has developed a microfluidic system to mimic lung capillaries. This work was motivated by acute respiratory distress syndrome (ARDS), in which blood cells become lodged in the microvasculature. The microfluidic system passes cells through small capillaries and allows quantification of the likelihood of aggregation and blockage. Use of devices such as this could provide an early warning for ARDS.

In a second project on blast-induced traumatic brain injury, Gabriele has developed a high-speed uniaxial stretcher that can simulate the effects of blast-induced injuries. This work has the potential to identify key events in blast-induced trauma at the tissue and cell level using model systems.

Gabriele's basic research is focused on understanding mechanical interactions of cells, such as the role of integrins in the propagation of strain-induced injury. Using magnetic tweezers, he has recently been able to apply and measure localized strains. His ongoing research uses micropatterning to control nuclear shape and evaluate its impact on chromatin organization and proliferation, as well as determining integrin-cadherin crosstalk (Fig. B.36).

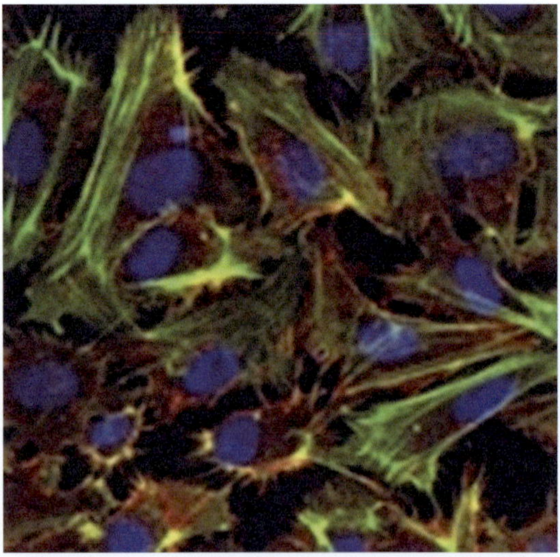

Fig. B.36 Cross talk between integrins and cadherins (Courtesy of Sylvain Gabriele, University of Mons)

Sources of Support

Funding is mainly provided by the Belgian National Research Fund and the University of Mons.

Collaborations and Possibilities

New technologies that may be useful for U.S. researchers are being developed by Gabriele's group, which is already in collaboration with Kit Parker's group at Harvard University.

Summary and Conclusions

Research at the intersection of physical sciences and biology is growing at the University of Mons thanks to the work of Gabriele. A multidisciplinary research institution, the Biosciences Institute, has been recently co-founded by Gabriele at the University of Mons to enhance Belgian collaborative research whereby physical and chemical sciences and engineering principles are being applied to diseases such as cancer research and oncology.

References

Gabriele, S., Benoliel, A.M., Bongrand, P., Theodoly, O.: Microfluidic investigation reveals distinct roles for actin cytoskeleton and myosin II activity in capillary leukocyte trafficking. Biophys. J. **96**, 4308-4318 (2009)

Gabriele, S., Versaevel, M., Preira, P., Theodoly, O.: A simple microfluidic method to select, isolate, and manipulate single-cells in mechanical and biochemical assays. Lab Chip. **10**, 1459–1467 (2010)

Versaevel, M., Grevesse, T., Gabriele, S.: Spatial coordination between cell and nuclear shape within micropatterned endothelial cells. Nat. Commun. **3**, 671 (2012)

http://portail.umons.ac.be/en2/pages/default.aspx

http://w3.umons.ac.be/perso/Gabriele.Sylvain/Home.html

University of Nürnberg-Erlangen

Site Address:	Nägelsbachstraße 49b, D-91052 Erlangen, GERMANY (Meeting took place at the Bavarian Academy of Sciences, Munich)
Date Visited:	9 May 2012
WTEC Attendees:	Sharon Gerecht, Parag Mallick, Owen McCarty (report author), Lance Munn, Hassan Ali
Host(s):	Prof. Ana-Sunčana Smith Tel.: +49 9131 85 20842 smith@physik.uni-erlangen.de

Overview

The University of Nürnberg-Erlangen is located in Erlangen, Germany, and is made up of over 600 professors, 24 clinics in the University Hospital, and nearly 33,500 students. The University has an Excellence Initiative graduate school in advanced optical techniques (SAOT) and an Excellence Initiative research cluster in engineering of advanced materials (EAM). These Excellence Initiatives are supported by the German Federal Ministry of Education and Research and aim to promote cutting-edge research and support the development and training of the next generation of scientists and researchers.

The EAM is a cluster of researchers focused on applying fundamental research towards to the creation and development of high-performance materials. The EAM is comprised of three interdisciplinary centers: (1) Functional Particle Systems; (2) Nano-analysis and Electron Microscopy; and (3) Multiscale Modeling and Simulation. The EAM focuses on the engineering of nanoelectronic materials, photonic and optical materials, catalytic materials, and lightweight materials. However, teams were built in the interdisciplinary environment of EAM to extend their activities into the field of biomaterials, with particular emphasis on artificial scaffolds for cell cultures.

Research and Development Activities

The WTEC panelists attended a presentation by Dr. Ana-Sunčana Smith, professor in the EAM Cluster and the department of physics. Her work centers on building and understanding biomimetic models for the cell adhesion process. This work requires the application of biophysical models to adhesive processes at both the molecular- and microscopic-length scales. Smith's work has demonstrated the emergence of characteristic length and time scales during nucleation. Figure B.37 represents a model of receptor-ligand interactions at the scale coupled to the deformations of the membrane at a micron scale (Reister-Gottfried et al. 2008).

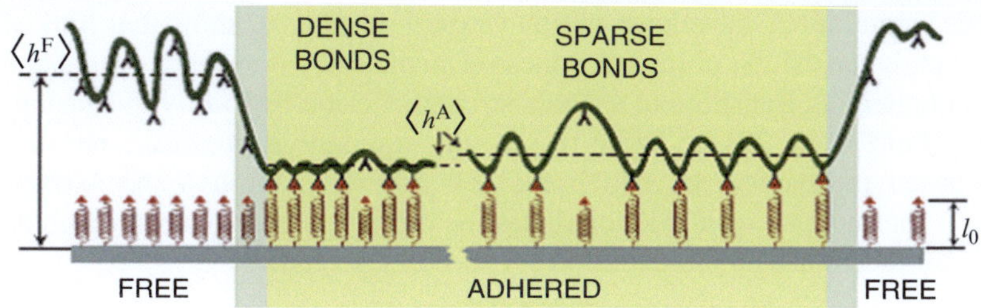

Fig. B.37 Model of adhering membrane. The effective binding affinity of a membrane for a surface is governed by the receptor density, separation distance (h), and receptor length (l) (Courtesy of Ana-Sunčana Smith)

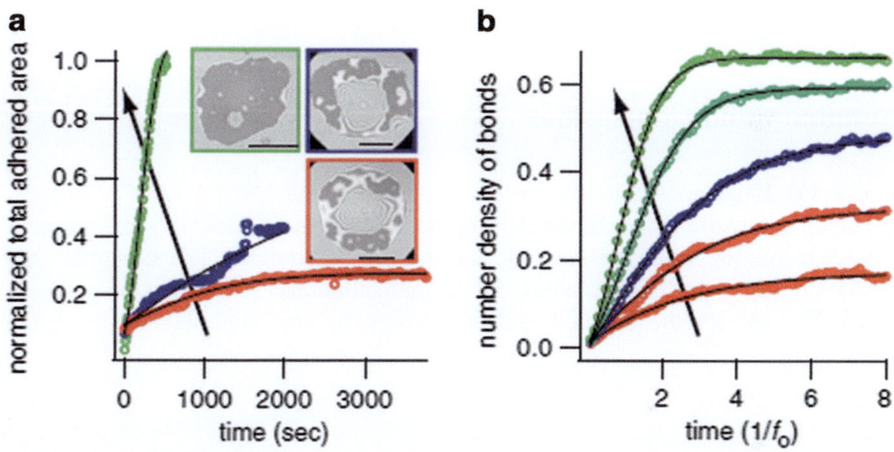

Fig. B.38 Modeling the dynamics of cell adhesion
(*a*) Adhered area (normalized by the equilibrium contact zone area) in time for vesicles on substrates with high, middle, and low E-selectin densities. For vesicles of comparable size, halving the concentration of E-selectin approximately doubles the equilibration time. The scale bar is 10 μm. (*b*) Average number of bonds in time in sets of 200 simulation runs for exp $(\upsilon_0/\kappa_B T) = 3.0, 3.25, 3.5, 3.9$, and 4.5. The directions of growing E-selectin density and υ_0 are shown with arrows. (Adapted from Reister-Gottfried et al. 2008)

Global adhesive forces are modeled as a series of individual bonds, which gives rise to the effective binding affinity of a cell membrane for a surface. The dynamics of cell adhesion can then be modeled by reaction-diffusion models that, depending on the density and strength of ligands and receptors, predict an exponentially saturating or power law growth (Fig. B.38). This model predicts that adhesion is a dynamical transition from nucleation of receptors in a free membrane to thermodynamically-regulated bond formation followed by saturation of receptor-ligand bonds.

These simulations have been verified experimentally to show that intrinsically strong bonds can exhibit ultraweak adhesion mediated by transiently bound domains and can undergo a transition to a stable strong adhesion by locally increasing bond density (Fenz et al. 2011). These results and simulations provide a physical sciences-based mechanism by which the cell migration through the extracellular matrix is thermodynamically favored, despite the fact that the bond strength of each individual receptor-ligand interaction is exceedingly high.

Sources of Support/Funding

This work is supported in part by the EAM; German Research Foundation; and Unity through Knowledge Fund from the Ministry of Science, Republic of Croatia.

Summary and Conclusions

The EAM Excellence Initiative research cluster at the University of Nürnberg-Erlangen is a world-leader in the development of advanced materials in communications technology, catalysis, energy, and transportation fields. The strength of the EAM is bridging the length scales from the molecular level (10^{-9} m) to macroscopic level (10^{-3} m) for the creation and application of new materials for emerging technologies.

References

Fenz, S.F., Bihr, T., Merkel, R., Seifert, U., Sengupta, K., Smith, A.S.: Switching from ultraweak to strong adhesion. Adv. Mater. **23**, 2622–2626 (2011)

Reister-Gottfried, E., Sengupta, K., Lorz, B., Sackmann, E., Seifert, U., Smith, A.S.: Dynamics of specific vesicle-substrate adhesion: from local events to global dynamics. Phys. Rev. Lett. **101**, 208103 (2008)

University of Paris Diderot

Site Address:	Material and Complex Systems Laboratory, UMR 7057 Centre National de la Recherche Scientifique and Université Paris-Diderot, CC 7056 10 Rue A. Domon et L. Duquet 75205 Paris Cedex 13, FRANCE www.msc.univ-paris-diderot.fr/?lang=en
Date Visited:	10 May 2012
WTEC Attendees:	Paul Janmey, Cindy Reinhart-King, Dan Fletcher, Lance Munn (report author), Jerry Lee, Nastaran Kuhn, Larry Nagahara, Hemant Sarin
Host(s):	François Gallet Head of Physics Department francois.gallet@univ-paris-diderot.fr Tel.: +33 1 57 27 61 03
	Loic Auvray Principal Investigator at CNRS, Biophysics Department loic.auvray@univ-paris-diderot.fr
	Atef Asnacios Group Leader Physics of Single Cell Mechanosensing Biophysics Department atef.asnacios@univ-paris-diderot.fr
	Sylvie Hénon Professor in Material and Complex Systems Laboratory sylvie.henon@univ-paris-diderot.fr Tel.: +33 1 57 27 62 15
	Florence Gazeau Professor in Material and Complex Systems Laboratory Biophysics Department Florence.gazeau@univ-paris-diderot.fr
	Jean-Francois Berret Professor in Material and Complex Systems Laboratory jean-francois.berret@univ-paris-diderot.fr Tel.: +33 1 44 27 44 97
	Vincent Fleury Professor in Material and Complex Systems Laboratory vincent.fleury@univ-paris-diderot.fr

Overview

The University Paris Diderot science sector teams mainly work with the French National Center for Scientific Research (CNRS) and National Institute of Health and Medical Research (INSERM). A large number of science sector teams also work with other universities or institutions, such as Paris Descartes, Pierre et Marie Curie, Paris Sud, Paris Est, the École Normale Supérieure Paris, the École Supérieure de Physique et de Chimie Industrielles de la Ville de Paris, and the Observatoire de Paris. Some of these teams are located on the sites of these institutions. These teams mainly work in mathematics, computer science, physics, chemistry, biology and earth science, with research on the latter being done in close partnership with the Paris Institute of Geophysics.

Research and Development Activities

Dr. François Gallet

Gallet's research focuses on the mechanics of cells and tissues (Fig. B.39). A major interest is the interplay between integrin and cadherin adhesive interactions, and how these are modulated and coordinated to allow appropriate cell-cell and cell-surface adhesions. He spreads cells on fibronectin (FN) patterns of different sizes and mimics cell-cell adhesion by attaching beads coated with E-cadherin fragments to the cell membrane. The bead-cells contacts have less stiffness for cells widely spread on FN.

Gallet is also actively collaborating with Prof. Benoit Ladoux (University of Paris-Diderot) and Prof. Sylvie Dufour (Institut Curie), studying how cells produce force on micropillars as they interact with other cells through cadherin binding. Their results show that cells bound to each other tend to produce more force than single cells.

Dr. Atef Asnacios

Asnacios uses sophisticated biophysical approaches to study tissue and cytoskeletal mechanics. It has been shown that substrate rigidity can be modulated to direct cell spreading, migration, and stem cell differentiation. He is investigating the

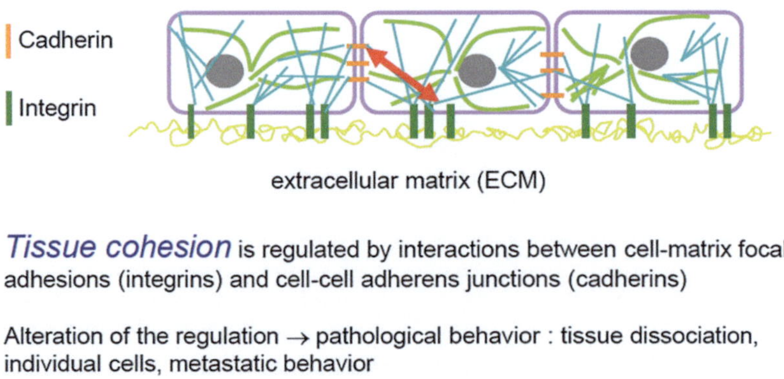

Fig. B.39 Gallet's research addresses epithelial microanatomy and the cross-talk between cadherin and integrin adhesions (Courtesy of Ana- Sunčana Smith)

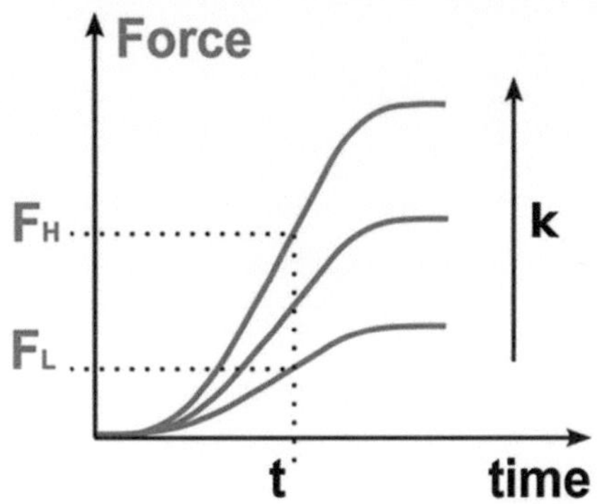

Fig. B.40 Applying high or low force (FH, FL, respectively) to a cell results in different dynamics of reciprocal force generation. This could explain why cells polarize on anisotropic substrates along the stiffest axis (From Fouchard et al. 2011)

mechanisms by which cells probe and detect local mechanical properties. To do this, he performs single cell traction force measurements and analyzes changes in cell structure (cell shape, force generation, stress fiber organization, and adhesion complexes). He has shown that cell contractility adapts to the local external stiffness, reflecting the force-dependent kinetics of myosin binding to actin. One interpretation of this work is that contractile acto-myosin units, by themselves, are sufficient to sense forces and respond to external mechanical perturbations.

His group is now combining traction force measurements with total internal reflection fluorescence microscopy of labeled proteins in adhesion complexes to simultaneously monitor the kinetics of force adaptation and adhesion complex remodeling (Fig. B.40).

Dr. Loïc Auvray

Auvray works on the dynamics of protein folding. He creates well-defined nanopores through which proteins can be extruded. The system mimics the translocation of biomacromolecules across a membrane, and can be used to track changes in tertiary structure during the process. Accurate measurements of protein melting temperatures are readily made with this methodology (Fig. B.41). The process is quantified by measuring conductance across the pore. An important finding is that the unfolding curve does not depend on the structure or net charge of the nanopore.

Dr. Jean-François Berret

Berret studies polyelectrolytes and copolymers in solutions. He presented work on magnetic rods, or nanowires, created by aggregation of small, 10 nm magnetic particles. The resulting nanowires are 200 nm in diameter and range in length from

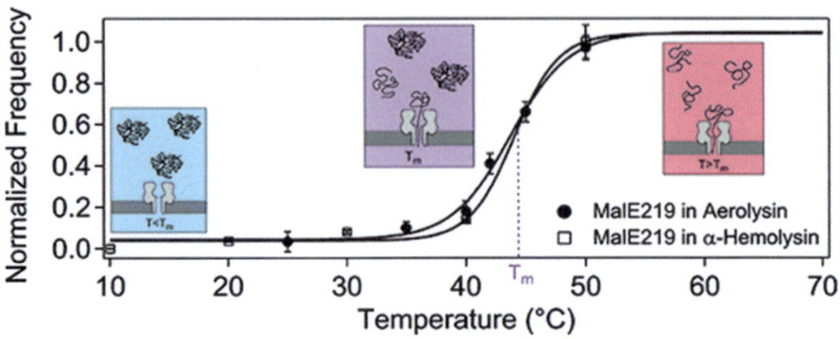

Fig. B.41 Macromolecule translocation through nanopores. As temperature increases, thermal unfolding allows passage through the pores (Courtesy of L. Auvray)

3-15 μm. When added to culture media, they are naturally internalized, and can be visualized, tracked, or manipulated. In one application, Berret is measuring the thermal rotational diffusion of the rods to estimate the internal viscosity of cells. Values are surprisingly high—around 200 cP. In future work, he plans to manipulate these rods to disrupt the cytoplasm of targeted cancer cells, inducing cell death.

Dr. Vincent Fleury

Fleury studies the role of tissue hydrodynamics in development and evolution. His hypothesis is that the topology of blood vessel networks and tissue morphology are determined by flow patterns. He uses many models, including chicken embryos and jellyfish, to study angiogenesis and branching morphogenesis (Fig. B.42). Using time-lapse video-microscopy, he tracks arterial-venous differentiation, and has demonstrated that flow shapes the global patterning of the arterial tree and regulates the activation of the arterial markers ephrinB2 and neuropilin 1.

Drs. Florence Gazeau and Claire Wilhelm

Gazeau and Wilhelm are developing magnetic nanoparticles for medical imaging and therapy. The particles are easily taken up by cells and can be imaged using standard technologies. Immediate applications include cell tracking via MRI, and MRI imaging of inflammation. It is also possible to manipulate the particles or particle-laden cells with a magnetic field, so cell therapy and tissue engineering might be directed or facilitated by externally-applied magnetic force. Using this

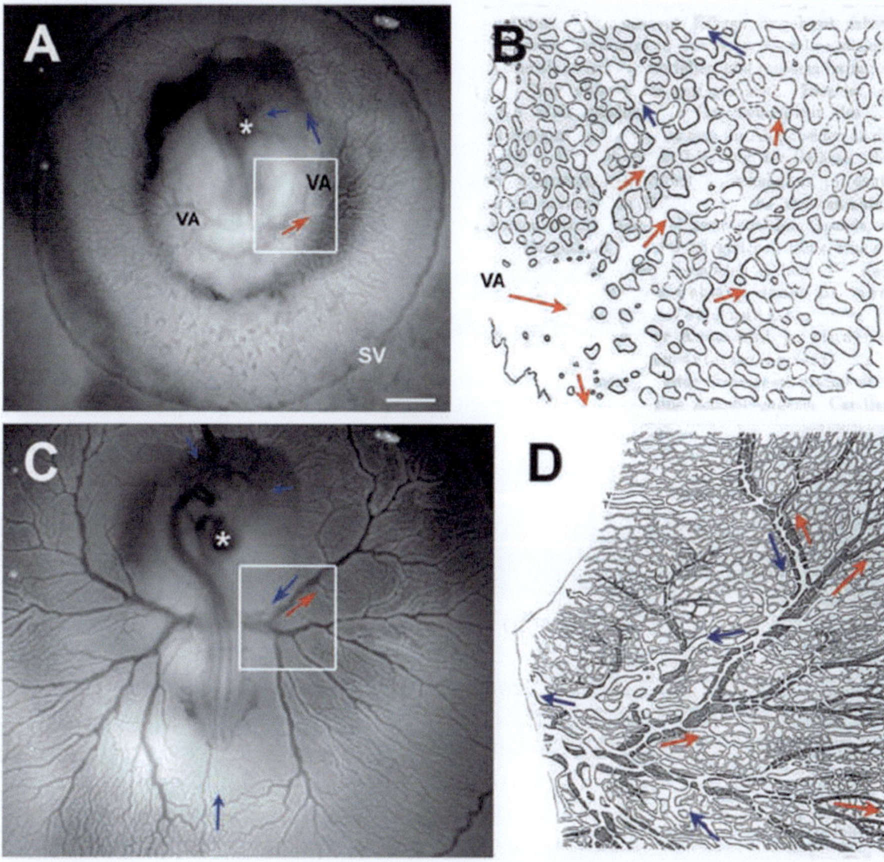

Fig. B.42 Development of the chick vasculature. (From Freund et al. 2012)
(*A*) Formation of the vitelline artery (VA). Red and blue arrows indicate arterial and venous flow, respectively. The heart is marked by the asterisk. (*B*) Outline of the vascular system in the boxed area in *A*. (*C*) After 26 h, significant remodeling has occurred. (*D*) Illustration of the vascular system in the boxed area in *C*. Veins are white, arteries are gray; at this stage, veins and arteries run in parallel. SV: sinus vein. Scale bar =1100 μm

approach, Gazeau and Wilhelm have shown directed localization of particles in tumors. This could be used to deliver photoactivatable drugs or therapeutic cells, or induce nanoparticle-mediated hyperthermia.

Another interesting application is the manipulation of cells in culture to force aggregation in pre-designated shapes, with defined kinetics.

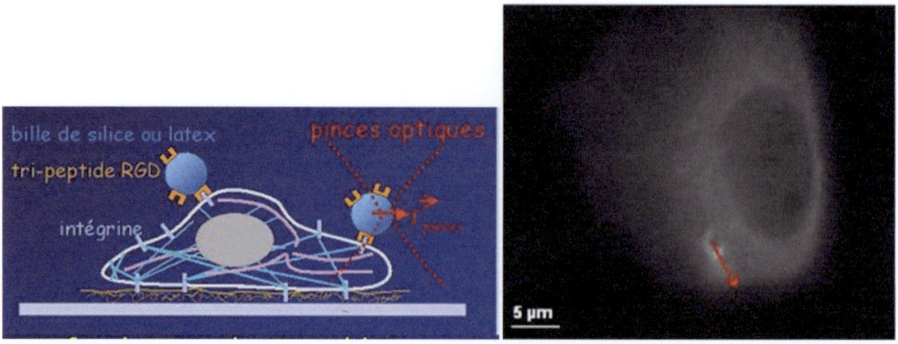

Fig. B.43 Applying cyclic forces to beads attached to integrins results in cell stiffening and concentration of actin at the bead focal adhesions (Courtesy of S. Hénon)

Dr. Sylvie Hénon

Hénon studies cell mechanics, focusing on the role of intermediate filaments. She attaches functionalized beads to cells and manipulates them with optical tweezers to measure cell mechanical properties. She applies relatively sophisticated methods to measure viscoelastic moduli and cell mechanical "creep" in addition to Young's moduli. Using these methods, she discovered that cells submitted to successive creep experiments become stiffer with time. The dynamics depend on cell type, but fibroblasts stiffen in approximately 10 min.

To investigate the roles of cytoskeletal components in the stiffening process, Hénon uses a GFP-actin fusion protein to visualize actin dynamics (Fig. B.43). By pulling on beads coated with RGD peptides, she is able to see actin recruitment at the location of the bead during the stiffening process. Hénon is also exploring clinical relevance of her concepts: cells responsible for myofibrillar myopathies with desmin mutations display quantitatively different mechanical properties in her assays.

Translational Efforts

Gazeau and Wilhelm plan to use their magnetic particles for clinical imaging, diagnosis, and therapy.

Sources of Support

External funding generally comes from French agencies and foundations including:

- Association pour la Recherche sur le Cancer
- Centre National de la Recherche Scientifique
- Association Française contre les Myopathies
- Ligue Contre le Cancer
- Agence Nationale de la Recherche
- Ministère de la Recherche

Summary and Conclusions

Diderot is more focused on basic research than other sites visited. Physics concepts and approaches are very much the central theme here, but many of the researchers are attempting to relate their work to the cancer field. In light of physics and oncology, highlights include a variety of sound studies from multiple groups on cell mechanics and force generation. Fleury's work on hydrodynamics and tissue morphogenesis is also likely relevant to cancer progression.

References

Allioux-Guerin, M., Icard-Arcizet, D., Durieux, C., Henon, S., Gallet, F., Mevel, J.C., Masse, M.J., Tramier, M., Coppey-Moisan, M.: Spatiotemporal analysis of cell response to a rigidity gradient: a quantitative study using multiple optical tweezers. Biophys. J. **96**, 238–247 (2009)

Boryskina, O.P., Al-Kilani, A., Fleury, V.: Body plan in tetrapods: is it patterned by a hyperbolic tissue flow? Cell Cycle. **10**, 3801–3802 (2011)

Breton, M., Prevel, G., Audibert, J.F., Pansu, R., Tauc, P., Le Pioufle, B., Francais, O., Fresnais, J., Berret, J.F., Ishow, E.: Solvatochromic dissociation of non-covalent fluorescent organic nanoparticles upon cell internalization. Phys. Chem. Chem. Phys. **13**, 13268–13276 (2011)

Fleury, V.: Clarifying tetrapod embryogenesis by a dorso-ventral analysis of the tissue flows during early stages of chicken development. Bio. Syst. **109**(3), 460–474 (2012)

Fleury, V., Boryskina, O.P., Al-Kilani, A.: Hyperbolic symmetry breaking and its role in the establishment of the body plan of vertebrates. C. R. Biol. **334**, 505–515 (2012)

Fouchard, J., Mitrossilis, D., Asnacios, A.: Acto-myosin based response to stiffness and rigidity sensing. Cell Adh. Migr. **5**, 16–19 (2011)

Freund, J.B., Goetz, J.G., Hill, K.L., Vermot, J.: Fluid flows and forces in development: functions, features and biophysical principles. Development. **139**, 1229–1245 (2012)

Galimard, A., Safi, M., Ould-Moussa, N., Montero, D., Conjeaud, H., Berret, J.F.: Thirty-femtogram detection of iron in Mammalian cells. Small. **8**, 2036–2044 (2012)

Gazeau, F., Wilhelm, C.: Magnetic labeling, imaging and manipulation of endothelial progenitor cells using iron oxide nanoparticles. Fut. Med. Chem. **2**, 397–408 (2010)

Guardia, P., Di Corato, R., Lartigue, L., Wilhelm, C., Espinosa, A., Garcia-Hernandez, M., Gazeau, F., Manna, L., Pellegrino, T.: Water-soluble iron oxide nanocubes with high values of specific absorption rate for cancer cell hyperthermia treatment. ACS Nano. **6**, 3080–3091 (2012)

Lartigue, L., Wilhelm, C., Servais, J., Factor, C., Dencausse, A., Bacri, J.C., Luciani, N., Gazeau, F.: Nanomagnetic sensing of blood plasma protein interactions with iron oxide nanoparticles: impact on macrophage uptake. ACS Nano. **6**, 2665–2678 (2012)

Levy, M., Wilhelm, C., Luciani, N., Deveaux, V., Gendron, F., Luciani, A., Devaud, M., Gazeau, F.: Nanomagnetism reveals the intracellular clustering of iron oxide nanoparticles in the organism. Nanoscale. **3**, 4402–4410 (2011)

Merstorf, C., Cressiot, B., Pastoriza-Gallego, M., Oukhaled, A.G., Bacri, L., Gierak, J., Pelta, J., Auvray, L., Mathe, J.: DNA unzipping and protein unfolding using nanopores. Meth. Mol. Biol. **870**, 55–75 (2012)

Oukhaled, A.G., Biance, A.L., Pelta, J., Auvray, L., Bacri, L.: Transport of long neutral polymers in the semidilute regime through a protein nanopore. Phys. Rev. Lett. **108**, 088104 (2012)

Payet, L., Martinho, M., Pastoriza-Gallego, M., Betton, J.M., Auvray, L., Pelta, J., Mathe, J.: Thermal unfolding of proteins probed at the single molecule level using nanopores. Anal. Chem. **84**, 4071–4076 (2012)

Qi, L., Chapel, J.P., Castaing, J.C., Fresnais, J., Berret, J.F.: Stability and adsorption properties of electrostatic complexes: design of hybrid nanostructures for coating applications. Langmuir ACS J. Surf. Coll. **23**, 11996–11998 (2007)

University of Rostock

Site Address:	(meeting at University of Leipzig) Linnéstrasse 5 04103 Leipzig, GERMANY www.uni-rostock.de/en/
Date Visited:	11 May 2012
WTEC Attendees:	Paul Janmey, Cindy Reinhart-King (report author), Dan Fletcher, Lance Munn, Hemant Sarin
Host(s):	Adelinde Uhrmacher, Professor of Modeling and Simulation Department of Computer Science Faculty of Computer Science and Electrical Engineering University of Rostock D-18051, Rostock, GERMANY adelinde.uhrmacher@uni-rostock.de Tel.: +49 381 498 7610

Overview

The University of Rostock has nine faculties divided into institutes and clinics. The working group on modeling and simulation in the Department of Computer Science at the University of Rostock interfaces modeling, simulation, and artificial intelligence. Their research centers on developing modeling and simulation methods and their application in different areas. The methodological developments refer to modeling formalisms, particularly to modeling formalisms supporting variable structure models, multilevel models, and spatial models, efficient execution of these models (in terms of approximate and parallel approaches), and on support for *in silico* experiments (e.g., by exploiting workflow technologies). The application areas of the methods range from demography, software development, to cell biology.

Research and Development Activities

While in Leipzig, the WTEC team met with Adelinde Uhrmacher from the Department of Computer Science. Her talk was titled: "Modeling and simulating spatial dynamics in cell biology: Computer scientist on systems biology." She emphasized approaches to spatial simulation—ML-Rules—a multilevel rule-based modeling method (Maus et al. 2011), and a spatial variant—ML-Space (Bittig et al. 2011). ML-Rules allows one to compactly describe and combine compartmentalized dynamics, including inter- and intra-cellular dynamics as well as processes at the cellular level such as proliferation of cells, apoptosis, and cell differentiation (Mazemondet et al. 2011). ML-Rules assumes well mixed solutions within the compartments. It does not capture phenomena that are induced by the molecular crowding within cells. Therefore, the language for ML-Space has been developed with decidedly spatial semantics. Here, species can be defined as individual particles that react due to collisions, or as a population of species residing in a small area. It inherits from ML-Rules the compact description and the ability to describe processes at different organizational levels however they adhere to spatial physical constraints. ML-Space has been used to investigate lipid rafts as compartments, with a focus on their movements and the activity of receptors in rafts.

Adelinde Uhrmacher, a computer scientist, emphasizes the importance of separation of concern in modeling and simulation, and thus to clearly distinguish between the model, its execution, and defining an *in silico* experiment with this model. The latter can mean parameter scanning and optimization. Similar to a wet-lab experiment, the *in silico* experiment needs to be carefully documented to be of any value—including how has the model been validated and how have the simulation results been achieved. Uhrmacher's group has designed a general purpose plug-in-based modeling and simulation framework which has already been applied to develop different modeling and simulation tools for cell biology. Currently, the framework includes more than 700 plugging and more than 100 plug-in types (e.g., different modeling formalisms, execution algorithms, steady state analyzers, etc.). It also provides intelligent support to configure suitable experiments on demand.

Figure B.44 shows an illustration of the hierarchical modeling concept. Different-shaped nodes correspond to different species names while attributes are color-coded. Stacking of identical nodes represents the amount of a certain species. In the figure, *A* is a graphical representation of a hierarchical model structure via nested

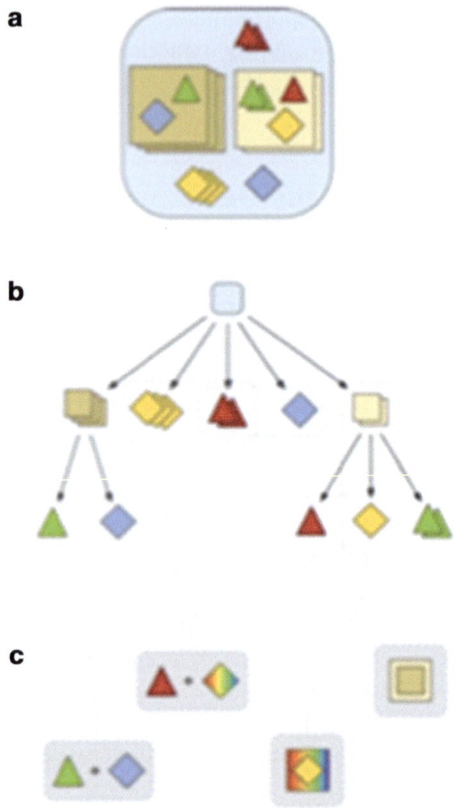

Fig. B.44 Nested model structure (From Maus et al. 2011)

nodes. *B* shows the same model structure alternatively depicted as a directed tree graph. Note that besides atomic species (triangles and diamonds), species containing a sub-solution (squares) might be attributed so that each species at each level might have its own state. In *C* are examples of matching different reactant patterns within the hierarchical model structure. The rainbow shadings in the second and third pattern illustrate variable instead of defined colors, i.e., attributes (Maus et al. 2011) (Fig. B.45).

Summary and Conclusions

A multi-scale simulation of cell processes should be useful in understanding biological processes.

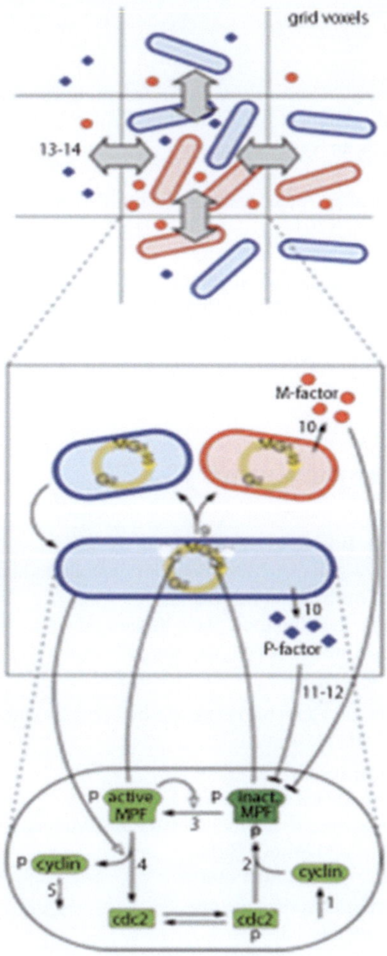

Fig. B.45 Schematic description of the example model. (From Bittig et al. 2011)
At the bottom level, interacting proteins describe the intracellular dynamics of a fission yeast cell (reactions *1-5*). The intermediate level describes dynamics of entire cell states, i.e., cell growth (*6*), cell cycle phase transitions (*7-9*), and division including mating type switching (*9*). In addition, cells may secrete pheromone molecules (P-factor and M-factor) to the extracellular medium (*10*). Various inter-level causalities between the intermediate and the bottom level influence processes both in an upward (*7-9*) and downward causation manner (*4,11-12*). The top level discretizes the environment of cells into multiple fictive compartments in order to study spatial dynamics of pheromone diffusion and displacement of cells (*13-14*). Although spatial dynamics referring to compartments and particle diffusion between cells can be modeled, excluded volume effects cannot be described in ML-Rules; therefore one has to move to ML-Space

References

Bittig, A.T., Haack, F., Maus, C., Uhrmacher, A.M.: Adapting rule-based model descriptions for simulating in continuous and hybrid space. In: Proceedings of the 9th international conference on computational methods in systems biology 2011, pp. 161–170, New York, NY, USA, ACM. CMSB '11 (2011)

Maus, C., Rybacki, S., Uhrmacher, A.M.: Rule-based multi-level modeling of cell biological systems. BMC Syst. Biol. **5**, 166 (2011)

Mazemondet, O., Hubner, R., Frahm, J., Koczan, D., Bader, B.M., Weiss, D.G., Uhrmacher, A.M., Frech, M.J., Rolfs, A., Luo, J.: Quantitative and kinetic profile of Wnt/beta-catenin signaling components during human neural progenitor cell differentiation. Cell. Mol. Biol. Lett. **16**, 515–538 (2011)

www.mosi.informatik.uni-rostock.de/mosi/

University of the Basque Country

Site Address:	P.O. Box 644, 48080 Bilbao, SPAIN www.ehu.es/p200-hmencont/en/contenidos/informacion/basic_facts/en_inf/basicfacts.html (The meeting took place at the University of Barcelona Institute for Bioengineering of Catalonia)
Date Visited:	7 May 2012
WTEC Attendees:	Sharon Gerecht, Parag Mallick (report author), Owen McCarty, Lance Munn, Hassan Ali
Host(s):	Prof. José M.G. Vilar Biophysics Unit (CSIC-UPV/EHU) Department of Biochemistry and Molecular Biology Tel.: +34-94 601 3450 Fax: +34-94 601 3360 j.vilar@ikerbasque.org www.vilarlab.org

Overview

The University of the Basque Country (UPV/EHU) has been certified as a Campus of International Excellence by the Spanish Ministry of Education. UPV/EHU and its partners stimulate research, creativity, and innovation in order to establish a new, sustainable economic and social model. Researchers at UPV/EHU seek to extend their international influence by working on a "brain-gain" basis and establishing cross-border campuses. It is clear that they are significantly connected to the broader international community.

Towards this, UPV/EHU hosts an unrestricted and multicultural community. Their international relations office manages over 900 exchange agreements with other Spanish, European, and international universities. Every year, the three campuses turn into lively melting pots where over 900 foreign students blend in with local trainees. The university takes part in SICUE, the Spanish universities students' exchange program; ERASMUS, the most extended European mobility program; and they have created their own exchange programs with Latin America (UPV/EHU-AMÉRICA LATINA) and other destinations (United States, Canada, The Philippine Islands, New Zealand, and Russia). A number of teaching staff travels to Central and South American universities.

The UPV/EHU and the National Research Council of Spain (CSIC) host together the Biophysics Unit (CSIC-UPV/EHU), a cross-disciplinary center at the interface between physics, biochemistry and molecular biology.

Research and Development Activities

Dr. Jose Vilar

The Vilar group is broadly interested in understanding and *accurately* predicting molecular, cellular, and cell-population behaviors in terms of the interactions of the components and vice versa.

Although traditional approaches have been successful at identifying cellular components and their interactions, the Vilar lab believes it is critical to piece back together all the genetic, biochemical, molecular, and structural information into a physiologically relevant description of the cell using "constructive" methods. Consequently, they use computational modeling as a tool for transforming molecular detail into a more integrated form of understanding complex behavior. They are interested, not only in the interactions between cellular components, but also in the resulting cellular behavior and its integration into the physiological context of an organism. Their main focus is investigating the role of long-range interactions in controlling biochemical and cellular processes across scales.

The group is currently working in several areas including:

- Gene regulation (RXR and other nuclear hormone receptors, Fig. B.46)
- Signal transduction (EGF and TGF-β pathways)
- Control of cell growth and death (Bcl-2/Bax in metabolism and apoptosis)
- Classification of Leukemia samples

Vilar presented some early work on the lac repressor, which binds to a primary operator O1 and prevents the RNA polymerase from transcribing the genes. If it is not bound, transcription proceeds at a given rate. Unfortunately, regulation of transcription is complex. In addition to O1 there are two sites outside the control region, the so-called auxiliary operators O2 and O3, which closely resemble O1 and where the repressor can also bind. However, they are much weaker than O1 (10 and 300 times less). Moreover, elimination of either one of them leaves the repression level practically unchanged. However, the role of O2 and O3 are actually quite

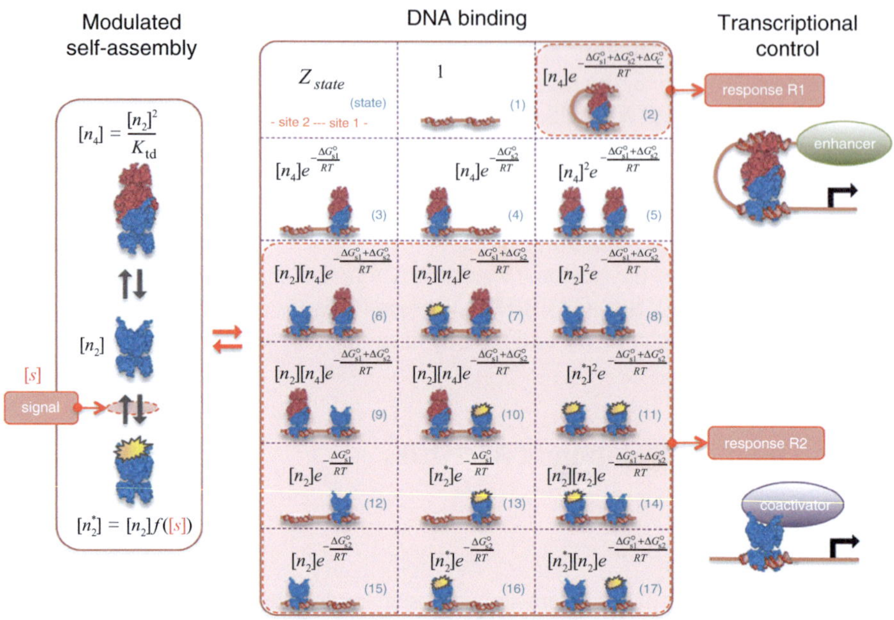

Fig. B.46 Quantitative modeling of control of gene expression by modulated self-assembly of the retinoid X receptor (RXR). (From Vilar and Saiz 2011)
The model considers how intracellular signals are processed through modulated self-assembly into populations of different RXR oligomeric species that upon DNA binding engage in transcriptional control

significant: simultaneous elimination of both of these operators reduced the repression level about 100 times. Deeper investigations and computational modeling were able to detail how DNA looping could explain this result. Vilar then proceeded to demonstrate how general processes of self-assembly could lead to these sorts of emergent behaviors in regulation by mechanisms previously thought to be distinct.

Vilar also used an entropy-based approach to develop the highest performing method in the DREAM 6 competition where the goal was to diagnose acute myeloid leukemia using flow cytometry.

Sources of Support/Funding

Vilar receives funding from Ikerbasque, the Basque Foundation for Science, for both research and endowing his professorship, and from the Spanish Ministry of Science and Innovation.

Summary and Conclusions

Vilar's research program at Ikerbasque and the Biophysics Unit (CSIC-UPV/EHU) is extremely exciting. His research focus is on an important intersection between biologically important problems and novel, highly formal approaches to attack them. In particular, he has been leading efforts to identify mechanisms of emergent behavior in cellular regulation and information transfer in cancer.

References

Vilar, J.M., Saiz, L.: Control of gene expression by modulated self-assembly. Nucleic Acids Res. **39**, 6854–6863 (2011)
www.ehu.es/biologiacomputacional/#Publications
www.ehu.es/biologiacomputacional/#Research
www.ikerbasque.net/
www.unidaddebiofisica.org/

Uppsala University and Science for Life Laboratory Uppsala

Site Address:	Meeting at Uppsala University Dag Hammarskjölds väg 20 752 37 Uppsala, SWEDEN www.uu.se/
Date Visited:	8 May 2012
WTEC Attendees:	Paul Janmey, Cindy Reinhart-King, Dan Fletcher (report author), Nastaran Kuhn, Nicole Moore, Hemant Sarin
Host(s):	Karin Forsberg Nilsson Professor of Stem Cell Research Department of Immunology, Genetics, and Pathology Vice Director, Science for Life Laboratory Uppsala Karin.nilsson@igp.uu.se Tel.: +46 18 471 41 58
	Fredrik Nikolajeff Associate Professor, Center Manager, Explorative Research Uppsala Berzelii Technology Centre for Neurodiagnostics fredrik.nikolajeff@angstrom.uu.se Tel.: +46 18 471 30 36
	Magnus Malmqvist Associate Professor, Biomedical Radiation Sciences Department of Radiology, Oncology, and Radiation Science magnus.malmqvist@bioventia.com Tel.: +46 18 55 53 00

Overview

Uppsala University is one of the world's oldest universities, with a rich academic and research history. Uppsala University, along with the Royal Institute of Technology (KTH), has some of Europe's most advanced laboratories for nanomaterial science with applications in pharmaceutics, biotechnology, and energy. Science for Life Laboratory (SciLifeLab) is a joint venture between four universities; Karolinska Institutet, KTH, Stockholm University, and Uppsala University. At Uppsala University, the SciLifeLab project is based at the Biomedical Center. SciLifeLab researchers study the molecular basis for human complex disease. Their methods include identifying genetic risk factors, biomarkers, and mechanisms by applying novel technologies to patient samples unique to the Nordic countries, as well as comparative biology proteomic and genomic approaches. The efforts of SciLifeLab in Uppsala are coordinated with SciLifeLab in Stockholm to provide a national infrastructure for molecular biosciences.

Research and Development Activities

Fredrik Nikolajeff described the activities of the Uppsala Berzelii Technology Centre for Neurodiagnostics, a unique collaboration of researchers in academic, medical, and commercial institutions that focuses on Alzheimer's disease, chronic pain, and the development of new technologies. The goal of the Centre is to identify new biomarkers and methods that can be used for early diagnosis of diseases. It draws on the strong medical and academic resources in the Uppsala-Stockholm area. For example, positron emission tomography (PET) was used with the tracer 11C-D-deprenyl (DDE) to identify inflammation in patients with enduring pain after rear-impact car accidents (Linnman et al. 2011). Twenty-two patients with whiplash associated disorder (grade II) and 14 healthy controls were investigated using the PET method and were found to have clear indications of sustained injury (Fig. B.47). The Centre continues advanced work on label-free infrared spectroscopy, biobanking of cerebrospinal fluid and neuronal tissue, and proximity ligation assays. Strong connections with industry enable direct translation of advances into commercial products.

Karin Forsberg Nilsson gave an overview of the SciLifeLab program and its activities, which were inspired by the Broad Institute in the United States, as well as related research activities at Uppsala University. The use of domestic animals is one such platform. She described a unique approach to the study of cancer using dogs as models. Dogs are an ideal model because they have almost the same genes as humans, experience the same environment as humans, and get similar diseases. Dogs are also ideal models because disease progression and treatment can be monitored over long periods of time by owners and veterinarians.

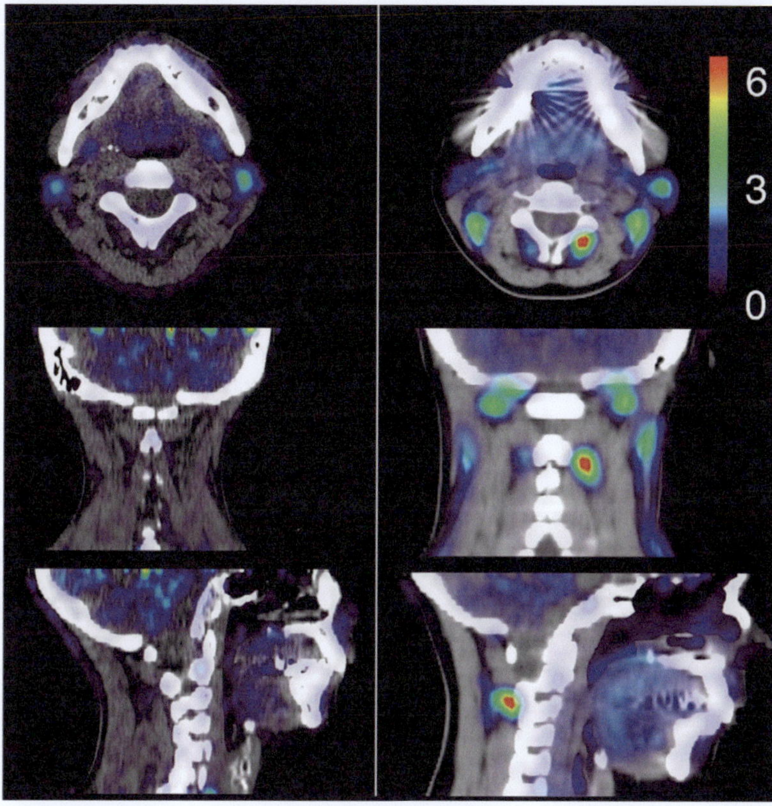

Fig. B.47 DDE uptake in a representative healthy control (*left*) and a whiplash patient (*right*). The patient displays high DDE uptake in the adipose tissue right of the spinous process of C2. PET images are overlaid on the subject's individual CT anatomy and tracer uptake is expressed as standardized uptake values (From Linnman et al. 2011)

Forsberg-Nilsson's research focuses on the study of neural stem cells and their role in neuronal tissue development and brain cancers (Fig. B.48). As part of the Comprehensive Cancer Consortium, Forsberg-Nilsson is carrying out a longitudinal collection of patient data and glioma samples to create a biobank resource of glioma initiating cells (also called cancer stem cells) for further studies on brain cancer development and treatment. She noted that there is a high degree of patient participation in biobanking efforts in Sweden due to trust in the health care system and a willingness to participate in research.

Magnus Malmqvist, co-inventor of the Biacore surface plasmon resonance method for studying biomolecular interactions in real time, discussed the development of new technologies at Uppsala University. Noting that "the workshop is as important as the science," Malmqvist described a new technology for measuring

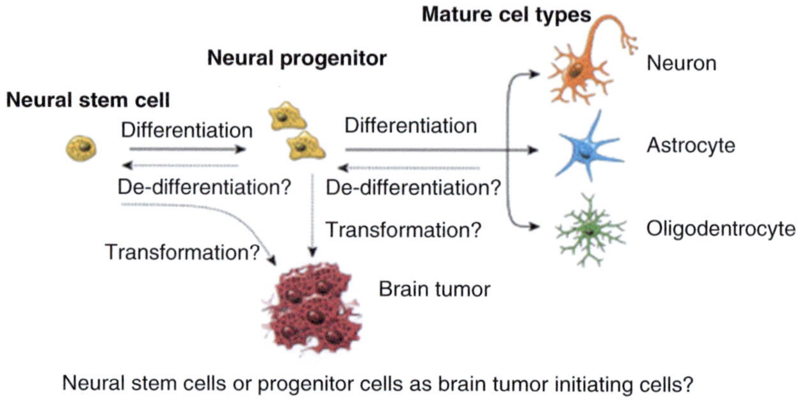

Fig. B.48 The role of neural stem cells on brain development and brain cancer (From www.igp.uu.se/Research/Cancer_and_vascular_biology/karin_forsberg_nilsson/)

ligand affinity to cells and their kinetics called LigandTracer. This technology uses cells whose surfaces are periodically exposed to labeled ligands to optically determine rates of adsorption. Data can be visualized by an interaction map relating association and dissociation constants, providing a new way to view ligand interactions with cells. This mapping software is available for use by other researchers to view their data.

Translational Efforts

Uppsala University is strongly connected to industry through collaborative research projects, enabling a direct way for new innovations to be turned into commercial products or clinical procedures. Furthermore, Sweden handles patents in a unique way. The rights to the invention stay with the inventors, not the university, as is the case in the United States. The inventors are then free to partner with companies or other organizations to pursue the patenting and development of their technology.

Sources of Support

The Swedish government has made a strong commitment to funding biomedical research through the Research and Innovation Bill, 2009-2012.

Collaborations and Possibilities

Sweden's outstanding collection of patient samples and other biobanking efforts provide a powerful resource for studies of physical science and oncology.

Summary and Conclusions

Uppsala University is conducting cutting-edge research at the intersection of physical sciences and oncology, with strong clinical and translational connections.

References

Forsberg, M., Holmborn, K., Kundu, S., Dagalv, A., Kjellen, L., Forsberg-Nilsson, K.: Undersulfation of heparan sulfate restricts differentiation potential of mouse embryonic stem cells. J. Biol. Chem. **287**, 10853–10862 (2012)

Linnman, C., Appel, L., Fredrikson, M., Gordh, T., Soderlund, A., Langstrom, B., Engler, H.: Elevated [11C]-D-deprenyl uptake in chronic Whiplash Associated Disorder suggests persistent musculoskeletal inflammation. PLoS One. **6**, e19182 (2011)

Niklasson, M., Bergstrom, T., Zhang, X.Q., Gustafsdottir, S.M., Sjogren, M., Edqvist, P.H., Vennstrom, B., Forsberg, M., Forsberg-Nilsson, K.: Enlarged lateral ventricles and aberrant behavior in mice overexpressing PDGF-B in embryonic neural stem cells. Exp. Cell Res. **316**, 2779–2789 (2010)

www.uu.se/

www.uu.se/en/research/

http://suls.se/

http://suls.se/research/how/research-infrastructure/

www.igp.uu.se/Research/Cancer_and_vascular_biology/karin_forsberg_nilsson/

www.igp.uu.se/Research/Cancer_and_vascular_biology/karin_forsberg_nilsson/more/

www.teknik.uu.se/material/staff/haen/index.html

www.teknik.uu.se/material/Tribomaterials/mim/index.html

www.nanobridge.eu

www.uuinnovation.uu.se/

www.berzelii.uu.se/

Venetian Institute of Molecular Medicine

Site Address:	University of Padua Via Orus 2, 35129 Padua, ITALY www.vimm.it
Date Visited:	11 May 2012
WTEC Attendees:	Sharon Gerecht (report author), Parag Mallick, Owen McCarty, Hassan Ali
Host(s):	Nicola Elvassore Venetian Institute of Molecular Medicine Via Orus 2, 35129 Padua Tel.: +39 049 7923 211 Fax: +39 049 7923 250 www.vimm.it/research/group.php?PI=172 Department of Chemical Engineering University of Padova via Marzolo 9, 35131 Padova Tel.: +39 049 827 5469 Fax: +39 049 827 5461 nicola.elvassore@unipd.it www.bioera.dipic.unipd.it/Bioera/
	Giacinto Scoles European Research Council Grantee at the Azienda Ospedaliera Universitaria dell'Universita' di Udine, Padiglione #13 Udine 3300, ITALY Tel.: +39 0432 559 413 Fax: +39 040 375 8565 gscoles@princeton.edu www.princeton.edu/~chemdept/Scoles/SCOLES%20NEW/Scolessite/index.html
	Barbara Molon Foundation for Advanced Biomedical Research Venetian Institute of Molecular Medicine – VIMM via Orus 2 - 35129 Padua, ITALY www.vimm.it, www.fondbiomed.it barbara.molon@unipd.it

Overview

The Venetian Institute of Molecular Medicine (VIMM) is a new center for advanced biomedical research. It was established 10 years ago by the Foundation for Advanced Biomedical Research (a private consortium) as a joint venture with the University of Padua and the Padua Health Authority.

VIMM is strategically located between the University Hospital and the Preclinical Science Campus. It occupies an area of about 3,000 m^2. Once the remaining 8,000 m^2 is finished, the complex will become the major center for research in molecular medicine in the Veneto region. Creation of VIMM, a modern building with facilities for advanced research, was made possible by private fundraising, which actively involved industries and financial institutions.

The scientific goal of the institute is to integrate basic and clinical research in order to achieve a faster transfer of the developments of molecular and cellular biology into the clinic. The basic research themes of the institute revolve around four basic areas: (1) structural biology; (2) cellular biology; (3) host-pathogen interactions and its implications for gene therapy; and (4) cellular and molecular oncology. Currently, there are 20 principal investigators, of which three are young investigators. Half of the investigators hold dual appointments with the University of Padua.

VIMM's heart is represented by newly recruited postdoctoral fellows and faculty members. A special effort is being made to attract promising scientists in Europe and the United States.

Dr. Nicola Elvassore

Sharing facilities in VIMM and University of Padua, research in Elvassore's laboratory applies engineering principles with biological approaches to rationally understand the mechanisms governing cell behavior. The main engineering tools to recapitulate the cell microenvironment include substrate engineering, micropatterning technology, and microfluidics technology. Utilizing substrates with stiffness ranging from 12 to 21 kPa, the group has shown that sarcomere formation during myoblast differentiation occurs on softer substrate (15 kPa) which in turn enables the maturation and functionality of myotubes (Serena et al. 2010). This model is now being used to study Muscular Dystrophy, where the soft surrounding tissue of the diseased muscle results in dysfunctional myotubes with impaired calcium release and upregulation in dystrophin expression. In a recent collaboration with Piccolo, the groups identified the YAP/TAZ [Yorkie-homologues YAP (Yes-associated protein) and TAZ (transcriptional coactivator with PDZ-binding motif)] as nuclear relays of mechanical signals exerted by extracellular rigidity and cell shape, independently from Hippo pathways (Dupont et al. 2011). Utilizing micropillar substrates with varying rigidity, the groups were able to decouple matrix stiffness and adhesion. The ease of use of the technology facilitated robust examination and in-depth understanding of the underlying mechanism. Piccolo counted 7,000 experiments in which the technology enabled discoveries to move forward. Control over substrate topography (using micropatterning) was shown to affect myoblast proliferation and differentiation (Zatti et al. 2012) and recently is being applied to study cardiac differentiation of human embryonic stem cells. Final examples were presented for the use of microfluidic technology to fabricate a miniature culture system for the robust study of cellular behaviors in response to a soluble factor. One device is comprised of culture chambers that allow control over the soluble environment with on-line monitoring of the cultured cells. Another microfluidic technology includes a gas exchanger for control over dissolved oxygen levels for the study of the hypoxic effect on calcium transients in response to electrical stimulation of muscle cell derivatives (Fig. B.49).

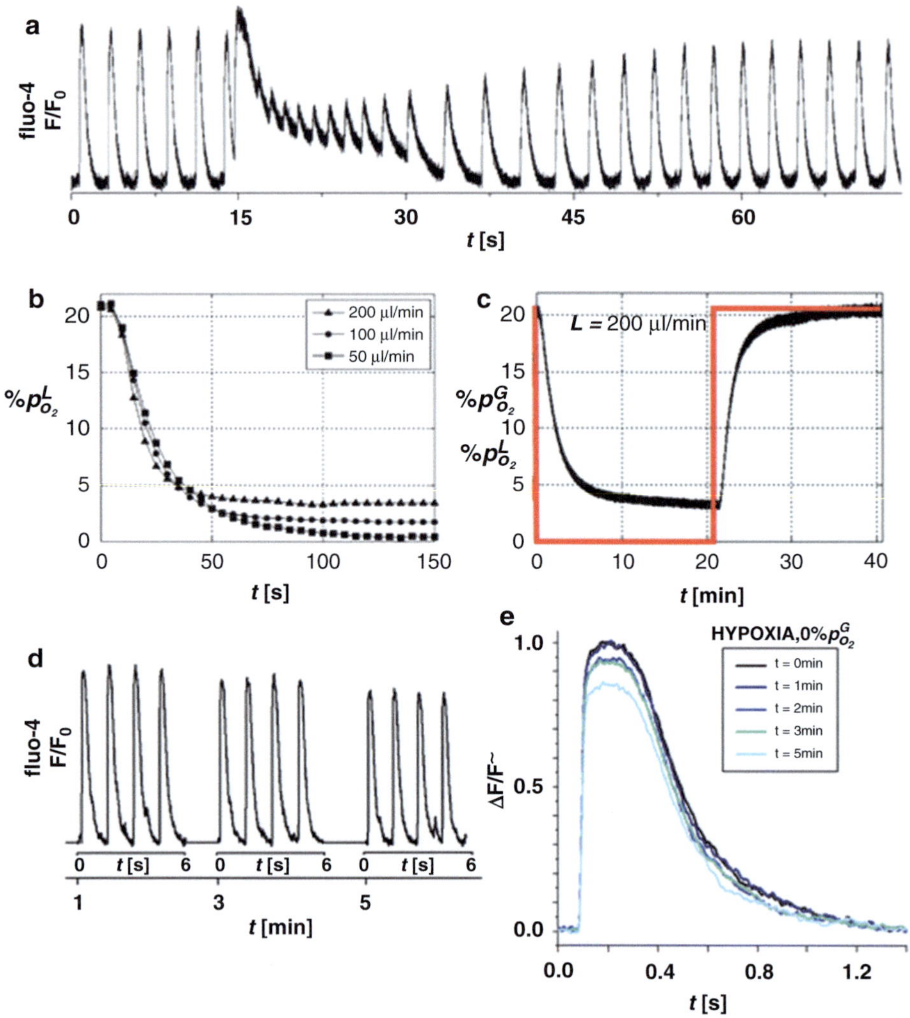

Fig. B.49 Effect of hypoxia on calcium transients (From Martewicz et al. 2012)
(*A*) Calcium dynamics in a cardiomyocyte under electrical stimulation revealed through Fluo-4 in the microfluidic culture chamber during perfusion of 50 µl of 10 mM caffeine solution; cardiomyocyte displays normal Ca^{2+} transients and response to caffeine with full recovery after wash out. (*B*, *C*) Liquid phase oxygen dynamics during step impulse of the oxygen partial pressure in the gas phase monitored by Ru(ddp) with three different flow rates (*B*) and by optic fiber sensor with = 200 µl min − 1 (*C*) at exchanger outlet. (*D*) Calcium transients sequence at different time points after hypoxic stimulus to the cell culture. (*E*) Comparison of single normalized calcium transients at different time points after hypoxic stimulus

Dr. Giacinto Scoles

Scoles is well-known in North America as a chemical physicist. He works at the University Hospital of the University of Udine, the largest city in the northeast region of Italy. He has recently obtained support through the ERC for building a network of laboratories in Italy's northwest region that comprises, in addition to Scoles' lab and Dr. C.A. Beltrami's pathology lab in Udine, the IOM microfabrication lab and the synchrotron light lab elettra in Basovizza, Trieste, and the oncopharmacology lab of Dr. G. Toffoli at the Center of Reference for Oncology in Aviano, Pordenone. The aim of this collaboration is to use MEMS- and AFM-based nanotechnologies to impact the diagnostics and eventually the cure of metastatic cancer.

During the past 10 years, Scoles shifted the center of his scientific activity to try to use nanotechnology tools for biomedical applications. He has determined the structure of self-assembled monolayers of long-chain alkyl sulfides on gold(111) (Cossaro et al. 2008). The focus and the goal of current research is the quantitative, high-throughput measurement of proteins and their interactions (interactomics) in samples produced by a very small number of cells or within single cells. As an example, the group had studied a new way to fabricate multiple protein nanosensors using DNA-directed immobilization (DDI) in combination with a nanografting (NG), an atomic force microscopy-based technique. Binding sites of protein with well defined local environment are designed by combining NG and DDI (Fig. B.50). The group now aims to make new inroads into quantitative diagnostics and disease monitoring starting from the study of allergies to small molecule drugs because of their importance for a more personalized tailoring of chemotherapy. Another example

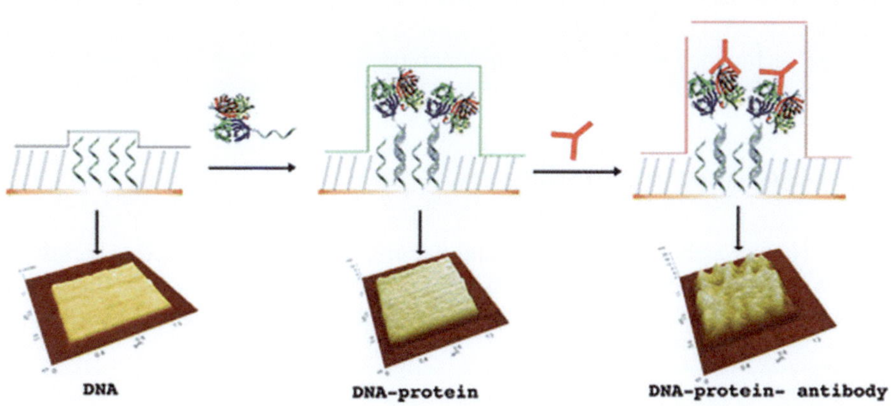

Fig. B.50 Atomic force microscopy nanografting was utilized to prepare DNA nanopatches of different sizes (200×200 to 1000×1000 nm^2) onto which DNA-protein conjugates can be anchored through DNA-directed immobilization. (From Bano et al. 2009)
Height measurements were used to assess the binding of the proteins as well as their subsequent interaction with other components, such as antibodies. The results indicate that nanografted patch arrays are well suited for application in biosensing and could enable the fabrication of multifeature protein nanoarrays

showed the immobilization of prion proteins onto nanostructured surfaces in two different orientations, as demonstrated by differential height (i.e., topographic) measurements. This approach allows the estimation of binding parameters and the full characterization of the nanoscale biorecognition process and thus opens the way to several high sensitivity diagnostic applications (Sanavio et al. 2010).

Dr. Barbara Molon

Molon presented her recent study on T lymphocyte trafficking to the tumor, unveiling an unexpected mechanism of tumor evasion. The group has described a novel reactive nitrogen species (RNS)-dependent posttranslational modification of chemokines that has a profound impact on leukocyte recruitment to mouse and human tumors. Intratumoral RNS production induces CCL2 chemokine nitration and hinders T cell infiltration, resulting in the trapping of tumor-specific T cells in the stroma that surrounds cancer cells. Reasoning on that, the group designed and developed a new compound—AT38—that efficiently interferes with RNS generation. A time-scheduled administration of AT38 in tumor-bearing mice caused a reduction in nitrotyrosine formation and the subsequent unmasking of tumor-infiltrating lymphocyte (TIL) chemo-attractant signals. Utilizing photocrosslinked hydrogel, unmodified CCL2 was placed within the mass of untreated tumors enabling T-cell infiltration into the tumor lesion (Fig. B.51). These data indicated that the mechanism by which AT38 improves TIL infiltration is based on unmodified chemokine (CCL2) bioavailability. AT38 administration in tumor-bearing mice could pre-condition the tumor microenvironment and thus support cancer elimination by tumor-specific CTLs (Molon et al. 2011). Current studies continue to utilize biomaterial systems, based on hydrogel technology for biomimetic immunology approach both *in vitro* and *in vivo*.

Sources of Support/Funding

VIMM is a privately owned institute and funding for research is raised from private sources, government via FIRB grants, and European Union grants via the FP7 program of applied grants and the more "free" approach of the program IDEAS of the ERC.

Summary and Conclusions

VIMM promotes promising translational research activities by providing extensive infrastructure to support high-level collaborative work in the Veneto region. Young investigators are recruited to enhance the integration of physical and engineering approaches for advancing biological discoveries and applications.

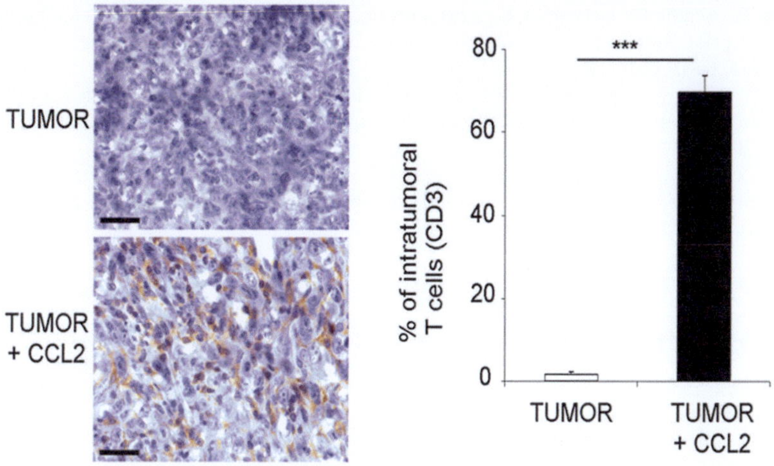

Fig. B.51 MCA-203 fibrosarcoma tumor samples from mice that received intratumoral injections of CCL2 (0.5 μg in hydrogel) were stained for CD3 by immunohistochemistry. The graphs represent the quantification of immunoreactive cells. Scale bar: 50 μm. (Courtesy of B. Molon)

Through two ERC grants (the first is the one mentioned above and the second was obtained by Dr. Maurizio Prato, an internationally recognized carbon nanostructures organic chemist), and through grants from AIRC and the Italian government, the eastern part of Italy was able to mount what can be considered one of the largest scientific/technological efforts towards understanding and defeating cancer.

References

Bano, F., Fruk, L., Sanavio, B., Glettenberg, M., Casalis, L., Niemeyer, C.M., Scoles, G.: Toward multiprotein nanoarrays using nanografting and DNA directed immobilization of proteins. Nano. Lett. **9**, 2614–2618 (2009)

Cossaro, A., Mazzarello, R., Rousseau, R., Casalis, L., Verdini, A., Kohlmeyer, A., Floreano, L., Scandolo, S., Morgante, A., Klein, M.L., Scoles, G.: X-ray diffraction and computation yield the structure of alkanethiols on gold(111). Science. **321**, 943–946 (2008)

Dupont, S., Morsut, L., Aragona, M., Enzo, E., Giulitti, S., Cordenonsi, M., Zanconato, F., Le Digabel, J., Forcato, M., Bicciato, S., Elvassore, N., Piccolo, S.: Role of YAP/TAZ in mechanotransduction. Nature. **474**, 179–183 (2011)

Martewicz, S., Michielin, F., Serena, E., Zambon, A., Mongillo, M., Elvassore, N.: Reversible alteration of calcium dynamics in cardiomyocytes during acute hypoxia transient in a microfluidic platform. Integr. Biol. (Camb.). **4**, 153–164 (2012)

Molon, B., Ugel, S., Del Pozzo, F., Soldani, C., Zilio, S., Avella, D., De Palma, A., Mauri, P., Monegal, A., Rescigno, M., Savino, B., Colombo, P., Jonjic, N., Pecanic, S., Lazzarato, L., Fruttero, R., Gasco, A., Bronte, V., Viola, A.: Chemokine nitration prevents intratumoral infiltration of antigen-specific T cells. J. Exp. Med. **208**, 1949–1962 (2011)

Sanavio, B., Scaini, D., Grunwald, C., Legname, G., Scoles, G., Casalis, L.: Oriented immobilization of prion protein demonstrated via precise interfacial nanostructure measurements. ACS Nano. **4**, 6607–6616 (2010)

Serena, E., Zatti, S., Reghelin, E., Pasut, A., Cimetta, E., Elvassore, N.: Soft substrates drive optimal differentiation of human healthy and dystrophic myotubes. Integr. Biol. (Camb.). **2**, 193–201 (2010)

Zatti, S., Zoso, A., Serena, E., Luni, C., Cimetta, E., Elvassore, N.: Micropatterning topology on soft substrates affects myoblast proliferation and differentiation. Langmuir. **28**, 2718–2726 (2012)

Weizmann Institute of Science

Site Address:	234 Herzl Street Rehovot 76100, ISRAEL Tel.: +972 8 934 3111 Fax: +972 8 934 4107 www.weizmann.ac.il/
Date Visited:	14 May 2012
WTEC Attendees:	Paul Janmey (report author) and Sharon Gerecht
Host(s):	Ronen Alon The Linda Jacobs Chair in Immune and Stem Cell Research Department of Immunology Tel.: +972 8 934 2482 Fax: +972 8 934 2724 ronen.alon@weizmann.ac.il www.weizmann.ac.il/immunology/AlonPage.html
	Tom Shemesh Post-doctoral fellow, Department of Molecular Cell Biology Laboratory of Professor Alexander Bershadsky Tel.: +972 8 934 2936, 4071 Fax: +972 8 934 4125 tom.shemesh@weizmann.ac.il
	Roy Bar-Ziv Faculty of Chemistry, Department of Materials and Interfaces Helen and Martin Kimmel Center for Nanoscale Science Perlman Building Rehovot 76100 Tel.: +972 8 934 2069 roy.bar-ziv@weizmann.ac.il www.weizmann.ac.il/materials/barziv/
	Sam Safran Fern and Manfred Steinfeld Professorial Chair Faculty of Chemistry Department of Materials and Interfaces Tel.: +972 8 934 3362 Fax +972 8 934 4138 sam.safran@weizmann.ac.il www.weizmann.ac.il/fluids/SafranGroup

Overview

The Weizmann Institute of Science (WIS) is one of the foremost research institutions in Israel and has a long tradition of interdisciplinary work and especially strong interactions between theorists and experimentalists in biophysics. The mission of WIS is to "address the most urgent challenges facing humanity." One of the institute's major focus areas is cancer research.

The Biological Physics group (www.weizmann.ac.il/Biological_Physics/) is part of the interdisciplinary research efforts at WIS. Interviewees Drs. Sam Safran, Roy Bar-Ziv, and Alexander Bershadsky are among the members of the Biological Physics Group, which coordinates and supports joint research in the departments. A recent example of this outreach to experimental physicists, cell biologists, and theorists is a Minisymposium on Biological Machines: Physics and Bioengineering (www.weizmann.ac.il/conferences/physbio/).

There is a long history of interactions between physicists, especially theorists and experimentalists, between WIS and other Israeli institutions. Examples include the combined efforts of theory by Safran (WIS) and Michael Kozlov (Tel Aviv University) with the laboratories of Benjamin Geiger, the Professor Erwin Neter Professorial Chair of Cell and Tumor Biology and Bershadsky, the Joseph Moss Professorial Chair of Biomedical Research in the Department of Molecular Cell Biology (WIS). Their work is devoted to explaining how application of force to focal adhesions induces their growth and remodeling and on defining the role of forces in regulating and organizing the cytoskeletal changes that drive cell motility. Such physical models are integrated with proteomic and genomic studies to identify the mechanisms underlying cancerous transformation, either due to deregulated growth or to failure to undergo apoptosis that can result from altered cell adhesion, morphology, or motility control.

Research and Development Activities

Dr. Ronen Alon

The Alon lab studies chemokine signaling to leukocyte integrins, with a special interest in how physical features such as fluid shear stress and spatial organization affect adhesion and activation at endothelial and extravascular contacts (Alon and Shulman 2011, Cinamon et al. 2004, and Schreiber et al. 2007; Shulman et al. 2011). A current focus is to define the interplay between biochemical and physical effects on how the leukocyte integrin LFA1 binds its cognate receptor ICAM-1 in endothelial cells when leukocytes move along the endothelium as blood flows through a vessel. Other projects concern similar integrin-mediated cell-cell adhesion involving lymphocytes adhering to antigen-presenting cells.

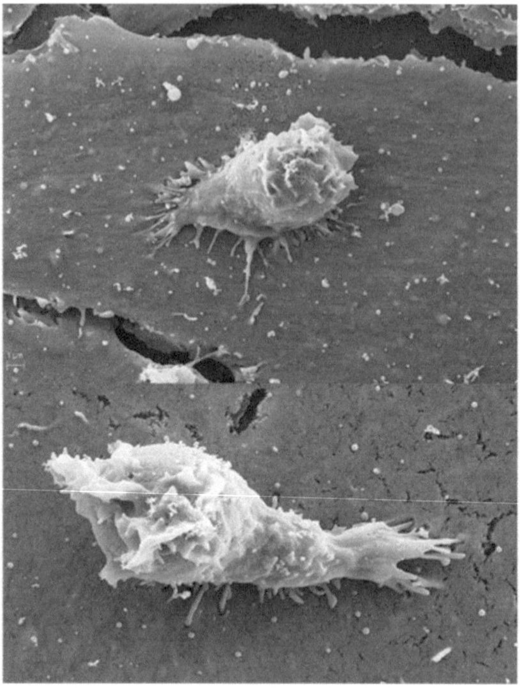

Fig. B.52 Two T lymphocytes sending invasive filopodia into the body of a cytokine stimulated endothelial cell in the presence of shear forces. (From www.weizmann.ac.il/immunology/AlonPage.html)
The density of these filopodia is increased by up to five fold in the presence of maximal stimulatory conditions (shear stress, high levels of apical chemokines)

From a physical perspective, an intriguing question is how the same pair of ligands (LFA1-ICAM-1) that is regulated by stresses that occur in shear flow for lymphocyte-endothelial cell adhesion also works in settings such as the immune synapse in the absence of similar fluid shear stresses.

An important issue raised by the host is the need for increased scanning electron microscopy (EM) and immune-EM imaging of cells to visualize and quantify interactions at length scales that cannot be captured by light microscopy. An example of the utility of these ultra-resolution methods is shown in Fig. B.52, where submicron scale contacts between lymphocytes and endothelial cells stimulated by both shear stress and chemokine signals trigger the protrusion and the subsequent migration of the lymphocytes though the endothelial barrier.

Similarly, signaling moieties need to be studied in the correct spatial orientation and amounts that occur *in vivo*, rather than by application of global stimulation with ligand concentrations and spatial distribution that may not correspond to the geometry and magnitudes *in vivo*. This need again emphasizes the importance of measuring the physical features of the microenvironmental landscape in order to understand how biochemical signaling impacts the cell.

Drs. Tom Shemesh and Alexander Bershadsky

The Bershadsky lab studies how cells move, and the mechanical forces necessary for cells to attach themselves to the substrate and to one another. In exploring the points of contact, which act as mechanical "sensors" that provide the cell with information about its environment and determine its behavior, he has learned that in cancer cells, the activity of these "sensors" is disrupted, which likely accounts for the cell's difficulty in adhering to substrates and, consequently, their greater mobility.

Work from the Bershadsky group in collaboration with theorists provided some of the first quantitative models for how force applied to cell adhesion sites could alter their growth (Bershadsky et al. 2006; Shemesh et al. 2005), and therefore how molecular changes in abnormal cells can alter mechanical control of cell growth, migration, and proliferation (Fig. B.53).

Recent work from this group provides a computational model for the protrusion of the leading edge of a motile cell by considering how membrane tension, frictional forces, and actin polymerization rates combine to control the rate of cell protrusion (Fig. B.54).

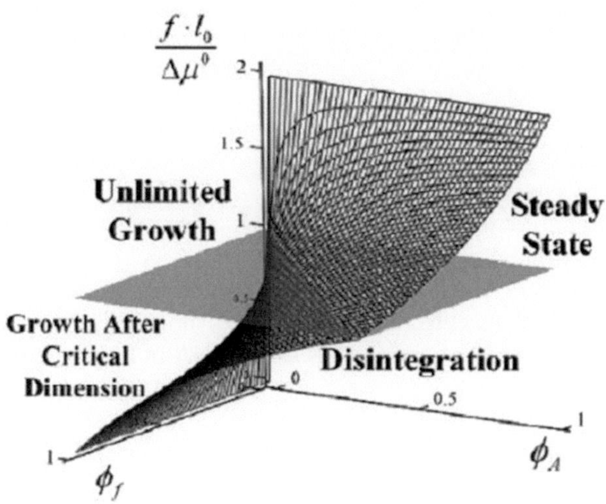

Fig. B.53 Phase diagram of the system representing the different regimes of focal adhesion aggregate assembly-disassembly corresponding to different ranges of three system parameters. (From Shemesh et al. 2005)
(*1*) the density of the points of force application along the aggregate length ϕ_f, the density of the points of the aggregate anchoring to the substrate ϕ_A; (*2*) the dimensionless parameter ($\chi = (f \cdot l\,0)/(\Delta\mu 0)$), characterizing the ratio between the molecular energy provided by an elementary pulling force, $f \cdot l\,0$; and (*3*) the difference of the protein standard chemical potentials in the aggregated and free states, $\Delta\mu 0$

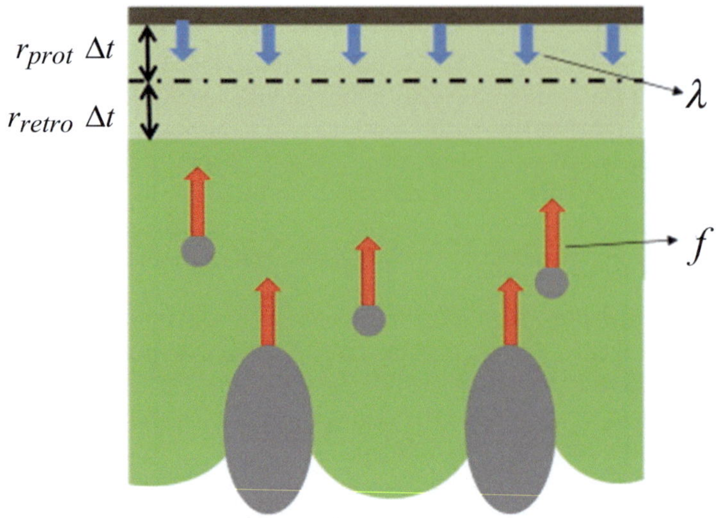

Fig. B.54 Schematic description of the model for actin polymerization-driven cell protrusion. (From Shemesh et al. 2012)
At a discrete time step Dt, the modeled components of the lamellipodium are the nascent and mature cell adhesions (gray circles and ellipsoids, respectively), the cell membrane (brown), and actin gel (green). Newly polymerized gel is shown in lighter green, with a dash-dot line indicating the position of the membrane before polymerization. Blue arrows designate force exerted on the gel by the cell membrane due to tension, whereas red arrows represent friction forces (*f*) applied by the adhesions to the gel. r_{prot} and r_{retro} denote the rates of membrane protrusion and retrograde actin flow, respectively

Dr. Roy Bar-Ziv

The Bar-Ziv group combines methods and concepts of soft matter physics with development of artificial biological systems. One project involves design and production of surface-tethered DNA arrays of controlled density to study transcription in crowded environments (Fig. B.55) that might more closely resemble those *in vitro*, compared to traditional approaches using DNA in free solution (Daube et al. 2010, Shemer et al. 2012).

A second project is aimed at improving methods to precisely identify and target a specific state of a living cell. Cell states could be better identified by the expression pattern of several genes rather than of a single gene. Therefore, autonomous identification can be achieved by a system that measures the expression of these genes and integrates their activities into a single output. The Bar-Ziv lab has constructed a system that diagnoses a unique state in yeast in which two independent pathways, methionine anabolism and galactose catabolism, are active. Their studies show that cells can autonomously report on their state, identify the state of interest, and inhibit their growth accordingly. Further work will determine if such systems could be applied to clinical problems such as identification of aberrant versus normal growth (Nissim and Bar-Ziv 2010).

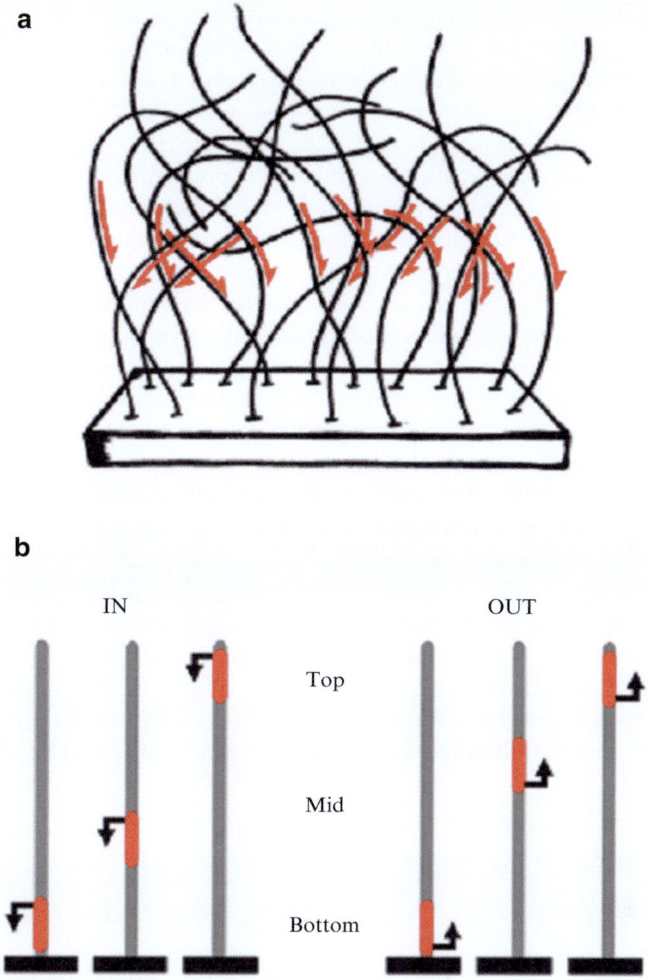

Fig. B.55 The synthetic DNA brush (From Daube et al. 2010)
(A) A scheme of a DNA brush (2,160 bp) with a transcriptional (TX) unit (300 bp) in between T7 promoter and terminator, oriented to the surface. (*B*) Depiction of six brush configurations with TX units (*Red*). The position of the promoter for OUT is at 60, 1,220, and 1,800 bp for *Bottom*, *Mid*, and *Top*, respectively

Dr. Sam Safran

The Safran group applies insights from theories of soft matter physics to the interface of physics and biology in several contexts, including the mechanics and organization of the cell membrane and the mechanisms by which cellular processes such as proliferation, differentiation, and tissue development are controlled by the mechanical properties of cells and their environment.

Theoretical models in which cells are treated in terms of active force dipoles have been important for interpreting experimental studies on cell mechanics (Schwarz and Safran 2002). The theory includes non-equilibrium cell activity, local mechanical equilibrium, and random forces to determine cell response to static and dynamic stress, as well as the curvature of the substrate (Biton and Safran 2009). To understand how substrate rigidity determines the polarization of cells, Safran and colleagues have generalized the treatment of elastic inclusions in solids to "living" inclusions whose active polarizability, analogous to that of non-living matter, results in feedback in response to matrix stresses (Zemel et al. 2006).

These models can explain recent observations of the non-monotonic dependence of stem cell polarization on matrix rigidity, and other morphogenetic effects such as the dependence of muscle striation on substrate properties (Freidrich et al. 2011). These findings provide a mechanical correlate for the existence of optimal substrate elasticity for cell differentiation and function. A recent extension of models that consider how cell-generated forces deform the substrates, and how substrate deformation feeds back to alter cell structure allows for substrates to exhibit non-linear elasticity, and therefore comes significantly closer to being able to model the deformations of natural matrices composed of fibrous polymers (Shokef and Safran 2012) (Fig. B.56).

An example of the utility of such models to explain cellular reorganization in response to substrate stiffness is seen in dependence of intracellular actin fiber alignment on the elastic modulus of the substrate. Figure B.57 shows agreement of theory with experimental measurements of liquid crystalline ordering of actin filament bundle for cells of different axial ratios that adhere to substrates with different stiffnesses.

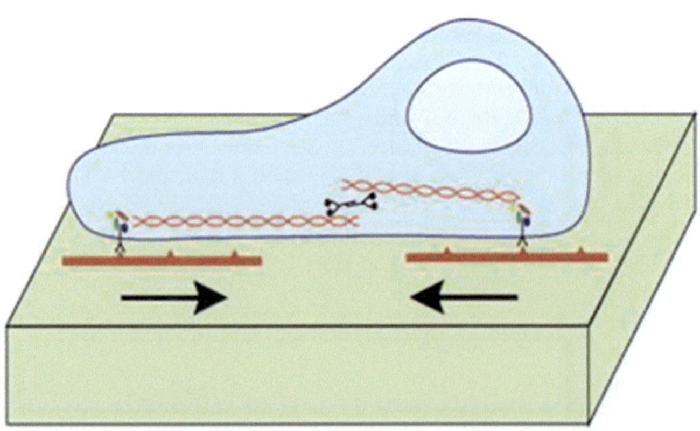

Fig. B.56 Schematic diagram by which cell contractility generated by actin-myosin interactions generates a fore dipole at the cell/substrate interface (From www.weizmann.ac.il/Biological_PhysicB.57s/photos/gallery.html)

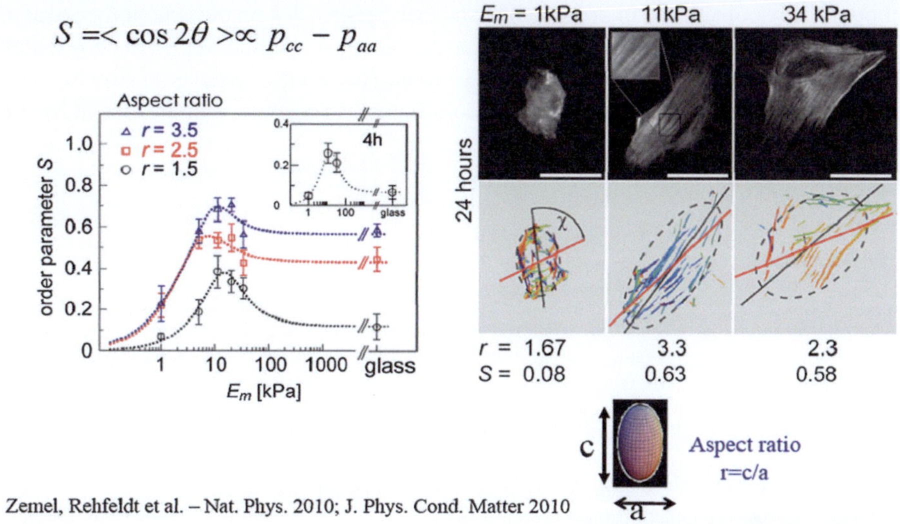

Fig. B.57 The effect of axial cell elongation on stress-fiber polarization and experimental values of the order parameter S for different elastic substrates. (From Zemel et al. 2010a)
The experimental values of the stress-fiber order parameter, S, 24 h after plating the cells, for the three groups of cells (of aspect ratios r =1:5;2:5;3:5) as a function of Young's modulus of the matrix, E_m. χ is the angle between each stress fiber in the cell and the long axis of the fitted ellipse. Within each of the different groups, S is maximal for E_m = 11 kPa and generally increases with aspect ratio r, in agreement with our theoretical predictions (Zemel et al. 2010a, b)

Sources of Support/Funding

Funding for interdisciplinary research in the host groups has come from institutional sources and grants from the Israel Science Foundation, Minerva Foundation, Clore Center for Biological Physics, Kimmelman Center for Biomolecular Structure and Assembly, German Academic Exchange Service, and the U.S.-Israel Binational Science Foundation, among other grant agencies. There are also close ties between researchers at WIS and the Mechanobiology Institute at the National University of Singapore.

Summary and Conclusions

WIS has an exceptionally strong history of productive collaboration between physicists and biologists, and a special strength in physical theories that have informed and motivated experimental biologists. Many of these collaborations have lasted a decade or more and argue strongly for a long-term commitment to the physical science/biology interface. The role of theory in biology and biomedicine has also been thoughtfully considered in order to have a clear perspective on the role of theory and computation as either an effort to primarily model a specific set of experimental results or provide a general framework to identify common aspects of diverse biological phenomena. The groups at WIS collaborate extensively with groups in North America, Europe, Asia, and elsewhere.

References

Alon, R., Shulman, Z.: Chemokine triggered integrin activation and actin remodeling events guiding lymphocyte migration across vascular barriers. Exp. Cell Res. **317**, 632–641 (2011)

Bershadsky, A., Kozlov, M., Geiger, B.: Adhesion-mediated mechanosensitivity: a time to experiment, and a time to theorize. Curr. Opin. Cell Biol. **18**, 472–481 (2006)

Biton, Y.Y., Safran, S.A.: The cellular response to curvature-induced stress. Phys. Biol. **6**, 046010 (2009)

Cinamon, G., Shinder, V., Shamri, R., Alon, R.: Chemoattractant signals and beta 2 integrin occupancy at apical endothelial contacts combine with shear stress signals to promote transendothelial neutrophil migration. J. Immunol. **173**, 7282–7291 (2004)

Daube, S.S., Bracha, D., Buxboim, A., Bar-Ziv, R.H.: Compartmentalization by directional gene expression. Proc. Natl. Acad. Sci. USA. **107**, 2836–2841 (2010)

Friedrich, B.M., Buxboim, A., Discher, D.E., Safran, S.A.: Striated acto-myosin fibers can reorganize and register in response to elastic interactions with the matrix. Biophys. J. **100**, 2706–2715 (2011)

Nissim, L., Bar-Ziv, R.H.: A tunable dual-promoter integrator for targeting of cancer cells. Mol. Syst. Biol. **6**, 444 (2010)

Schreiber, T.H., Shinder, V., Cain, D.W., Alon, R., Sackstein, R.: Shear flow-dependent integration of apical and subendothelial chemokines in T-cell transmigration: implications for locomotion and the multistep paradigm. Blood. **109**, 1381–1386 (2007)

Schwarz, U.S., Safran, S.A.: Elastic interactions of cells. Phys. Rev. Lett. **88**, 048102 (2002)

Shemer, G., Atsmon, Y., Karzbrun, E., Bar-Ziv, R.H.: Collective conformations of DNA polymers assembled on surface density gradients. J. Am. Chem. Soc. **134**, 3954–3956 (2012)

Shemesh, T., Geiger, B., Bershadsky, A.D., Kozlov, M.M.: Focal adhesions as mechanosensors: a physical mechanism. Proc. Natl. Acad. Sci. USA. **102**, 12383–12388 (2005)

Shemesh, T., Bershadsky, A.D., Kozlov, M.M.: Physical model for self-organizing of actin cytoskeleton and adhesion complexes at the cell front. Biophys. J. **102**, 1746–1756.

Shokef, Y., Safran, S.A.: Scaling laws for the response of nonlinear elastic media with implications for cell mechanics. Phys. Rev. Lett. **108**, 178103 (2012)

Shulman, Z., Cohen, S.J., Roediger, B., Kalchenko, V., Jain, R., Grabovsky, V., Klein, E., Shinder, V., Stoler-Barak, L., Feigelson, S.W., Meshel, T., Nurmi, S.M., Goldstein, I., Hartley, O., Gahmberg, C.G., Etzioni, A., Weninger, W., Ben-Baruch, A., Alon, R.: Transendothelial migration of lymphocytes mediated by intraendothelial vesicle stores rather than by extracellular chemokine depots. Nat. Immunol. **13**, 67–76 (2011)

Zemel, A., Bischofs, I.B., Safran, S.A.: Active elasticity of gels with contractile cells. Phys. Rev. Lett. **97**, 128103 (2006)

Zemel, A., Rehfeldt, F., Brown, A.E., Discher, D.E., Safran, S.A.: Cell shape, spreading symmetry and the polarization of stress-fibers in cells. J. Phys. Condens. Matter. **22**, 194110 (2010a)

Zemel, A., Rehfeldt, F., Brown, A.E., Discher, D.E., Safran, S.A.: Optimal matrix rigidity for stress fiber polarization in stem cells. Nat. Phys. **6**, 468–473 (2010b)

Appendix C. Site Visit Reports – Asia

Site visit reports are arranged in alphabetical order by organization name.

P. Janmey et al. (eds.), *Physical Sciences and Engineering Advances in Life Sciences and Oncology*, Science Policy Reports,
DOI 10.1007/978-3-319-17930-8

Academia Sinica [Workshop: Center for Nanomedicine Research, National Health Research Institutes]

Site Address:	128, Academia Road, Sec. 2 Nangang District, Taipei Taiwan
Date Visited:	June 6, 2013
WTEC Attendees:	Paul Janmey, Daniel Fletcher, Cynthia Reinhart-King (report author), Nastaran Kuhn, Patricia Foland
Host(s):	Dr. Chung-Shi Yang Center for Nanomedicine Research National Health Research Institutes Tel.: 886-37-246166 Fax: 886-37-586447 cyang@nhri.org.tw; cyang@ncnu.edu.tw

Overview

The National Health Research Institutes was founded in 1995 by the Taiwanese government as a non-profit foundation under the Department of Health. Its goal is to enhance medical research and improve healthcare in Taiwan. It helps to plan the overall direction of biomedical research in Taiwan, train scientists and physicians, support research activities across Taiwan, and help facilitate domestic and international collaboration. Researchers there are focused on both basic and disease-specific research. It has also set up several research resources, including a cell bank, a research database, the Health Research Information Network (HINT) and a bioinformatics core. It also supports several training programs for undergraduate, postdoctoral associates, and physicians. The NHRI contains several units, including a National Institute of Cancer Research and a Center for Nanomedicine Research, among others.

Research and Development Activities

Dr. Chung-Shi Yang is the Director of the Nanomedicine Center at NHRI. He has a strong research focus on Cancer Theranostics using radiation on nanoparticles. In one project, he is using synchrotron x-rays to synthesize conjugated nanoparticles of linear-like PEI and PEG-co-PEI for multiple applications, including tumor detection and imaging (Lin et al. 2012). The advantage of this method is that it is rapid without the use of various chemicals that can be pollutants. This method has been patented.

Dr. Yang's lab has used nanoparticles in concert with tomographically-reconstructed views of blood vessels to investigate size and shape of vessels in tumor versus normal tissue before and after chemotherapeutic agents (Fig. C.1, Chien

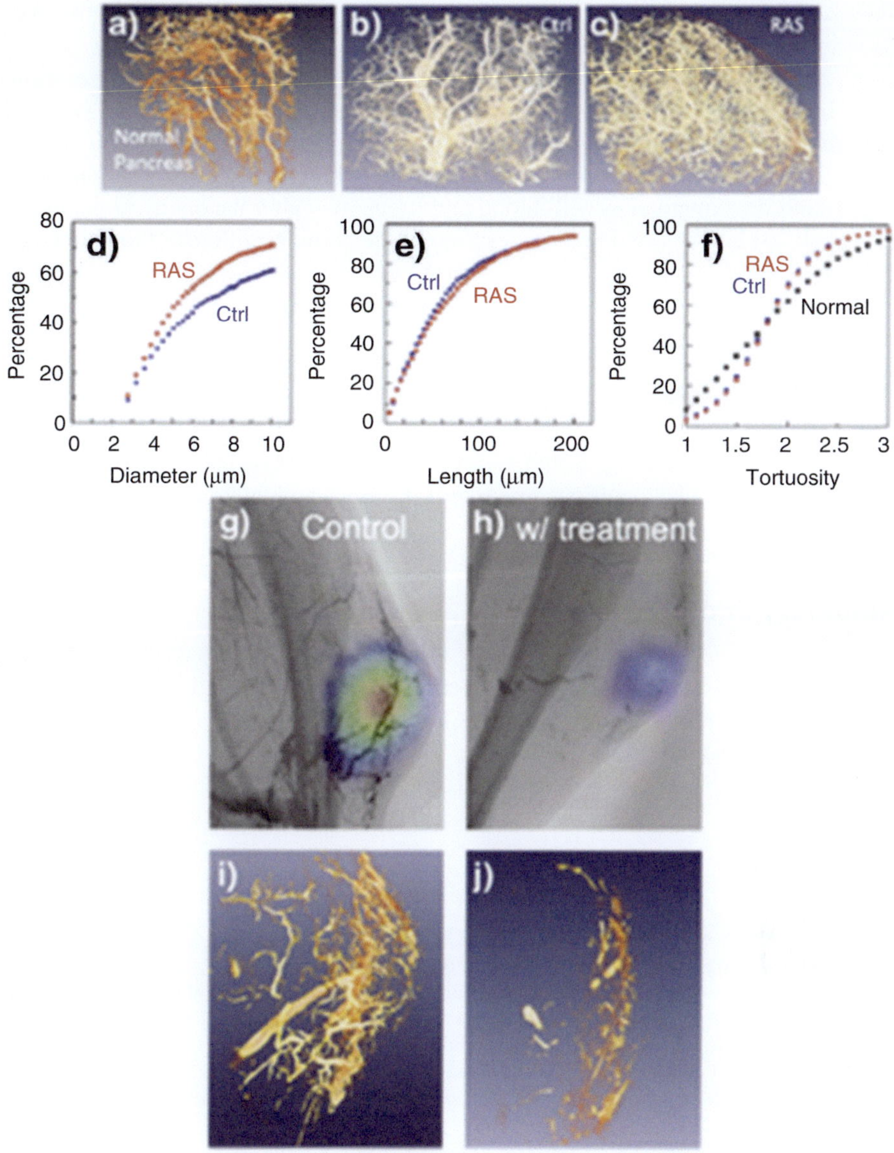

Fig. C.1 Stroma-mediated tumor microangiogenesis and its changes by bevacizumab. (From Chien et al. 2013)
(*a*)–(*c*): Tomographically reconstructed microvasculature images for orthotopic pancreatic cancer: (*a*) normal pancreas; (*b*) and (*c*): 3 days after implanting PANC1 cells into the pancreatic parenchyma of NOD-SCID mice without (Ctrl) or with (RAS) radiation activation. Field of view (reconstructed box): 924 μm-684 μm. (*d*) Percentage of vessels in (*b*) and (*c*) with diameter smaller than the horizontal scale value. (*e*) Percentage of vessels with length between adjacent branches smaller than the horizontal scale value. (*f*) "Tortuosity," measured as the ratio of the vessel length to the distance between branch points; the plot shows the percentage of cases in which the ratio is smaller than the horizontal scale value. (*g*), (*h*): 6 mm—8.8 mm images of xenografted NSCLC tumor treated with bevacizumab (10 mg/kg per day for 3 days) (*h*) or vehicle (*g*). Bioluminescence maps showing the tumor location are superimposed. (*i*), (*j*): Tomographically reconstructed 0.75 mm—1 mm images corresponding to (*g*) and (*h*)

et al. 2013). This method provides both excellent spatial and temporal resolution. They have also developed gold nanoparticles that are neutron-activated. These are both functional for imaging and treatment. The same group also has a strong interest in the use of electron microscopy for nanoparticle characterization. The goal is to understand how interaction with blood and tissue may alter nanoparticle geometry (Tai et al. 2012). They find that aggregation of the particles differs in blood in comparison to water.

Sources of Support

The NHRI receives funding from the government and public and private sectors. It distributes this money to various research institutions and internally.

Summary and Conclusions

The NHRI is an excellent "internal" research facility with strength across a number of areas. Ample funding to investigators allows flexibility and collaborations with the many close neighboring institutions.

References

Chien, C.C., Kempson, I.M., Wang, C.L., Chen, H.H., Hwu, Y., Chen, N.Y., Lee, T.K., Tsai, K.K., Liu, M.S., Chang, K.Y., Yang, C.S., Margaritondo, G.: Complete microscale profiling of tumor microangiogenesis, a microradiological methodology reveals fundamental aspects of tumor angiogenesis and yields an array of quantitative parameters for its characterization. Biotechnol. Adv. **31**(3), 396–401 (2013) doi:10.1016/j.biotechadv.2011.12.001

Lin, F.S., Chien, C.T., Chiu, W.C., Lin, S.Y., Tseng, F.G., Hwu, Y., Yang, C.S.: Chemical auxiliary-free polymerization yielding non-linear PEG for protein-resistant application. RSC Adv. **2**(18), 7174–7179 (2012)

Tai, L.A., Kang, Y.T., Chen, Y.C., Wang, Y.C., Wang, Y.J., Wu, Y.T., Liu, K.L., Wang, C.Y., Ko, Y.F., Chen, C.Y., Huang, N.C., Chen, J.K., Hsieh, Y.F., Yew, T.R., Yang, C.S.: Quantitative characterization of nanoparticles in blood by transmission electron microscopy with a window-type microchip nanopipet. Anal. Chem. **84**(15), 6312–6316 (2012)

Academia Sinica [Workshop: Genomic Research Center]

Site Address:	128, Academia Road, Sec. 2 Nangang District, Taipei Taiwan
Date Visited:	June 6, 2013
WTEC Attendees:	Paul Janmey (report author), Daniel Fletcher, Cynthia Reinhart-King, Nastaran Kuhn, Patricia Foland
Host(s):	Dr. Ying-chih Chang Genomic Research Center Academia Sinica Tel.: +886-2-27871277 yingchih@gate.sinica.edu.tw

Overview

The Chang laboratory specializes in development and implementation of biomimetic smart materials and interfaces to control the interaction of surfaces with specific cell types. One application of this research is the design of devices capable of isolated rare cells from a complex mixed population and capable, for example, or isolating and characterizing circulating cancer stem cells.

Research and Development Activities

The research focus of this lab is directed at the design and construction of a supramolecular architecture consisting of biomolecules or biomimetic materials that have specific interactions with cell or other biomaterial of interest. One application of devices using this strategy is to isolate viable cancer stem cells or other rare cells from a complex suspension under conditions that main cell viability. One approach taken is to construct an antibody-functionalized supported lipid bilayer (SLB), as a biomimetic and non-fouling membrane coating using antibodies specific or cancer stem cells (CTC). Figure C.2 shows the strategy for constructing such membranes. A recent report demonstrates their use with human blood spiked with two pre-stained colorectal cancer cell lines, HCT116 and colo205, to capture, purification and maintaining the viability of these cells. Over 97 % of HCT116, and 72 % of colo205 were captured, with the overall purity of cancer cells exceeding 95 %. Relates strategies have also been developed to optimize growth and differentiation of liver stem/progenitor cell (Tsai et al. 2012) and for controlled release of drugs from vesicles (Tseng and Chang 2012).

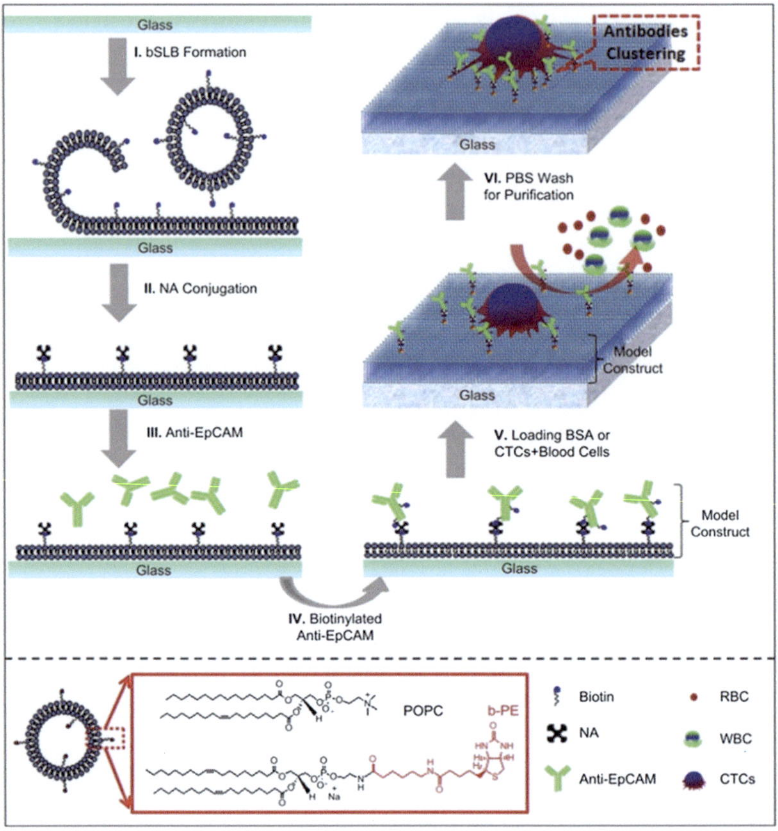

Fig. C.2 Schematic illustrations of sequential surface modification, cell capture, and purification. (From Wu et al. 2013)
The chemical structures of POPC and b-PE are shown at the bottom. The anti-EpCAM functionalized SLB glass surface is resistant to adhesion of protein (BSA) and normal blood cells, while adhering targeted cells such as HCT116 and colo205 (cancer cells)

Translational Efforts

The basic research on surface biophysics and biomimetic material chemistry are well suited for design of instrument with applications for diagnostics or cell separation.

Sources of Support

Academia Sinica, Taipei, Taiwan
Genomics Research Center, Academia Sinica

Summary and Conclusions

Fundamental principles of surface biophysics and biochemistry with emphasis on the properties of supported fluid lipid bilayers and their conjugation to biopolymers and antibodies have been used, often in combination with microfluidics, to design a number of devices with applications to specific cell types or biological processes. There is strong collaboration with cell and developmental biology to develop these devices for potential practical implementation.

References

Tsai, H.A., Shen, C.N., Chang, Y.C.: Use of surface properties to control the growth and differentiation of mouse fetal liver stem/progenitor cell colonies. Biomacromolecules. **13**, 3483–3493 (2012)

Tseng, P.Y., Chang, Y.C.: Tethered fibronectin liposomes on supported lipid bilayers as a prepackaged controlled-release platform for cell-based assays. Biomacromolecules. **13**, 2254–2262 (2012)

Wu, J.C., Tseng, P.Y., Tsai, W.S., Liao, M.Y., Lu, S.H., Frank, C.W., Chen, J.S., Wu, H.C., Chang, Y.C.: Antibody conjugated supported lipid bilayer for capturing and purification of viable tumor cells in blood for subsequent cell culture. Biomaterials. **34**, 5191–5199 (2013)

Academia Sinica [Workshop: Institute of Biochemistry]

Site Address:	128, Academia Road, Sec2 Nangang District, Taipei Taiwan
Date Visited:	June 6, 2013
WTEC Attendees:	Paul Janmey (report author), Daniel Fletcher, Cynthia Reinhart-King, Nastaran Kuhn, Patricia Foland
Host(s):	Kuan Wang Institute of Biochemistry Academia Sinica Tel.: +866-2-2785-5696 Ext. 3050 wangk@gate.sinica.edu.tw

Overview

The Kwan lab pioneered applications of biophysical chemistry to characterize two prominent muscle protein, titin and filamin. This lab also applies similar approaches to study the structure and function of intrinsically disordered proteins and extends their studies to define the potential of nanomedicine to improve methods for targeted drug delivery.

Research and Development Activities

A set of proteins that are inherently disordered under native conditions have unique physical properties that are important for the biological functions as elastomers within contractile cells and in mediation mechanotransduction. The most abundant and largest of these proteins, titin, has been extensively characterized in the Wang lab. One approach in this lab is to test the hypothesis that protein elasticity is an innate property of intrinsically disordered proteins and is manifested as a force-induced exposure of pre-formed protein binding sites in intrinsically disordered proteins. In the case titin, the intrinsically disordered region (PEVK) functions as a hybrid sensor and transducer of mechanical force to regulate protein/protein interactions in signaling pathways. A schematic diagram of the mechanisms by which the titin homolog TTN-1in c elegans is proposed to regulate elasticity of striated muscle, is shown in Fig. C.3.

For other disordered proteins such as nebulin, the compressive force of a stretched nebulin is required for it to interact and assemble with actin to form thin filaments (Yadavalli et al. 2009). This concept may be particularly relevant in the emerging field of mechanical biology and the effect of microenvironment on cell motility, cell proliferation and cell differentiation. Defining the physical aspects of intrinsically disordered proteins can be used to guide the engineering of smart elastomers (Tsai et al. 2012), develop methods for controlled drug discovery, and define the pathobiology of intrinsically disordered proteins in cardiovascular diseases and regeneration.

A second major effort of the Wang lab is the study of combination drugs and targeted drug delivery within the Nanomedicine Program of the Academia Sinica, as described on their web site:

> The short term goal of Academia Sinica's Nanomedicine Program is to develop a multiscale drug development and targeted delivery program that utilizes the Feedback Control System approach to rapidly optimize the formulation of drug cocktails for treating infectious diseases, cancer, and cardiovascular diseases. The Feedback [Control System] applies the well-known engineering feedback control principle to biological and clinical systems, and is particular powerful in optimizing the combination of three or more drugs to formulate "drug cocktails" at a significantly lower cost and faster pace per evaluation of a drug com-

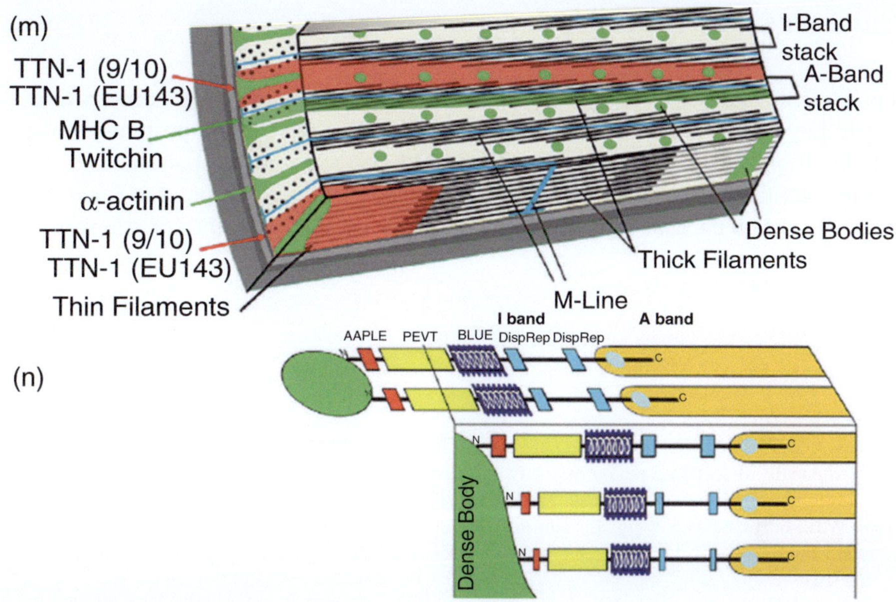

Fig. C.3 Regulation of elasticity in muscle tissue. (From Forbes et al. 2010)
(*m*) Drawing of a portion of body wall muscle indicating the organization of the myofilament lattice and antibody staining in this obliquely striated muscle. The drawing is oriented such that the edge on view of the obliquely striated I-band and A-band stacks have the same orientation as in the immunofluorescence images. Dense bodies are analogous to Z-disks in mammalian cross striated muscle. (*n*) Proposed model of TTN-1 orientation, maximal span and variable environment in the sarcomere. The TTN-1 molecules span from the dense bodies toward the A-band in an N-to-C orientation, with some reaching as far as the edge of the thick filaments

bination. The midterm goal is to deliver combination drugs to target tissues with functioning nanoparticles possessing homing agents on the surface. (From www.ibc.sinica.edu.tw/PI_DetailE.asp?Auto=47).

Translational Efforts

The work on protein biophysics and biochemistry is basic research and its implementation for novel bioengineered elastomers has many potential applications in smart materials and biomimetics. The work on drug delivery has a clear applied focus.

Sources of Support

Academia Sinica

Summary and Conclusions

Fundamental studies of the properties of intrinsically disordered proteins, originating in large part from work from the Wang group on the muscle proteins titin and nebulin, represent a large and increasing field of interest in mechanobiology, with many potential applications for defining how cells sense and respond to forces. The parallel work in nanomedicine and drug delivery also has potential for major contributions to biomedicine.

References

Forbes, J.G., Flaherty, D.B., Ma, K., Qadota, H., Benian, G.M., Wang, K.: Extensive and modular intrinsically disordered segments in C. elegans TTN-1 and implications in filament binding, elasticity and oblique striation. J. Mol. Biol. **398**, 672-689 (2010)

Tsai, W.L., Forbes, J.G., Wang, K.: Engineering of an elastic scaffolding polyprotein based on an SH3-binding intrinsically disordered titin PEVK module. Protein Expr. Purif. **85**, 187–199 (2012)

Yadavalli, V.K., Forbes, J.G., Wang, K.: Nanomechanics of full-length nebulin: an elastic strain gauge in the skeletal muscle sarcomere. Langmuir. **25**(13), 7496–7505 (2009)

Academia Sinica [Workshop: Institute of Biophysics, National Yang-Ming University]

Site Address:	128, Academia Road, Sec. 2 Nangang District, Taipei Taiwan http://www.ym.edu.tw/biophotonics/eng_version/index.html
Date Visited:	June 6, 2013
WTEC Attendees:	Paul Janmey, Daniel Fletcher (report author), Cynthia Reinhart-King, Nastaran Kuhn, Patricia Foland
Host(s):	Prof. Chau-Hwang Lee Chairman, Institute of Biophotonics National Yang-Ming University Tel.: 886-2-27898000 ext. 18 or 53 clee@gate.sinica.edu.tw http://www.rcas.sinica.edu.tw/~clee/index.html

Overview

Dr. Chau-Hwang Lee is Professor in the Institute of Biophotonics at National Ying-Ming University and Research Fellow at the Research Center for Applied Sciences at Academia Sinica, where he leads the High-Resolution Optical Microscopy and Applications (HiROMA) Laboratory. Dr. Lee is an expert in optical microscopy and microfabrication who applies novel technologies to study cell behavior and its implications for cancer. By developing microfluidic devices for culturing and observing cells in controlled microenvironments, Dr. Lee is isolating the effects of different parameters on cell growth and movement, with a particular focus on fibroblasts in co-cultures. This work illustrates the new perspectives that can be generated by combining physical science methods with questions in cancer biology.

Research and Development Activities

Dr. Lee and his research group are motivated by the question of how the microenvironment of a tumor influences the behavior of tumor cells. It has become clear that mechanical properties and cyclic stresses are important modulators of cell behavior, and stromal cells are key players in the tumor microenvironment. To address the question of how stromal cells affect tumor cell behavior, Dr. Lee and colleagues have develop a microfluidic cell culture chip that can support the growth of multiple types of cells at a time, such as breast cancer cells, glioblastoma cells, smooth muscle cells, microvascular endothelial cells, and fibroblast cells. These cell culture chips allow control of solutions and substrate interactions while at the same time enabling high resolution imaging of cell organization and behavior, including cell motility.

Fibroblasts are of increasing interest in cancer research as therapeutic targets. In particular, tumor-associated fibroblasts, which produce collagen type I that affects microenvironmental properties and cancer drug uptake, may play an important role in cancer biology and the development of new treatment strategies. Dr. Lee has recently developed a cell co-culture chip for lung cancer cells and lung fibroblasts (Fig. C.4).

By tracking the movements of each cell type, Dr. Lee and colleagues have found that the presence of fibroblasts in the culture affect the migration speeds of the cancer cells, presumably through a paracrine loop. This work suggests that preventing the fibroblasts from being activated could be essential for controlling cancer metastasis (Hsu 2011). Dr. Lee also showed that cyclic tensile stress of myofibroblasts can affect the migration speeds of cancer cells.

In a further advancement of the co-culture system, Dr. Lee demonstrated that it is possible to co-culture myofibroblasts, macrophages, and lung cancer cells. Interestingly, this study indicates that macrophages can suppress the effects of the myofibroblast cells, though this suppression is reversed by treating the macrophages with the antibody of TNF-alpha.

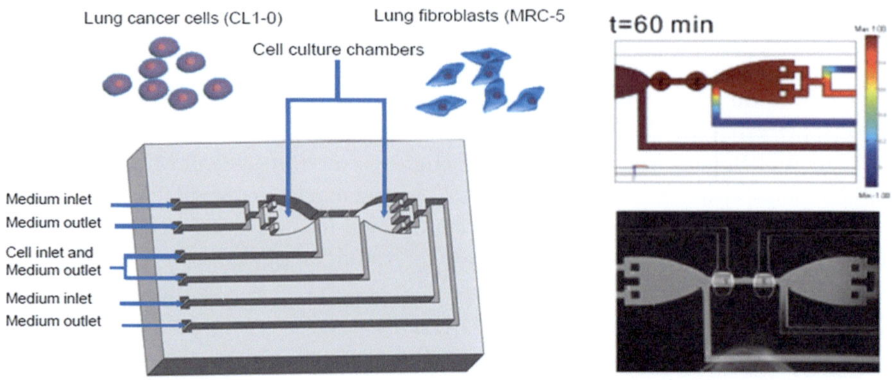

Fig. C.4 Microfluidic co-culture chip for studying the effect of lung fibroblasts on lung cancer cell behavior. The co-culture chip was designed to ensure that all the cells sense similar conditioned medium concentration from the other chamber (Courtesy of C.-H. Lee)

Translational Efforts

Understanding how the microenvironment of tumor cells affects their behavior is an important goal for both cancer biology and clinical treatments. The co-culture systems developed by Dr. Lee take a useful step forward in dissecting the complex interactions of different cell types. In particular, studies showing the effects of fibroblasts on tumor cell motility, and the effects of macrophages on myofibroblasts, point to potential targets for new pharmaceuticals. Importantly, the co-culture system will allow researchers to study the effects of different cell types on tumor cell behavior and may one day be turned into a personalized medicine technology.

Sources of Support

Funding was provided in part by Academia Sinica.

Summary and Conclusions

Microfluidic devices, like the co-culture devices developed by Dr. Lee, have the potential to play an increasingly important role in understanding the role of the microenvironment on tumor cell behavior. By isolating and modulating key variables such as cell type and cyclic strain, these microfluidic devices can be used together with high resolution optical imaging methods to track cell movements and other behaviors. Microfluidic methods for assaying cell behavior will continue to be important for basic studies and may in the future become basic clinical tools for evaluating patient response to chemotherapy or other personalize medicine applications.

References

Hsu, T.-H., Kao, Y.-L., Lin, W.-L., Xiao, J.-L., Kuo, P.-L., Wu, C.-W., Liao, W.-Y., Lee, C.-H.: Migration speed of cancer cells influenced by macrophages and myofibroblasts co-cultured in a microfluidic chip. Integr. Biol. **4**, 177 (2012)

Huang, J.-W., Pan, H.-J., Yao, W.-Y., Tsao, Y.-W., Liao, W.-Y., Wu, C.-W., Tung, Y.-C., Lee, C.-H.: Interaction between lung cancer cell and myofibroblast influenced by cyclic tensile strain. Lab Chip. **13**, 1114 (2013)

Xiao, J.-L. Pan, H.-J., Lee, C.-H.: Optically micropatterned culture of adherent cells. J. Biomed. Opt. **17**, 075004 (2012)

Xiao, J.-L., Lu, D.-H., Lee, C.-H.: Guiding the migration of adherent cells by using optical micropatterns. Appl. Phys. Lett. **102**, 123703 (2013)

Academia Sinica [Workshop: Institute of Physics]

Site Address:	128, Academia Road, Sec. 2 Nangang District, Taipei Taiwan
Date Visited:	June 6, 2013
WTEC Attendees:	Paul Janmey (report author), Daniel Fletcher, Cynthia Reinhart-King, Nastaran Kuhn, Patricia Foland
Host(s):	Dr. Keng-hui Lin Associate Research Fellow Institute of Physics Academia Sinica Tel.: +886-2-2789-6763 khlin@phys.sinica.edu.tw

Overview

The expertise of the Lin laboratory is in soft matter physics and its applications to problems in biology. A current interest is the application of methods using principles of colloid physics to produce three dimensional scaffolds from biocompatible materials that have biochemical and physical characteristic mimicking those of the extracellular matrix. Such material when seeded with cells can provide a test system for drug delivery, cell mechanics, and effect of therapeutic agents.

Research and Development Activities

Microfluidic systems are used to create foam-like structures with a crystalline ordering of spheres containing solutions of gelling agents such as alginate, gelatin, and other biopolymers. A process of drying and solvent exchange creates 3D scaffolds with curved surfaces and tunable dimension suitable for implantation with cells. A schematic of the method for producing these scaffolds is shown in Fig. C.5.

Imaging single cells and groups of cells within the spherical cavities of the scaffold is feasible using confocal fluorescence microscopy, and initial studies have shown that cells appear to sense the local stiffness of gel on a scale of tens of microns, and not the global stiffness of macroscopic scaffold. Elongation of cells depends on the scaffold mesh size relative to the cell volume (Lin et al. 2011).

Scaffolds can also be used to perform 3D traction force measurements and local forces can be measured by deformation of the initially spherical scaffold pore. Another novel use of this material is to measure the movement of cancer cells in response to an electric field applied to the scaffold (Sun et al. 2012).

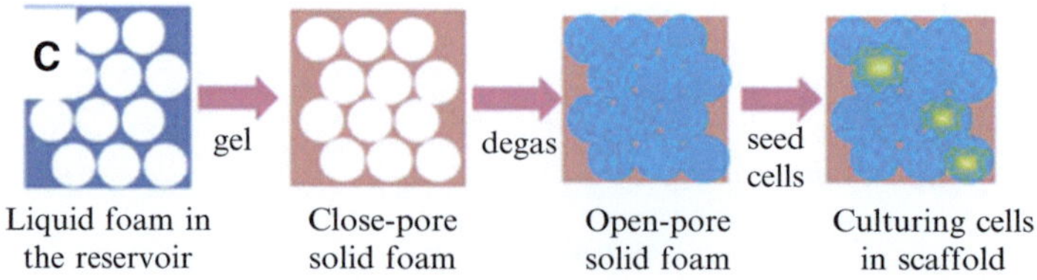

Fig. C.5 Construction of 3D ordered cellular solids. (From Lin et al. 2011)
Monodisperse liquid foam self-assembles into crystalline order. The liquid foam is gelled, at which point the bubbles became topologically closed pores in the solid foam. Finally, the closed-pore solid foam is transformed to an open-pore solid foam with some cavities by degassing under vacuum while immersed in liquid crosslinking solution. After exchange of medium, cells can be seeded within the scaffold

Translational Efforts

Developing the methods for creating 3D scaffolds involves basic research of soft matter and hydrodynamics, and the products of this work have potential applications in materials science, bioengineering, and diagnostics. Collaborations between this group in a physics institute and medical researchers have tested the viability of the scaffolds for implantation *in vivo* in the context of cartilage regeneration (Wang et al. 2012, Fig. C.6.)

Sources of Support

Academia Sinica Nano-Bio funding
National Science Council

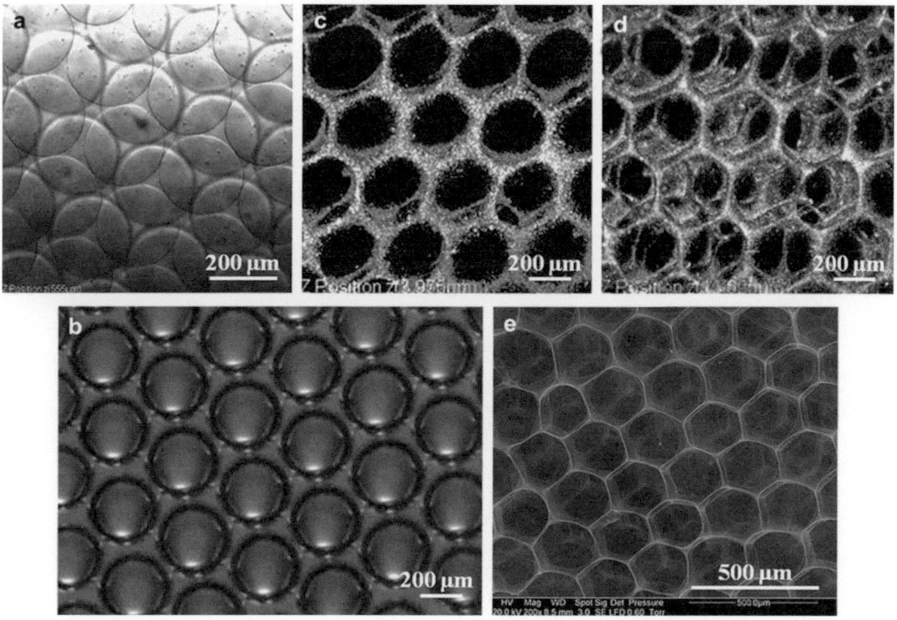

Fig. C.6 Uniform alginate bubbles generated by microfluidics. (From Wang et al. 2011) (*a*). The droplets formed a honeycomb structure after gelatin (*b*). Confocal microscope shows a 3D ordered array structure (*c* and *d*). After vacuum degassing process, the alginate forms a highly interconnecting porous structure with uniform pore size (*e*)

Summary and Conclusions

Principles of colloid chemistry and physics are combined with the strong tradition of microfluidic research throughout Taiwan to create novel soft biocompatible scaffolds for cell research and tissue engineering. Extensive collaboration between basic physics researchers and biomedical research groups is also a strength of this research effort.

References

Lin, J.Y., Lin, W.J., Hong, W.H., Hung, W.C., Nowotarski, S.H., Gouveia, S.M., Cristo, I., Lin, K.H.: Morphology and organization of tissue cells in 3D microenvironment of monodisperse foam scaffolds. Soft Matt. **7**, 10010–10016 (2011)

Sun, Y.S., Peng, S.W., Lin, K.H., Cheng, J.Y.: Electrotaxis of lung cancer cells in ordered three-dimensional scaffolds. Biomicrofluidics. **6**(1), 014102–014102-14 (2012) doi:10.1063/1.3671399

Wang, C.C., Yang, K.C., Lin, K.H., Liu, H.C., Lin, F.H.: A highly organized three-dimensional alginate scaffold for cartilage tissue engineering prepared by microfluidic technology. Biomaterials. **32**, 7118–7126 (2011)

Wang, C.C., Yang, K.C., Lin, K.H., Liu, Y.L., Liu, H.C., Lin, F.H.: Cartilage regeneration in SCID mice using a highly organized three-dimensional alginate scaffold. Biomaterials. **33**, 120–127 (2012)

Academia Sinica [Workshop: Institute of Statistical Science]

Site Address:	128, Academia Road, Sec. 2 Nangang District, Taipei Taiwan http://www.stat.sinica.edu.tw/statnewsite/
Date Visited:	June 6, 2013
WTEC Attendees:	Paul Janmey, Daniel Fletcher (report author), Cynthia Reinhart-King, Nastaran Kuhn, Patricia Foland
Host(s):	Dr. Chen-Hsiang Yeang Assistant Research Fellow Institute of Statistical Science Academia Sinica Tel.: 886-2-2783-5611 #310, 6614-5630 Fax: 886-2-2783-1523 chyeang@stat.sinica.edu.tw http://www.stat.sinica.edu.tw/chyeang/

Overview

Dr. Chen-Hsiang Yeang is an Assistant Research Fellow in the Institute of Statistical Science at Academia Sinica where he works in the field of computational biology. His work involves development of algorithms, statistical methods, and models to advance two main areas of computational biology: cancer genomics and molecular evolution. A major goal of Dr. Yeang's work is to identify drivers of tumor phenotypes by analyzing various types of molecular aberrations and single cell heterogeneities. This work shows how careful mathematical modeling can reveal novel strategies to understand and control evolving populations at multiple scales.

Research and Development Activities

Dr. Yeang approaches the problem of identifying drivers of tumor phenotypes by developing mathematical models to capture aberrations at the molecular level. One approach is to define association modules of the aberrations, such as mutation and copy number. The association modules are then integrated into a logistic regression model, and layered model selection is used to determine the associations. Dr. Yeang is in the process of applying this framework to glioblastoma multiforme (GBM), making use of cancer genomic data from the Cancer Genome Atlas. Preliminary results show that a small number of association modules can predict patient survival. These results validate the general approach and suggest that the association module framework may be a powerful way to both make sense of cancer genomic data and predict outcomes for cancer patients. Further work with an expanded data set and additional cancer genomes will help to build confidence in the methods and the predictions.

Dr. Yeang is also using mathematical modeling to understand the role that single-cell heterogeneity has on personalized medicine treatments. Tumors are notoriously heterogeneous at the single cell level, but most cancer therapies can't or don't take that heterogeneity into account. In a recent publication, Dr. Yeang and colleagues showed through modeling that accounting for this heterogeneity would motivate an alternative approach to personalized cancer therapy, which they termed "nonstandard" personalized medicine (Fig. C.7, Beckman 2012). This work is another promising example of how statistical methods informed by cancer biology can lead to new treatment strategies.

Translational Efforts

The statistical models developed by Dr. Yeang have important and direct implications for cancer therapies. His models, both those that are aimed at tracking association modules from cancer genomes and those that analyze the effects of single cell

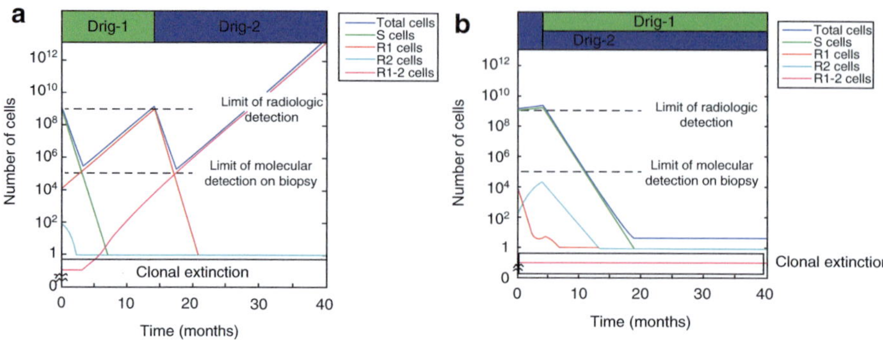

Fig. C.7 Example of how cancer treatment with conventional personalized medicine (**a**) differs from the "nonstandard" personalized medicine (**b**) proposed by Dr. Yeang and colleagues showing how treatments that incorporate single-cell heterogeneity have improved outcomes (See Beckman, 2012 for full details of the "nonstandard" personalized medicine)

heterogeneity on cancer treatment, show that detailed analysis of data can help to guide thinking about cancer progression and therapy. With further clinical interactions, this work has the potential to guide new approaches to cancer treatment that would benefit many patients.

Sources of Support

Dr. Yeang is supported by Academia Sinica.

Summary and Conclusions

Computational biologists like Dr. Yeang are showing that mathematical models and statistical analysis can reveal new associations and novel approaches to treatment. This approach will continue to be vitally important not only to make sense of vast quantities of data but also to predict have evolution will affect the growth and progression of tumor cells.

References

Beckman, R.A., Schemmann, G.S., Yeang, C.H.: Impact of genetic dynamics and single-cell heterogeneity on development of nonstandard personalized medicine strategies for cancer. Proc. Natl. Acad. Sci. USA. **109**(36), 14586–14591 (2012)

Chen, D.H., Chang, A., Liao, B.Y., Yeang, C.H.: Functional characterization of motif sequences under purifying selection. Nucleic Acids Res. **41**(4), 2105–2120 (2013)

Suen, S., Lu, H.S., Yeang, C.H.: Evolution of domain architectures and catalytic functions of enzymes in metabolic systems. Genome Biol. Evol. **4**(9), 852–869 (2012)

Academia Sinica [Workshop: National Tsing Hua University]

Site Address:	128, Academia Road, Sec. 2 Nangang District, Taipei Taiwan http://www2.ess.nthu.edu.tw/en/faculty.php
Date Visited:	June 6, 2013
WTEC Attendees:	Paul Janmey, Daniel Fletcher, Cynthia Reinhart-King (report author), Nastaran Kuhn, Patricia Foland
Host(s):	Fan-Gang Tseng Department of Engineering and System Science, Rm. 418 National Tsing Hua University 2 Kuang Fu Road, Hsinchu 300, Taiwan Tel.: +886-03-5715131 34270 fangang@ess.nthu.edu.tw

Overview

National Tsing Hua University is a research university originally founded in 1956 with a focus on nuclear science studies, expanding into engineering and now encompassing virtually all disciplines. It is ranked among the best universities in Taiwan and ranked 107th in the world (Time Higher Education, 2010). It has many research centers to promote interdisciplinary work, in addition to divisions focused on supporting government and industrial collaborations and technology transfer.

The Engineering and System Science Department has three major focus areas: Nanotechnology and MEMS, Nuclear and Energy Engineering and Engineering Physics. They report that the department exceeds $3 M USD in research expenditures annually. They collaborate with other academic partner institutions and industry.

Research and Development Activities

Dr. Tseng's lab works in the area of micro and nanotechnologies, bringing together biology, physics and engineering into innovative applications. He has approximately 10–20 people in his lab working in applications related to cancer. As indicated on his website (http://www2.ess.nthu.edu.tw/~fangang/eg_about_us.htm). More than 80 million NT dollars have been invested to develop seven major current state-of-the-art and well-equipped laboratories, which include 10 K clean room, chemistry laboratory, nanoscale thin film deposition laboratory, MEMS testing and packaging laboratory, nano-/micro-fluidics system and nano-optics testing laboratory, atomic force microscopy and nanoindentation laboratory, and nano-/micro-fluidics simulation laboratory which occupy about 330-square- meter of space.

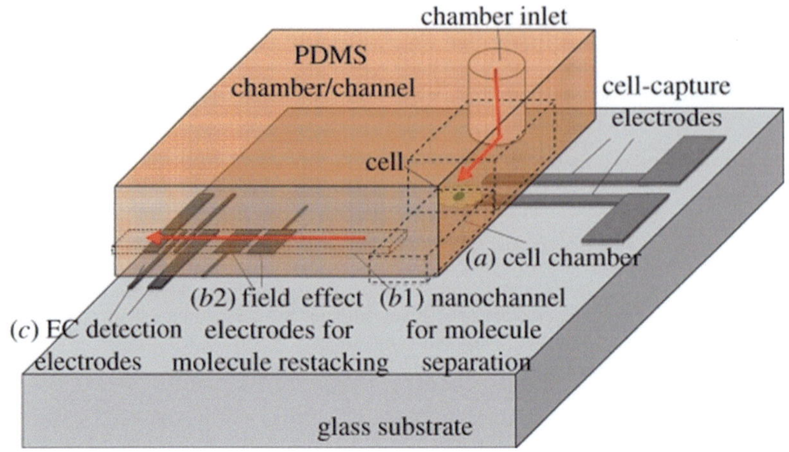

Fig. C.8 Schematic of a Nano-CEEC chip designed for living single-cell analysis (From Wu et al. 2011)
Step 1: (*a*) cell loading, capturing and culturing in the cell chamber. Step 2: (*b1, b2*) DAEKF separation, including sample collection, nanoseparation and restacking in the nanochannel. Step 3: (*c*) amperometric detection by the EC detection electrodes

Dr. Tseng's laboratory has eight major research thrusts (1) ultrasensitive single molecule protein nanoarray, (2) fast single molecule DNA sequencing, (3) ultrasensitive and high-speed detection of biochemical release agent at a single cell or multiple cells level, (4) ultrasensitive and high-speed dynamic monitoring and detection of neurotransmitter, (5) high efficiency and non-toxic nanoparticles for biomolecules and cellular study, (6) integrated optoelectronic microsystem, (7) integrated circuit design for micro-/nano-system, and (8) surface tension dominant nano-/micro-system.

One project of interest is his work isolating CTCs for both early detection and for studies of drug resistance. His lab is developing a device that runs whole blood samples through a three-stage process in approximately 30 min to isolate CTCs (Fig. C.8). In the first step, cells are isolated using hydrodynamics and inertial forces based on size exclusion. In the second stage, silicon nanowires entangle white blood cells. In the third stage, self-assembled arrays allow for the formation of a single cells layer for washing. Integration of this device with a dual asymmetry, electrokinetic flow focusing device will allow for the isolation of specific markers within the cell. They have also developed a Nanocapillary electrophoretic electrochemical chip (Nano-CEEC chip), which can trap individual cells, wash away the others, and lyse that cell for further analysis (Wu et al. 2011).

In other work, Chen-Hsien Liu is using DEP-based technologies to create liver-like tissue constructs with heterogeneous subpopulations (Fig. C.9). Shape, size and cell placement can all be controlled.

Gwo-Bin Vincent Lee is developing microfluidic systems for cancer cell analysis (Fig. C.10, Wang et al. 2013). Specifically, he is interested in gene-based detection methods using on-chip multiplex PCR with a focus on ovarian and lung cancers.

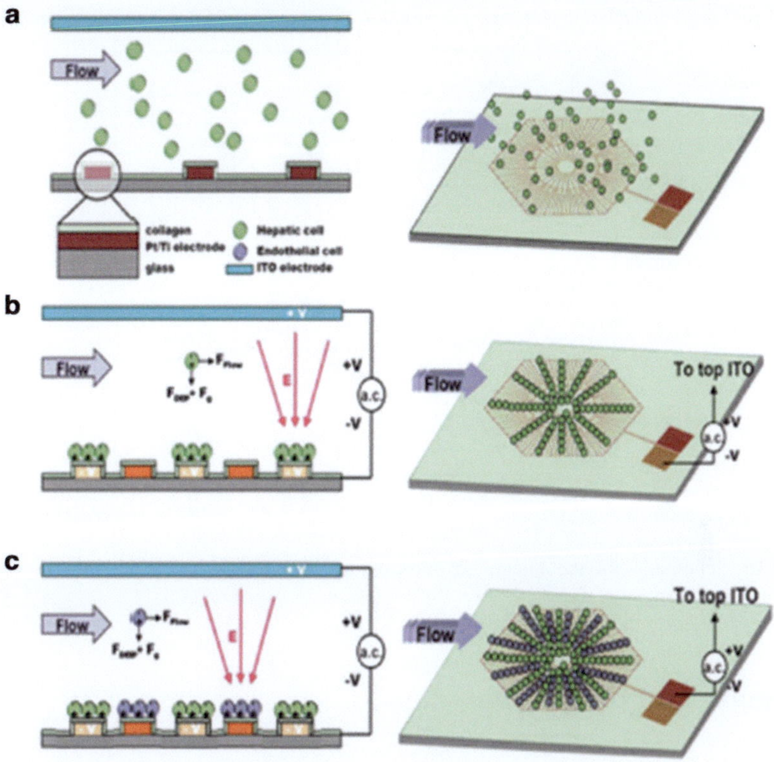

Fig. C.9 The configuration and operation principles of DEP-based heterogeneous lobule-mimetic cell patterning (Adapted from Ho et al. 2013)
(*a*) Randomly distributed hepatic cells are loaded into the microfluidic chamber. (*b*) The hepatic cells are captured and patterned onto the first DEP patterning electrodes to form the radial cell-string pattern after the vertical positive DEP voltage is applied. (*c*) The endothelial cells are, then, loaded, guided and positioned in-between the patterned hepatic cells via applying the positive DEP voltage on the second DEP patterning electrode. The heterogeneous integration of hepatic and endothelial cells is performed to mimic the hexagonal lobule of liver tissue

Translational Efforts

Several biomedical related companies in Taiwan have contacted Dr. Tseng's lab regarding cooperation in the fields of CTCs diagnosis, sperm separation, and DNA micro array for early virus detection. Prof. GB Lee has initiated a start-up company for chip based RT-PCR diagnosis.

Sources of Support

National Sciences Council of Taiwan

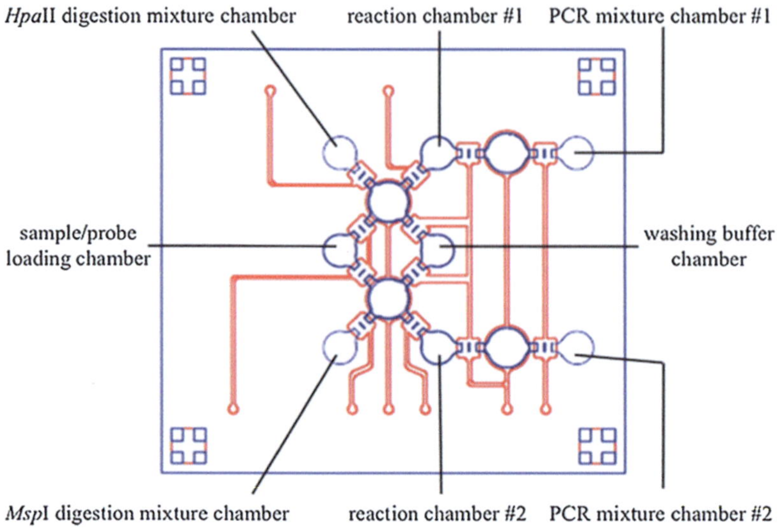

Fig. C.10 Schematic for the integrated microfluidic chip for detection of DNA methylation (Adapted from Wang et al. 2013)

Summary and Conclusions

Overall, National Tsing Hua University has clear demonstrated strength in MEMS/microfabrication for cell manipulation and analysis. They recruit highly qualified graduate students and in some cases support very large laboratories. There is extensive infrastructure to support research.

References

Ho, C.T., Lin, R.Z., Chen, R.J., Chin, C.K., Gong, S.E., Chang, H.Y., Peng, H.L., Hsu, L., Yew, T.R., Chang, S.F., Liu, C.H.: Liver-cell patterning Lab Chip, mimicking the morphology of liver lobule tissue. Lab Chip. **13**(18), 3578–3587 (2013) doi:10.1039/c3lc50402f

Wang, C.H., Lai, H.C., Liou, T.M., Hsu, K.F., Chou, C.Y., Lee, G.B.: A DNA methylation assay for detection of ovarian cancer cells using a HpaII/MspI digestion-based PCR assay in an integrated microfluidic system. Microfluidics Nanofluidics (in press) (2013) doi:10.1007/s10404-013-1179-8. (Published online April 2, 2013)

Wu, R.G., Yang, C.S., Cheing, C.C., Tseng, F.G.: Nanocapillary electrophoretic electrochemical chip, towards analysis of biochemicals released by single cells. Interf. Focus. **1**(5), 744–753 (2011)

Academia Sinica [Workshop: Research Center for Applied Sciences]

Site Address:	128, Academia Road, Sec. 2 Nangang District, Taipei Taiwan http://rcas.sinica.edu.tw/
Date Visited:	June 6, 2013
Host(s):	Dr. Yi-chung Tung Assistant Research Fellow Research Center for Applied Sciences, Academia Sinica Tel.: 886-2-2789-8000 ext 67 Fax: 886-2-2782-6680 tungy@gate.sinica.edu.tw http://www.rcas.sinica.edu.tw/~tungy/RCAS.TungLab/Welcome.html

Overview

Dr. Yi-Chung Tung is a Research Fellow in the Research Center for Applied Sciences at Academia Sinica, where he leads a laboratory that focuses on the development of integrated biomedical microdevices. Dr. Tung's interdisciplinary research involves the use of microdevices to control the fluid environments around cells and to analyze small numbers of cells. Recent work from Dr. Tung's laboratory has succeeded in controlling the gaseous microenvironment around cells in a cell culture device, in particular the levels of oxygen (O_2) and nitrous oxide (NO), and has shown that gas levels and gas gradients can have profound effects on cells. Dr. Tung's results highlight the importance of controlling not only the physical microenvironment of cells but the gaseous microenvironment.

Research and Development Activities

The microenvironment around cells influences their growth and behavior through soluble factors, physical cues, and dissolved gasses. It is well known that certain tissues and tumors are associated with changes in levels of dissolved gasses, but this property has been difficult to study *in vitro* due to the challenge of creating controlled gaseous microenvironments. Dr. Tung has addressed this issue by developing a microfluidic cell culture device with spatial control of dissolved O_2 (Fig. C.11, Peng 2013). The device works by confining oxygen scavenging chemical reactions underneath the cell culture chambers, causing the local oxygen levels to vary spatially. Using the device, Dr. Tung and colleagues tested the effect of the anti-cancer drug triapazamine (TPZ) on an adenocarcinomic human alveolar basal epithelial cell line at three different oxygen levels. They found that the toxicity of TPZ is hypoxia induced, providing a new method for investigating the role of the gaseous microenvironment on the mechanism of this important anti-cancer drug.

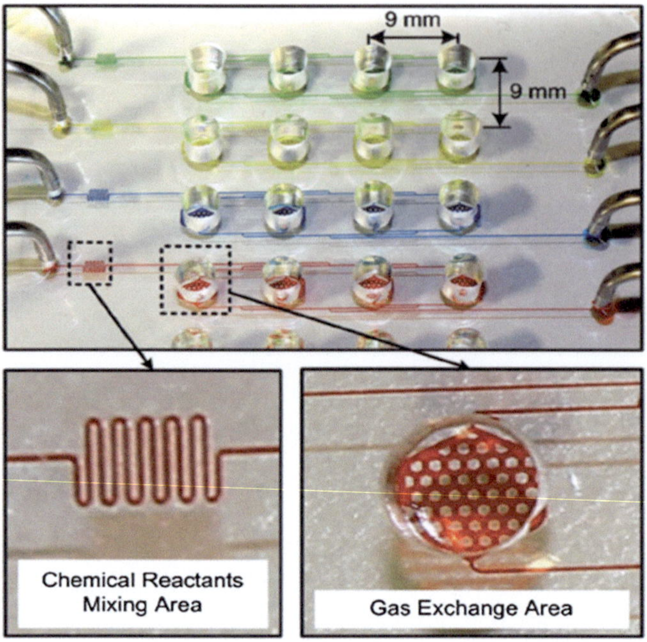

Fig. C.11 Integrated microfluidic culture device with controlled gaseous microenvironments. The device enables variation in oxygen levels and establishment of oxygen gradients (From Peng 2013)

Dr. Tung is extending his group's ability to control the gaseous microenvironment by developing a device to control NO that is being used to test the role of NO levels and gradients on smooth muscle cells and other cells.

Translational Efforts

Expanded control of cell microenvironments to include gas levels provides researchers with new tools to explore tumor behavior and screen pharmaceuticals for activity that is affected by gas levels. Dr. Tung has demonstrated how his device can be used to understand the hypoxia-dependent activity of TPZ, and further development of the device will enable more extensive evaluations of the gas-dependent behavior of normal and tumorigenic cells.

Sources of Support

Dr. Tung's research is supported by the National Health Research Institutes (NHRI) in Taiwan, National Science Council (NSC) in Taiwan, and the Academia Sinica Research Program in Nanoscience and Nanotechnology.

Summary and Conclusions

Microfluidic devices with integrated functions such as control of the gaseous micro-environment are poised to play more significant roles as research tools and potentially as clinical screening tools in the future.

References

Chen, Y.-H., Peng, C.-C., Cheng, Y.-J., Wu, J.-G., Tung, Y.-C.: Generation of nitric oxide (NO) gradients in microfluidic devices for cell culture using spatially controlled chemical reactions. Biomicrofluidics **7**(6), 064104 (10 pp.). (2013) doi:10.1063/1.4829775

Peng, C.C., Liao, W.H., Chen, Y.H., Wu, C.Y., Tung, Y.C. (2013). A microfluidic cell culture array with various oxygen tensions. Lab Chip. **13**(16), 3239–3245

Academia Sinica [Workshop: Tunghai University]

Site Address:	128, Academia Road, Sec. 2 Nangang District, Taipei Taiwan	
Date Visited:	June 6, 2013	
WTEC Attendees:	Paul Janmey, Daniel Fletcher, Cynthia Reinhart-King (report author), Nastaran Kuhn, Patricia Foland	
Host(s):	Ming-Jer Tang, M.D., Ph.D. Tunghai University 1 University Road, Tainan, 70101 Taiwan Tel.: (+866) 6 235 3535 ext 5420-5422 Fax: (+886) 6 236 2780 mjtang1@mail.ncku.edu.tw	

Overview

Tunghai University was founded in 1955 and enrolls 17,000 students. It offers 35 Master's programs and 14 Ph.D. programs. It encourages industrial and academic collaborations and cross-affiliations with other institutions and sister schools. It has a number of research centers that provide a mechanism to bring faculty together across departments to focus on questions in a common theme. These include the Life Sciences Research Center and Nanotechnology Center. Within these centers, there are multiple core laboratories including, Electron Microscopy, Confocal Microscopy, Genomics and Proteomics and Cell Culture.

Functional Focus

The overall focus of Tunghai University for growth, as stated on their website, is in the areas of green science, digital cultural creative industry, and corporate social responsibility. However, there are separate focuses within the university including in the Life Sciences Research Center, which focuses on Developmental Biology and Disease.

Research and Development Activities

Ming-Jer Tang's research efforts are in the area of cancer mechanobiology. He collaborates with a number of groups including Yang Kao Wang at Taipei Medical University. He has found that transformed cells evade soft-substrate induced apoptosis (Wang et al. 2007, Fig. C.12). His lab has used AFM to probe the mechanical properties of a number of cells from breast, bladder, pancreatic, and cervical cancers. They show that cancer cells are softer than their normal counterparts and have less adaptability to substrate rigidity. Caveolin-1 plays a key role in mechanosensing potentially by regulating the formation of the actin cap which affects the stiffness of the cell as measured by AFM. In epithelial cells however caveolin-1 affects cell-cell junctions which may in turn affects cell stiffness. His lab has found that, using the mPAD system, traction forces decrease in transformed cells.

Translational Efforts

The Central Taiwan Science Parks are being heavily invested in as an "R&D hotspot" with a focus on the use of nanotechnology in industry. The goal of the Center for Nanoscience and Technology is to integrate the nanotechnology efforts at Tunghai with those of the Science Parks.

Sources of Support

National Science Council, National Health Research Institute, Ministry of Education, and other local and overseas sources.

Summary and Conclusions

Tunghai University is well-positioned to be involved in both training and research in nanotechnology. Their plan to integrate research efforts across the university and through the science park will position them well to stay at the cutting edge of research developments and translation into industrial applications.

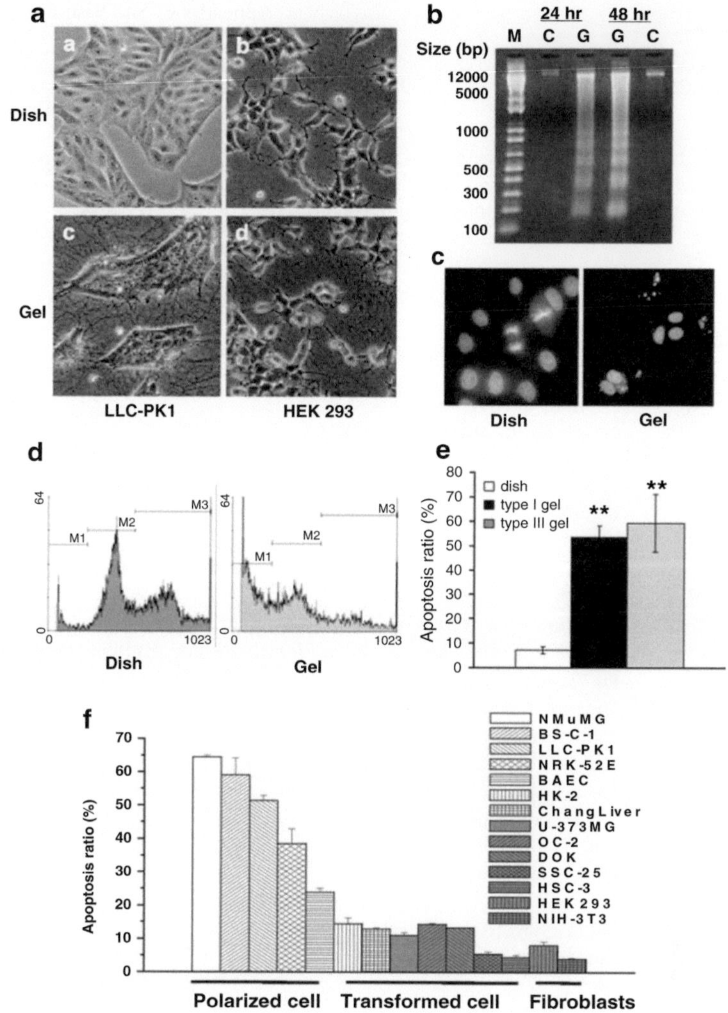

Fig. C.12 Characterization of collagen gel-induced cell death. (From Wang et al. 2007)
(*A*) LLC-PK1 epithelial cells (*left*) and HEK 293 fibroblasts (*right*) were cultured on culture dishes (*upper*) or type I collagen gel (*lower*) for 24 h. Cell morphology was observed under the same power field of phase-contrast microscopy. (*B*) small molecular weight DNA extracted from LLC-PK1 cells cultured on culture dishes (lanes 1 and 4) or type I collagen gel (lanes 2 and 3) for 24 or 48 h was resolved using 2 % agarose gel electrophoresis and stained with ethidium bromide. Lanes 2 and 3 show 200-bp-based DNA laddering patterns. (*C*) Hoechst 33258 staining (5 μg/ml) of LLC-PK1 cells cultured on culture dishes or type I collagen gel for 24 h. Immunofluorescence micrographs were taken under the same power field after UV light excitation. Nuclear condensation and fragmentation were seen in LLC-PK1 cells cultured on collagen gel. (*D*) cell cycle analysis was assessed using flow cytometry (FACScan) of propidium iodide-stained LLC-PK1 cells cultured on culture dishes or type I collagen gel for 48 h. M1 indicates the sub-G0 phase, M2 indicates the G0/G1 phase, and M3 indicates the S/G2/M phase of the cell cycle. (*E*) the apoptotic ratio (%) in LLC-PK1 cells cultured on culture dishes, type I collagen gel, or type III collagen gel for 48 h was assessed using FACScan analysis. (*F*) the apoptotic ratio (%) of different cell lines cultured on type I collagen gel for 48 h, including epithelial (NMuMG, BS-C-1, LLC-PK1, NRK-52E, and MDCK), endothelial (BAECs), mesenchymal (HEK 293 and NIH-3 T3), and tumor cells (HK-2, Chang liver, U-373 MG, OC-2, DOK, SSC-25, and HSC-3). Only polarized cells developed apoptosis on collagen gel

References

Wang, Y.H., Chiu, W.T., Wang, Y.K., Wu, C.C., Chen, T.L., Teng, C.F., Chang, W.T., Chang, H.C., Tang, M.J.: Deregulation of AP-1 proteins in collagen gel-induced epithelial cell apoptosis mediated by low substratum rigidity. J. Biol. Chem. **282**(1), 752–763 (2007)

Cancer Science Institute of Singapore, National University of Singapore

Site Address:	14 Medical Drive No. 12-01, Centre for translational Medicine Singapore 117599 http://www.csi.nus.edu.sg	
Date Visited:	June 3, 2013	
WTEC Attendees:	Paul Janmey, Owen McCarty (report author), Cynthia Reinhart-King, Sharon Gerecht, Nastaran Kuhn, Larry Nagahara, Patricia Foland	
Host(s):	Dr. Daniel Tenen[1] Director, Cancer Science Institute of Singapore, NUS dtenen@bidmc.harvard.edu csidgt@nus.edu.sg	
	Dr. Jean-Paul Thiery Senior Principal Investigator csitjp@nus.edu.sg	
	Dr. Richie Soong Research Associate Professor, Senior Principal Investigator Tel.: (65) 6516 8055 csirs@nus.edu.sg	

Overview

The Cancer Science Institute of Singapore (CSI Singapore) focuses on the development of basic and translational studies to gain an understanding of the pathogenesis of cancer and identify novel treatment strategies to combat this disease. The mission of is to catalyze world-class research in cancer sciences through collaboration between research strengths and outstanding international collaboration.

The research mission of CSI Singapore is focused in two areas of expertise: Cancer Biology & Stem Cells, and Experimental Therapeutics. Research is focused on key cancer diseases which are endemic in Asian populations, namely leukemia and gastric, liver, and lung cancer. Additionally, efforts are underway to develop a research focus in breast cancer phenotypes that are specific to the Asian population.

---

[1] Dr. Tenen was not present during the WTEC panel's visit.

CSI Singapore is located on three floors at the Center for Translational Medicine, premises within the Yong Loo Lin School of Medicine, and is situated opposite to the approximately 1000-bed National University Hospital. Dr. Daniel Tenen serves as the Director.

Research and Development Activities

Dr. Daniel Tenen

Dr. Tenen is a leader in the field of transcriptional regulation, hematopoiesis and cancer. His research efforts have focused on transcription factors, gene regulation, and their relationship to normal differentiation and cancer. He has successfully characterized transcription factors that play a role in the differentiation of hematopoietic stem cells into specific lineages; isolated and characterized regulatory elements of genes which are expressed at different stages of myeloid differentiation, including CD34, and two master transcription factors which are regulators of myeloid development: PU.1 and C/EBP alpha. He currently places focus on understanding regulation, signal transduction pathways; and interacting partners of transcription factors and their role in stem cells. Other current projects in his laboratory include targeting specific genes in hematopoietic stem cells, analysis of cancer mutations, analysis of transcription factor targets and gene signatures in cancer using genomic approaches; and elucidating the role of microRNAs in leukemias.

Dr. Jean-Paul Thiery

Prof Jean-Paul Thiery's focus of research is to unravel the mechanisms of invasion and metastasis of carcinoma cells. His seminal contributions in the study of bladder carcinoma have led to the discovery of molecular mechanisms governing the formation of superficial or invasive tumors. His current studies involve oncogenomics and functional approaches to characterize breast carcinoma basal subtypes and ovarian carcinoma. Prof Thiery is also known for his studies on mechanochemistry of cell adhesion and migration and for his work on the ontogeny of the neural crest.

Dr. Richie Soong

The Soong group is focused on understanding the differing genetic, environmental and cultural drivers in cancer in Asian populations. In particular, his group utilizes clinical samples from cancers that are endemic in Asian populations, such as gastric, liver, and lung cancer, to optimize and develop novel diagnostics and treatment platforms. His group has five main interests: (1) development of a link between pharmacogenetics and Asian treatment outcomes, (2) development of approaches to characterize the cancer genome and methylome via next generation sequencing,

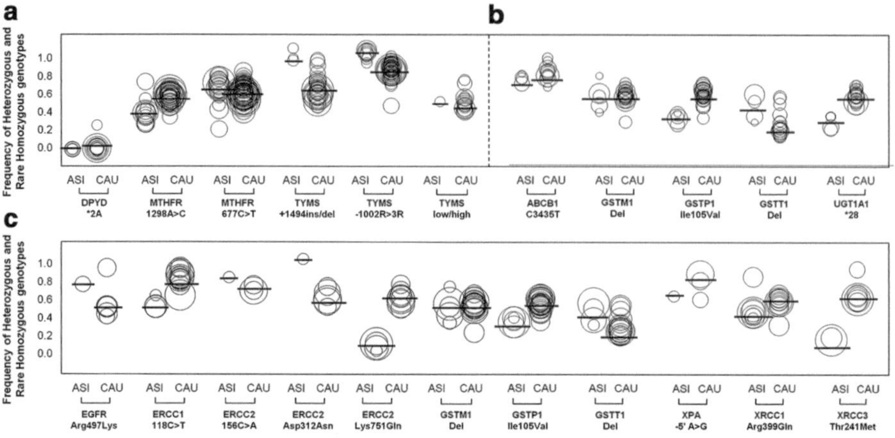

Fig. C.13 Population differences in chemotherapy outcomes. (Adapted from Loh et al. 2012) Genotype frequencies (sum of heterozygous and rare homozygous) for pharmacogenetic variants with respect to (*a*) 5-fluorouracil, (*b*) irinotecan, and (*c*) oxaliplatin. The center of the circle indicates the reported frequency and the size of the circle indicates the size of the study. Horizontal lines indicate the weighted average genotype frequency for the variant. Only variants previously investigated in more than one study in either of the ethnic populations were graphed. CA, Caucasian; EA, East Asian

(3) creation of microRNA biomarkers as cancer diagnostics, (4) optimization of circulating tumor cell assays for personalized medicine, and (5) discovery and validation of novel drug biomarkers in Asian cancers. His group has recently performed a meta-analysis and characterization of genotype frequency study to demonstrate that East Asian and Caucasian colorectal cancer patients differ significantly in the frequency of many pharmacogenetic variants and in related chemotherapy outcomes (Fig. C.13, Loh et al. 2012).

Translational Efforts

CSI Singapore has established a multidisciplinary program in Experimental Therapeutics in order to facilitate the development of targeted therapies in patients. The clinical faculty within the Centre of Translational Medicine have expertise in the design and administration of clinical trials, and are actively involved in partnering with pharmaceutical companies in order to pioneer clinical trials in East Asian patient populations.

Sources of Support

CSI Singapore was founded in 2008, made possible in part through a $172 million "Research Center of Excellence" grant. Physician-scientist teams are supported through the Ministry of Education.

Summary and Conclusions

CSI Singapore is ideally located in Southeast Asia to be the world leader in translational research focused on development of diagnostic platforms and therapeutic strategies for cancers that endemic in the Asian population, in particular gastric and liver cancer and certain leukemias.

References

Loh, M., Chua, D., Yao, Y., Soo, R. A., Garrett, K., Zeps, N., Platell, C., Minamoto, T., Kawakami, K., Iacopetta, B., Soong, R.: Can population differences in chemotherapy outcomes be inferred from differences in pharmacogenetic frequencies? Pharmacogenomics J. (2012) (Epub ahead of print June 26, 2012) doi:10.1038/tpj.2012.26

Pabst, T., Mueller, B.U., Zhang, P., Radomska, H.S., Narravula, S., Schnittger, S., Behre, G., Hiddemann, W., Tenen, D.G.: Dominant-negative mutations of CEBPA, encoding CCAAT/enhancer binding protein-alpha (C/EBPalpha), in acute myeloid leukemia. Nat. Genet. **27**(3), 263–270 (2001)

Rosenbauer, F., Tenen, D.G.: Transcription factors in myeloid development: balancing differentiation with transformation. Nat. Rev. Immunol. **7**, 105–117 (2007) doi:10.1038/nri2024

Thiery, J.P.: Epithelial-mesenchymal transitions in tumour progression. Nat. Rev. Cancer. **2**(6), 442–454 (2002)

East China University of Science and Technology (ECUST) [Workshop: Institute of Mechanics, CAS]

Site Address:	Meilong Road 130 200237 Shanghai, China Prof. Yongsheng Li http://www.ecust.edu.cn/s/2/t/31/main.htm
Date Visited:	June 5, 2013
WTEC Attendees:	Paul Janmey (report author), Daniel Fletcher, Cynthia Reinhart-King, Nastaran Kuhn, Xiaofeng Zhu, Patricia Foland
Host(s):	Mian Long Institute of Mechanics, Chinese Academy of Sciences China mlong@imech.ac.cn

Overview

The Long laboratory studies several aspects of mechanobiology including the mechanisms of adhesion mediated by transmembrane proteins that respond to shear stress and the influence of subphase topography and mechanics on mesenchymal cells.

Research and Development Activities

One project involves production of micropatterned surfaces with controlled mechanical properties to interrogate how mesenchymal cells react to these physical cues. An advance in this laboratory is to produce micro-fabricated polyacrylamide hydrogel substrates with different elastic moduli, and distinct topographies, in three dimensions to systematically test proliferation, morphology, adherent area, differentiation, and cytoskeletal re-organization of bone marrow-derived mesenchymal stem cells. The studies show that substrate stiffness or dimension is predominant in regulating cell proliferation by fostering cell growth on stiff, unevenly dimensioned substrates.

Controlling substrate elastic moduli promotes osteogenic or neuronal differentiation of rat bone marrow mesenchymal stem cells on stiff or soft substrates, respectively, whereas topography or dimension also plays a lesser role in directing cell differentiation. The experimental design enabling the separation of mechanical and topographical signals is shown in Fig. C.14 (Li et al. 2013b).

A second research interest involves elucidating the interplay between forces and biochemical kinetics in the control of neutrophil adhesion mediated by two different integrins, Mac-1 and LFA-1 expressed by neutrophils. Kinetic analysis shows different rates of engagement and dissociation of these integrin receptors (Li et al. 2013a), and molecular dynamics simulations show the effects of shear stress an stretch on the dissociation of the P-selectin/PSGL-1 complex (Kang et al. 2012).

Translational Efforts

Fundamental studies of mechanobiology and responses of cells to substrate stiffness are still relatively early in their development but the construction of novel soft substrate with defined [topography has potential application in improve methods for drug screening and other *in vitro* cell analysis.

Sources of Support

National Natural Science Foundation of China
National High-Technology Research and Development Program of China
National Key Basic Research Foundation of China
Knowledge Innovation Project of the Chinese Academy of Sciences

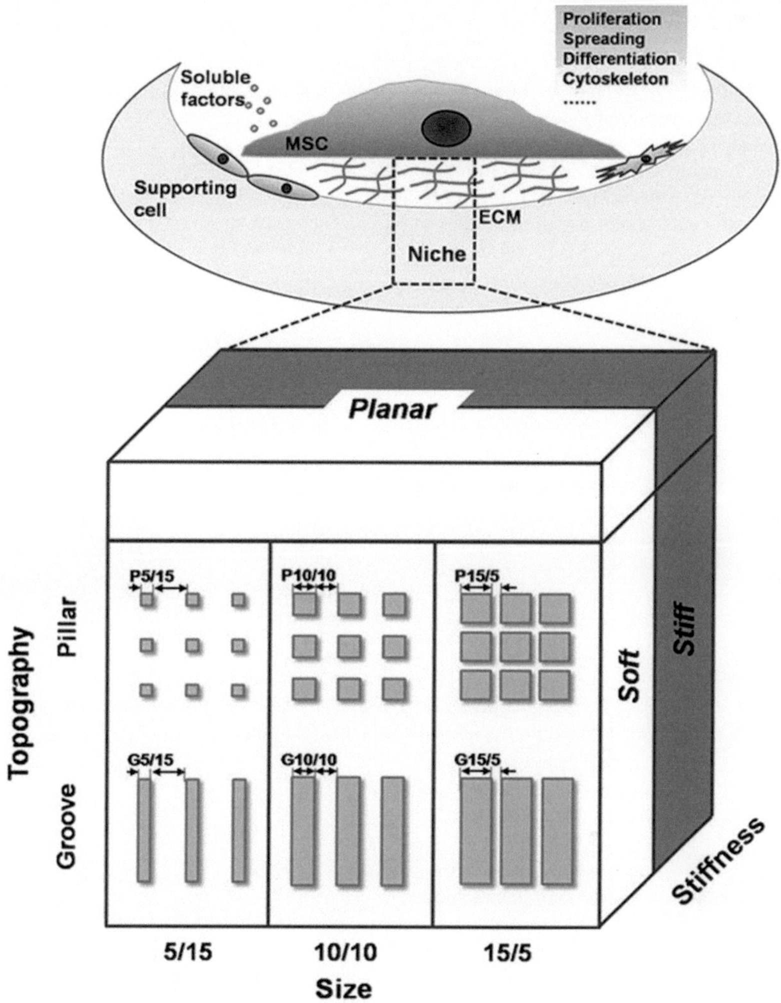

Fig. C.14 Schematic of experimental set-up for separation of mechanical and topographical signals (From Li et al. 2013b)
Micropatterning and altering formulation of hydrogels enables systematic testing of three regulating factors—stiffness, topography, and dimension. Planar substrates of varied stiffness are used as control

Summary and Conclusions

The work of this lab is developing new methods to study mechanobiology in a range of contexts and has produced important evidence for the relevance of such studies to understand normal and pathological ell response.

References

Kang, Y., Lu, S., Ren, P., Huo, B., Long, M.: Molecular dynamics simulation of shear- and stretch-induced dissociation of P-selectin/PSGL-1 complex. Biophys. J. **102**, 112–120 (2012)

Li, N., Mao, D. Lu, S., Tong, C., Zhang, Y., Long, M.: Distinct binding affinities of Mac-1 and LFA-1 in neutrophil activation. J. Immunol. **190**, 4371–4381 (2013a)

Li, Z., Gong, Y., Sun, S., Du, Y., Lu, D., Liu, X., Long, M.: Differential regulation of stiffness, topography, and dimension of substrates in rat mesenchymal stem cells. Biomaterials. **34**(31), 7616–7625 (2013b) doi:10.1016/j.biomaterials.2013.06.059. [Epub Jul 14, 2013]

East China University of Science and Technology (ECUST) [Workshop: National Cheng Kung University]

Site Address:	Meilong Road 130 200237 Shanghai China http://www.ecust.edu.cn/s/2/t/31/main.htm
Date Visited:	June 5, 2013
WTEC Attendees:	Paul Janmey (report author), Daniel Fletcher, Cynthia Reinhart-King, Nastaran Kuhn, Xiaofeng Zhu, Patricia Foland
Host(s):	Dr. Dar-Bin Shieh National Cheng Kung University Tainan 701, Taiwan Tel.: +886-6- 2353535, ext. 5410 Fax: +886-6- 2766626 dshieh@mail.ncku.edu.tw

Overview

The Shieh lab employs novel methods in biophotonics and nanoparticle delivery to improve methods of cancer cell detection *in vivo* and to design novel strategies for selective killing of malignant cells, with an emphasis on oral cancers.

Research and Development Activities

Several projects using nanomaterials and optical excitation explore potential applications of these technologies to cancer biology. One research interest is the development of in-cell gene scission at pre-designed sequence sites specific for mutations using Artificial Targeting Light Activated Nano Scissors (ATLANS) and a custom build photonic device (Tsai et al. 2010). Another project has demonstrate to the potential of multi-harmonic generation microscopy (Tsai et al. 2011) and Infrared microspectroscopy (Chiu et al. 2013) to aid in diagnosis of oral cancers. A third set of studies explores to potential of iron oxide nanoparticles for improved imaging (Rosen et al. 2012) and of iron-core gold shell nanoparticles to limit cancer-cell proliferation by a mechanism that involves selective mitochondrial damage to cancer cells, as shown in Fig. C.15 (Wu et al. 2011).

Translational Efforts

The research of the Shieh lab is highly translational in terms of applications to diagnostic and potential therapeutic treatments for oral cancers. Extensive collaborations between research and clinical groups have led to multiple patents.

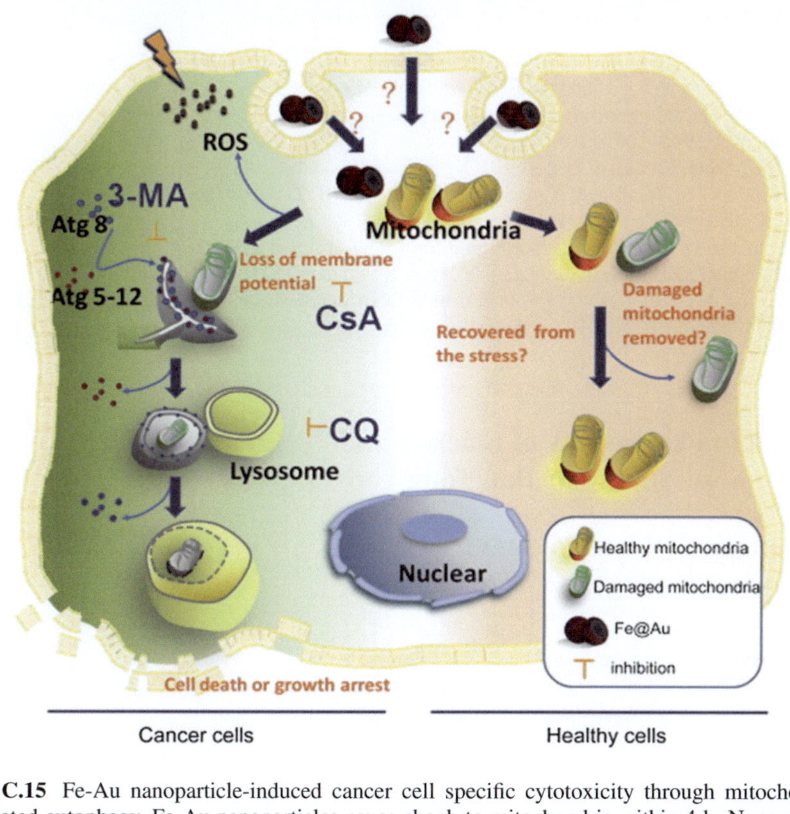

Fig. C.15 Fe-Au nanoparticle-induced cancer cell specific cytotoxicity through mitochondria mediated autophagy. Fe-Au nanoparticles cause shock to mitochondria within 4 h. Normal cells recover from the mitochondrial damage, but cancer cells undergo autophagy (From Wu et al. 2011)

Sources of Support

National Health Research Institute of Taiwan

National Taiwan University Research Center for Medical Excellence

National Nano Science and Technology Program funded by National Science Council (NSC), Taiwan.

Summary and Conclusions

This lab is actively involved in collaborations with several groups both within Taiwan and internationally to develop nanoparticle methods and optical spectroscopy and imaging directed at diagnosis and treatment of oral cancers. Combination of physical and material sciences with clinical medicine is an essential aspect of the research program.

References

Chiu, L.F., Huang, P.Y., Chiang, W.F., Wong, T.Y., Lin, S.H., Lee, Y.C., Shieh, D.B.: Oral cancer diagnostics based on infrared spectral markers and wax physisorption kinetics. Anal. Bioanal. Chem. **405**, 1995–2007 (2013)

Rosen, J.E., Chan, L., Shieh, D.B., Gu, F.X.: Iron oxide nanoparticles for targeted cancer imaging and diagnostics. Nanomed. Nanotechnol. Biol. Med. **8**, 275–290 (2012)

Tsai, M.R., Chen, S.Y., Shieh, D.B., Lou, P.J., Sun, C.K.: *In vivo* optical virtual biopsy of human oral mucosa with harmonic generation microscopy. Biomed. Opt. Exp. **2**, 2317–2328 (2011)

Tsai, T.L., Shieh, D.B., Yeh, C.S., Tzeng, Y., Htet, K., Chuang, K.S, Hwu, J.R., Su, W.C.: The down regulation of target genes by photo activated DNA nanoscissors. Biomaterials. **31**, 6545–6554 (2010)

East China University of Science and Technology (ECUST) [Workshop: Shanghai Jiao Tong University]

Site Address:	Meilong Road 130, 200237 Shanghai China http://www.ecust.edu.cn/s/2/t/31/main.htm
Date Visited:	June 5, 2013
WTEC Attendees:	Paul Janmey, Daniel Fletcher (report author), Cynthia Reinhart-King, Nastaran Kuhn, Xiaofeng Zhu, Patricia Foland

(continued)

Host(s):	Dr. Ping Ao Chang Jiang Distinguished Professor Room C203, Systems Biomedicine Building Shanghai Jiao Tong University Shanghai, 200240, China aoping@sjtu.edu.cn
	Prof. Yongsheng Li Key Laboratory for Ultrafine Materials of Ministry of Education School of Materials Science and Engineering, ECUST ysli@ecust.edu.cn

Overview

Dr. Ping Ao is the Chang Jiang Distinguished Professor at Shanghai Jiao Tong University, where he runs the Systems Biology Lab in the Shanghai Center for Systems Biomedicine. Trained as a theoretical physicist, Dr. Ao develops computational methods and theoretical frameworks for system biology. He brings a mathematical perspective to the complex behavior of biological systems, with particular attention to cancer and metabolism. Use of theories and approaches from non-equilibrium statistical dynamics are leading Dr. Ao to device a new description of biological systems called the endogenous molecular-cellular network hypothesis.

Research and Development Activities

Biological systems defy simple prediction, motivating extensive efforts to develop predictive models. Dr. Ao is tackling the challenge of developing predictive models of cancer progression, beginning with the formalisms of theoretical physics. Dr. Ao's efforts are aimed at creating the theoretical and computational tools that provide a framework for organizing, explaining, and predicting the behavior of tumors and other complex diseases. Based on experimental work that has identified oncogenes, tumor suppressors, and of other molecular components that are known to be linked to cancer progression, Dr. Ao and colleagues have put forward a concept they refer to as the endogenous molecular-cellular network hypothesis.

Dr. Ao's hypothesis is based on interactions between sets of modules of activity in cells (e.g., cell cycle, cell division) that can operate autonomously but interact with each other. According to Dr. Ao, "The interactions among these agents form an autonomous, nonlinear, stochastic, and collective dynamical network. We have tentatively named it as the endogenous molecular-cellular network." (Wang et al. 2013). An example of how this concept can be applied to genetic switching in Phage lambda is discussed in the paper and illustrated in Fig. C.16.

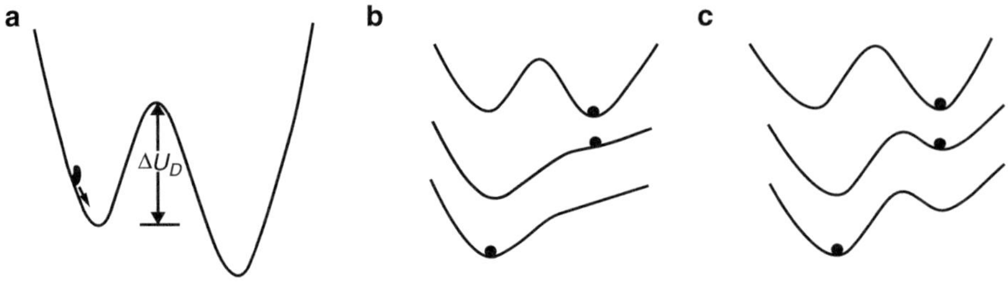

Fig. C.16 Schematic diagram of adaptive landscape of the phage lambda genetic switch, where the dynamic state of the biological system is represented as a *black dot* (From Wang et al. 2013) As the system progresses through different configurations, the dynamic state of the system may change or remain in a local minimum of the adaptive landscape. Wang et al. (2013) provide full details of the endogenous molecular-cellular network hypothesis and its applications

Translational Efforts

Dr. Ao has begun to apply this theoretical framework to prostate cancer and leukemia and initial results are encouraging. A method for predicting the behavior of cancers based on molecular details obtained from genetic or proteomic analysis of patient samples would have enormous impact, although the minimal input parameters needed for predictions based on the endogenous molecular-cellular network hypothesis are still to be determined.

Sources of Support

Dr. Ao is supported by the Shanghai Center for Systems Biomedicine.

Summary and Conclusions

Theoretical frameworks based on physical principles are needed to guide thinking about molecular and cellular systems and to make sense of the deluge of disease data now available. Dr. Ao's perspective on cancer as a non-equilibrium dynamic system is a useful step toward a more comprehensive model of cancer progression.

References

Ao, P., Galas, D., Hood, L., Zhu, X.: Cancer as robust intrinsic state of endogenous molecular-cellular network shaped by evolution. Med. Hypotheses. **70**, 678–684 (2008)

Wang, G.-W., Zhu, X.-M., Hood, L, Ao, P.: From Phage lambda to human cancer: endogenous molecular-cellular network hypothesis. Quant. Biol. **1**(1), 32–49, (2013). doi:10.1007/s40484-013-0007-1

Hong Kong Baptist University, Institute of Computational and Theoretical Studies (ICTS)

Site Address:	Kowloon Tong Kowloon Hong Kong http://www.icts.hkbu.edu.hk/	香港浸會大學 HONG KONG BAPTIST UNIVERSITY
Date Visited:	June 5, 2013	
WTEC Attendees:	Sharon Gerecht, Owen McCarty, Parag Mallick (report author), Sean Hanlon, Hassan Ali	
Host(s):	Professor Lei-Han Tang Director, Center for Quantitative Systems Biology Hong Kong Baptist University Tel.: (852) 3411-7031 lhtang@hkbu.edu.hk	
	Dr. Jian-Dong Huang University of Hong Kong Tel.: (852) 2819 2810 jdhuang@hku.hk	
	Dr. Quan Li Chinese University of Hong Kong Tel.: (852) 2609 6323 liquan@phy.cuhk.edu.hk	
	Dr. Jade Jue Shi Associate Director, Center for Quantitative Systems Biology Tel.: (852) 3411-7037 jshi@hkbu.edu.hk	
	Dr. Xue-Fei Li Post-doctoral fellow Center for Quantitative Systems Biology xuefeili@hkbu.edu.hk	
	Prof. Michel Van Hove Director and Chair Professor Institute of Computational and Theoretical Studies Hong Kong Baptist University Tel.: (852) 3411-2766 vanhove@hkbu.edu.hk	

Overview

Established in 1956, Hong Kong Baptist University has over 50 years of experience in providing broad-based and creativity-inspiring education. Within HKBU is the Institute of Computational and Theoretical Studies. The ICTS was established in

2011 to enhance inter-disciplinary research on computational and theoretical studies and attract local and overseas research groups for interdisciplinary and collaborative research. The objective was to create a critical mass that can conduct research of higher impact. Prior to the creation of the ICTS, HKBU had a number of smaller research centers and institutes studying everything from computer architectures to complex systems.

Functional Focus

The ICTS, together with its collaborating labs in sister institutions in Hong Kong, has a wide range of expertise spanning novel therapeutic strategies in cancer to investigations of nanoparticles to mathematical modeling.

Research and Development Activities

Jian-Dong Huang

The Huang lab at the Medical School, University of Hong Kong, is developing a novel therapeutic strategy in which synthetic biology approaches are used to create a novel tumor-targeting bacterium. Built upon a *Salmonella* backbone, they attempted to develop a bacterium that effectively targeted the tumor microenvironment, could deliver a toxic payload. To accomplish this targeting they put an essential gene under the control of oxygen sensitive promoter. Consequently, in presence of oxygen the gene is not expressed and the bacteria die. Particularly notable, because the bacteria was created by design and not by evolution, they were able to create bacteria titrated to survival at particular microenvironments. The bacteria showed excellent targeting and led to significant decreases in tumor burden. As part of the team, Dr. Xue-fei Li at the ICTS worked out a theoretical scheme that enables quantitative assessment of the environmental parameters that affect the efficacy of the treatment. This knowledge is used to fine-tune the synthetic circuits inside the engineered bacterium.

Quan Li

The Li group, in the Physics Department, The Chinese University of Hong Kong, is focused on nanoparticle-carrier based nanomedicine. The purposes of these studies were to enhance cellular uptake and to enable multifunctionality for both imaging and ablation. Some of the critical challenges faced include controlling of drug release, safety, and improvement of efficacy. To attack these challenges the Li group developed a novel silicon dioxide-therapeutic composite nanoparticle (Fig. C.17).

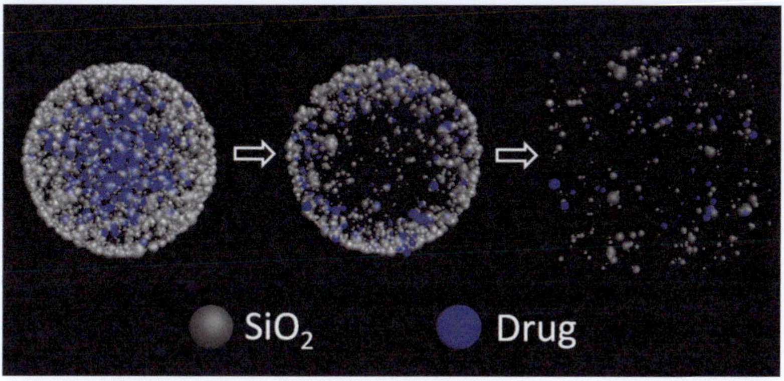

Fig. C.17 Schematic showing the drug release and carrier decomposition process of drug-loaded SiO_2 nanoparticles. (From Zhang et al. 2013)

These particles are natively decomposing and able to carry a therapeutic payload. Notably, the decomposition rate is controllable as a function of payload to SiO_2 ratio.

Jade Shi

The Shi lab aims to investigate drug response dynamics using a combination of live-cell imaging and mathematical modeling of pathway dynamics. One area that they looked at was cell type-dependent responses to anti-mitotic therapies (Fig. C.18). Notably, some cells die, whereas others escape arrest and continue to grow, but not divide. To investigate and overcome these challenges they investigated both mitotic arrest and apoptosis activation. They further attempted to interrogate where there is variance in apoptosis. They observed that BCL-XL and environment are major contributors. In an emerging area they are focused on cell-immune interactions.

Translational Efforts

Translational studies are done both locally and in collaboration with the mainland. It is clear that the focus of the studies is on clinically important problems.

Sources of Support

RGC, GRF and CRF grants; HKBU SDF grant for ICTS; RGC and HKBU grants for ICTS's High Performance Cluster Computing Centre.

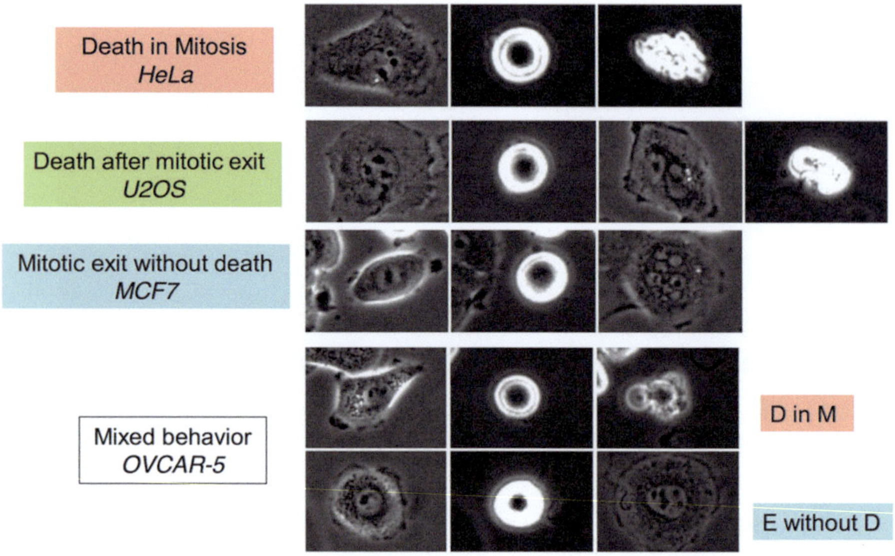

Fig. C.18 Cell type variation in response to anti-mitotic chemotherapeutics (Courtesy of J. Shi, HKBU)

Summary and Conclusions

Despite being under-resourced, ICTS is doing extremely exciting interdisciplinary research. They significantly interact with research groups from all over the world. Their close connection between mathematical modeling and experiment has them poised to have significant impact in the coming years.

References

Zhang, S., Chu, Z., Yin, C., Zhang, C., Lin, G., Li, Q.: Controllable drug release and simultaneously carrier decomposition of SiO_2-drug composite nanoparticles. J. Am. Chem. Soc. **135**(15), 5709–5716 (2013). doi:10.1021/ja3123015

Yu, B., Yang, M., Shi, L. Yao, Y., Jiang, Q., Li, X., Tang, L.-H., Zheng, B.-J., Yuen, K.-Y., Smith, D. K., Song, E., Huang, J.-D.: Explicit hypoxia targeting with tumor suppression by creating an "obligate" anaerobic Salmonella Typhimurium strain. Sci. Rep. **2**, 436 (2012). doi:10.1038/srep00436

Shi J, Zhou Y, Huang H.C., Mitchison T.J.: Navitoclax (ABT-263) accelerates apoptosis during drug-induced mitotic arrest by antagonizing Bcl-xL. Cancer Res. **71**(13), 4518–4526 (2011)

Institute of Molecular and Cell Biology, A*STAR

Site Address:	Proteos Building 61 Biopolis Drive Singapore, 138673 http://imcb.a-star.edu.sg/php/main.php
Date Visited:	June 3, 2013
WTEC Attendees:	Paul Janmey, Owen McCarthy, Cynthia Reinhart-King, Sharon Gerecht (report author), Nastaran Kuhn, Larry Nagahara, Patricia Foland
Host(s):	Prof. Wanjin Hong Executive Director, IMCB Protein Trafficking and Cancer Cell Biology mcbhwj@imcb.a-star.edu.sg
	Prof. Vinay Tergaonkar Senior Principal Investigator NFkB Signaling in Human Ailments vinayt@imcb.a-star.edu.sg
	Prof. Philipp Kaldis Senior Principal Investigator Cell Division and Cancer Research kaldis@imcb.a-star.edu.sg
	Prof. Uttam Surana Research Director Regulation of the Cell Division Cycle mcbucs@imcb.a-star.edu.sg
	Dr. Frederic Bard Senior Principal Investigator Regulation of Membrane Traffic fbard@imcb.a-star.edu.sg
	Dr. Thomas Leung Principal Investigator Targeting the RhoGTPase Effector kinases Unit mcbthoml@imcb.a-star.edu.sg
	Dr. Ivan Khang Nyee Tan Senior Research Fellow Targeting the RhoGTPase Effector kinases Unit kntan@imcb.a-star.edu.sg

Overview

The Institute of Molecular and Cell Biology (IMCB) was established in 1987 at the National University of Singapore (NUS) before becoming an autonomous research institute (RI) of the Agency for Science Technology and Research (A*STAR) and moving to Biopolis in 2004. Its mission is to develop and foster a vibrant research culture for cutting-edge biomedical sciences and for training high-quality Ph.D. students for the flourishing biotechnology and pharmaceutical industries in Singapore.

IMCB's research activities focus on four major fields:

- Animal Models of Development and Disease
- Cancer Genetics and Therapeutics
- Cell Biology in Health and Disease
- Structural Biology and Drug Discovery.

IMCB has numerous collaborations with industrial, translational clinical and academic partners both in Singapore and worldwide. The organizational structure of the IMCB is shown on the next page.

Research and Development Activities

Prof. Wanjin Hong, Executive Director

Professor Hong gave an overview about A*STAR and IMCB, research disciplines, and funding resources. From 2001 to 2004, Professor Hong was Acting Director of IMCB. During his directorship he managed the merger of IMCB with the Institute of Molecular Agrobiology (IMA) and oversaw the relocation of an expanded IMCB from the NUS campus to the Biopolis. Professor Hong was re-appointed the Executive Director of IMCB in November 2011. His current focus is to reposition IMCB by sustaining mechanistic basic research for novel discoveries in the area of cancer, infectious, metabolic, and neurological diseases as well as by increasing targeted translational research to enhance the value of IMCB's discoveries. We met with several investigators in IMCB and heard about their interest areas for research.

Vinay Tergaonkar, Ph.D., Senior Principal Investigator, Cancer Genetics and Therapeutics

Dr. Tergaonkar's work focuses on NFkB Signaling in Human Ailments using mainly the mouse model. There is some interest to look into the activation of NFkB by shear flow. Quantitative measurements focus mainly on genomics where they measured the 3D interactions of transcription binding (through 3C or ChipExo).

Philipp Kaldis, Ph.D., Senior Principal Investigator, Cancer Genetics and Therapeutics

The Kaldis laboratory studies cancer from the perspective of genetics using knockout mice specifically for cell cycle regulation (e.g., cell cycle dependent kinases). Interdisciplinary research in Kaldis's lab uses MRI and NMR technologies to look at liver metabolism in mice. Towards this, the Kaldis group is working with biophysicists at SBIC with expertise in metabolism and flux measurements.

Uttam Surana, Research Director, Cancer Genetics and Therapeutics and Chairman of the Graduate Program at IMCB

Dr. Surana studies the checkpoint controls related to genomic stability. Specifically, the group is interested in the mechanism underlying recovery after cell cycle arrest following DNA damage or the disruption of spindle-chromosome attachment. Another interest is antimitotic drugs in cancer treatment and looking at checkpoint controls in this context. Currently, they are also designing genetic screens to identify small molecule inhibitors against cell cycle regulators. Dr. Surana's interdisciplinary team is internal to IMCB.

Institute of Molecular and Cell Biology (IMCB)

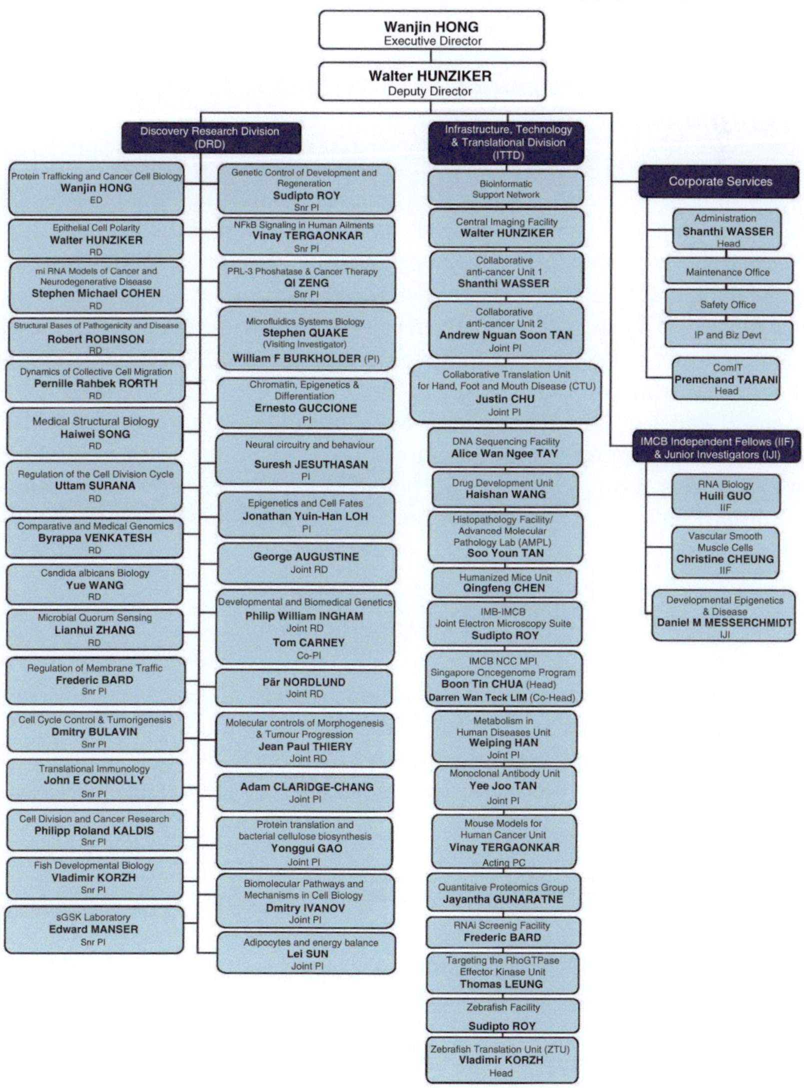

Frederic Bard, Senior Principal Investigator, Cell Biology in Health and Disease

The goal of Dr. Bard's group is to understand how trafficking of membrane-bound structures is regulated to mediate various cellular functions. Specifically, the group focuses on studying how trafficking regulation at the Golgi complex affects glycosylation in health and disease and how intracellular trafficking is exploited by pathogens and toxins. In a recent paper, the group has developed a quantitative morphological assay using three different Golgi compartment markers and quantitative image analysis, combined with a kinome- and phosphatome-wide RNAi screen (Fig. C.19). Using this approach they have provided a genetic overview of the signaling pathways that control the Golgi apparatus in human cells.

Thomas Leung, Ph.D., Principal Investigator, Technology and Translational Unit

We met with Dr. Leung and a Senior Research Fellow in his group, Dr. Ivan Khang NeeTan. The main interest of the group is the identification of the "targets" for Rho proteins and the understanding of their specific roles. Two of these Rho targets, Rho kinase ROK and Cdc42/Rac kinase MRCK, play distinct roles in regulating the actomyosin assembly in adherent cells in culture. In a recent publication, the group has identified of chelerythrine chloride as a specific MRCK inhibitor (Tan et al. 2011). Chelerythrine chloride's ability to block cellular activity of MRCK resulted in the specific loss of NM II-associated MLC phosphorylation in the lamella, and the consequential suppression of cell migration. Overall, the work in Dr. Leung's group is aimed at setting the stage for potential therapeutic interventions in such diverse diseases as cancer and neurodegenerative disorders.

Translational Efforts

Within Singapore there are no apparent barriers between collaborating with investigators outside of A*STAR (e.g., at the MBI). However, clinicians have very little time to do research, which makes translational research more challenging. A recent trend to fund translational work may help move research toward this direction.

Sources of Support

IMCB researcher apply for money through A*STAR—an intramural research program through the Ministry of Trade and Industry. As such, they are encouraged to undertake interdisciplinary research with other A*STAR institutions. For example, Dr. Kaldis's interdisciplinary work is funded by the Joint Council Office ($1 M over 3 years). In addition, researchers apply for external grants.

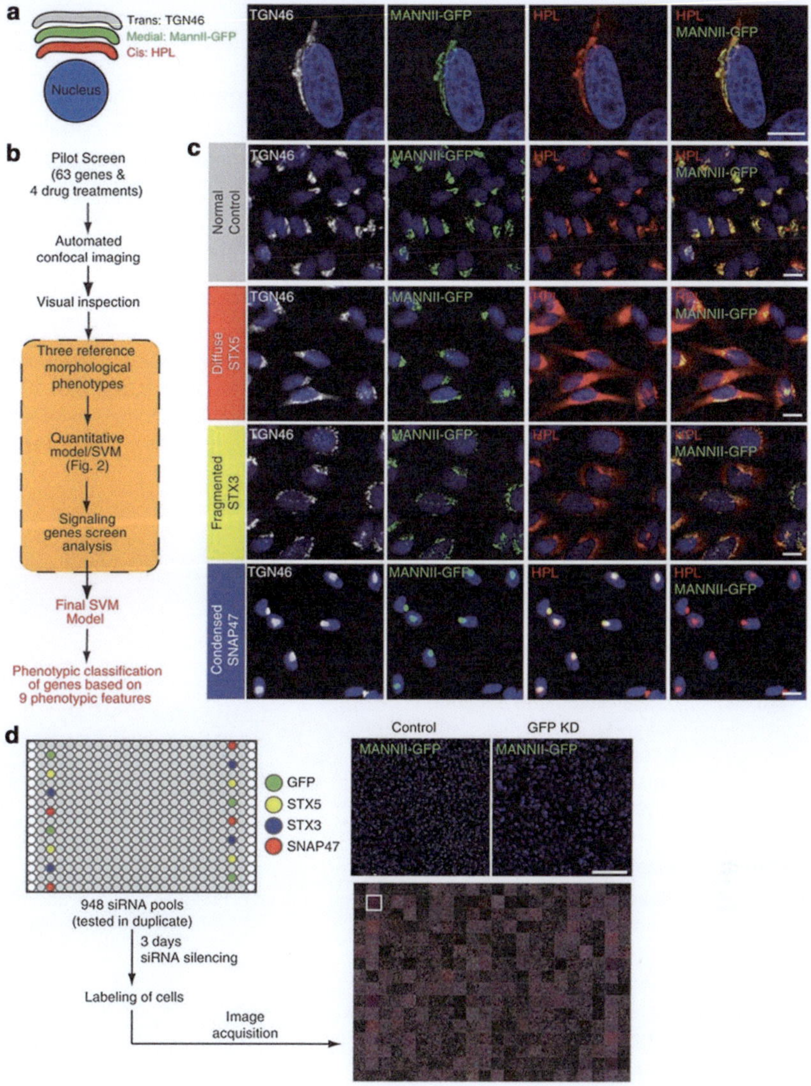

Fig. C.19 An imaging-based screen to identify Golgi organization phenotypes (From Chia et al. 2012)

(*A*) HeLa MannII-GFP (medial Golgi) cells were stained with cis Golgi marker HPL and trans Golgi marker TGN46. Compartments co-localized extensively even at × 100 magnification. Scale bar: 10 μm. (*B*) Schematic overview of the screening process. A pilot screen of 63 genes and 4 drug treatments was imaged using a × 20 objective and visually screened for changes in Golgi organization. Three Golgi phenotypes (diffuse, fragmented and condensed) were identified and used to train a preliminary Support Vector Machine (SVM) for quantitative scoring of treatments. Images of selected genes from the signaling genes screen were used to refine SVM training and obtain a final score. (*C*) Examples of the three reference phenotypes. STX5 knockdown induces a diffuse phenotype specifically for the cis Golgi while STX3 and SNAP47 knockdown induces a fragmented and condensed Golgi in all three compartments, respectively. Scale bar: 30 μm. (*D*) Workflow of the siRNA screen. Screen plates were loaded with controls for the three phenotypes for quality control in each plate. GFP knockdown and STX5 knockdown demonstrate homogeneous gene depletions in all wells seeded with the siRNAs. Scale bar: 200 μm

Summary and Conclusions

IMCB is the oldest institute in Singapore that focuses on doing the best science without a uniform mission. Recently, administration is encouraging collaboration among the different entities under A*STAR as well as translational research, which helps IMCB researchers expand their programs.

References

Chia, J., Goh, G., Racine, V., Ng, S., Kumar, P., Bard, F.: RNAi screening reveals a large signaling network controlling the Golgi apparatus in human cells. Mol. Syst. Biol. **8**, 629 (2012)

Tan, I., Lai, J., Yong, J., Li, S., Leung, T.: Chelerythrine perturbs lamellar actomyosin by selective inhibition of Myotonic Dystrophy kinase-related Cdc42-binding kinase. FEBS Lett. **585**, 1260–1268 (2011)

http://www.imcb.a-star.edu.sg/php/main.php

Kyoto University, Center for iPS Cell Research and Application

Site Address:	53 Kawahara-cho Shogoin, Sakyo-ku Kyoto 606-8507 Japan http://www.cira.kyoto-u.ac.jp/e/index.html	Center for iPS Cell Research and Application CiRA Kyoto University
Date Visited:	June 7, 2013	
WTEC Attendees:	Sharon Gerecht (report author), Owen McCarty, Parag Mallick, Sean Hanlon, Hassan Ali	
Host(s):	Koji Eto, M.D., Ph.D. Principal Investigator Department of Clinical Applications eto-g@cira.kyoto-u.ac.jp	
	Dr. Kazutoshi Takahashi Principal Investigator Department of Reprogramming Science ktakahashi-g@cira.kyoto-u.ac.jp	

Overview

The Center for iPS Cell Research and Application (CiRA) was established in 2008 under the auspices of the Institute of Integrated Cell-Materials Science (ICeMS). In 2009 Kyoto University created the iPS cell Research Fund. In 2010 the new building was completed and CiRA became an independent institute at Kyoto University led by Nobel Laureate Shinya Yamanaka.

The mission of the CiRA is to: (1) serve as the world-first core institute dedicated to leading iPS cell research; (2) pursue the possibilities of iPS cell through both fundamental and applied research with the goal of contributing to the development of regenerative medicine; and (3) maintain close ties with Kyoto University's Institute for Integrated Cell-Material Sciences, Graduate School of Medicine, and University Hospital to promote research collaborations and the cultivation and exchange of young scientists.

There are four departments and one division within the Institute:

- Department of Reprogramming Science
- Department of Cell growth and Differentiation
- Department of Clinical Application
- Department of Fundamental Cell Technology
- Uehiro Research Division for iPS Cell of Ethics.

The organizational structure of the center (as of July 2013) is shown below:

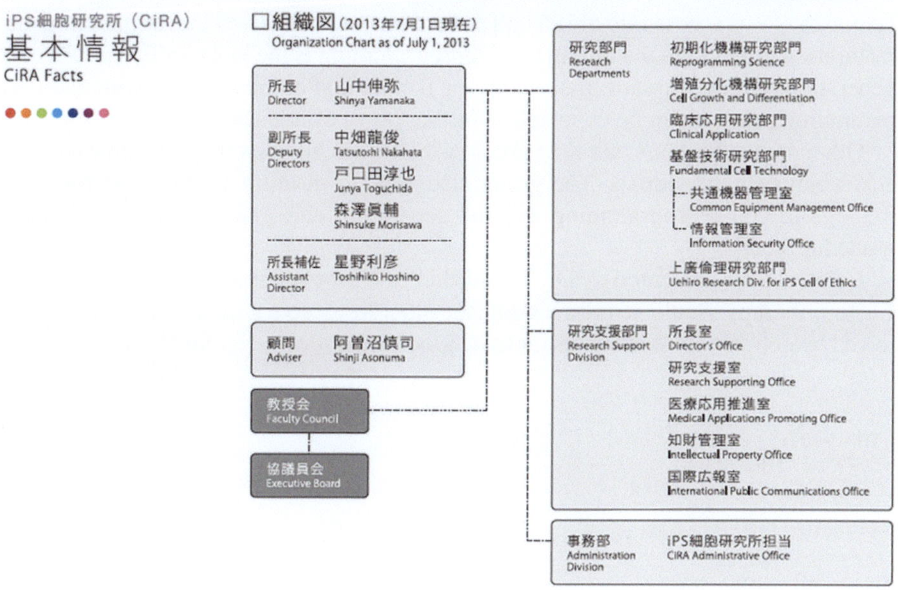

Research and Development Activities

Dr. Kazutoshi Takahashi, Department of Reprogramming Science

Dr. Takahashi group uses molecular and cell biological techniques to gain a better understanding of mechanisms underlying the pluripotent state. In a recent study, the group has demonstrated that a major obstacle to the creation of iPS cells lies in the maturation stage of the reprogramming process (Tanabe et al. 2013). Specifically, it was found that the success rate of cells in which reprogramming had been initiated (i.e., TRA-1-60+ cells) is only around 0.2 %. A detailed analysis of TRA-1-60-positive cells indicated that 75 % or more of cells had reverted to their state as of before the initiation of reprogramming and thus failed to become iPS cells. It was also found that, LIN28, which is known to improve reprogramming efficiency, prevents this reversion to the somatic cell state and improves iPS cell generation efficiency. Figure C.20 shows a model of the reprogramming process.

The findings of this research elucidated an important area of the mechanism behind the low success rate of iPS cell generation and suggested that stimulating the maturation rather than the initiation of reprogramming is the key to efficient iPS cell generation. Going forward, the search for factors and compounds to stimulate the maturation process can be expected to further boost efficiency.

Other areas of interest are the variations between cell lines that arise from differences between individuals. The group also aims to develop applications for their findings in cell reprogramming and the generation and maintenance of clinical–grade hiPS cells.

Related issue presented by Dr. Takahashi is the time required to generate therapy. Custom therapy would require significant time: to receive cells from a patient followed by generation of iPS cells (at least 3 months) followed by hiPS cell expansion

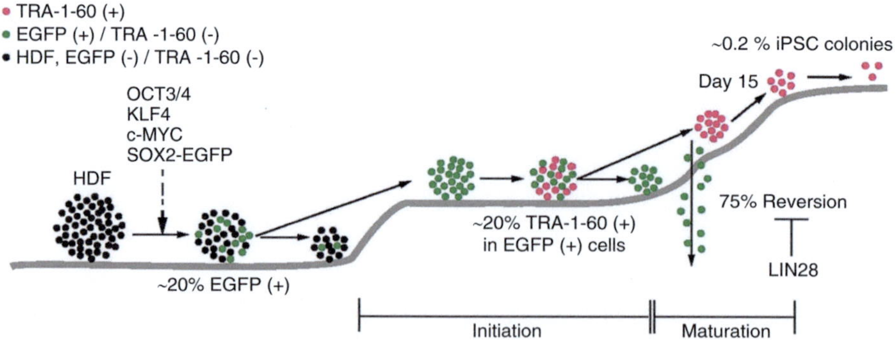

Fig. C.20 Model of the reprogramming process (From Tanabe et al. 2013).
Black dots, fibroblast or nontransduced cells; green dots, transduced cells with SOX2-IRES-EGFP; magenta dots, TRA-1-60 (+) cells

and differentiation (at least 3 months). How can we overcome this time requirement (6+ months), which is critical for patients? One approach is to generate a bank of hiPS cells of a variety of genetic backgrounds using HLA matching. Five HLA-homozygous donors will cover 30 % of Japanese people; ≈140 HLA-homozygous donors will cover 90 % of Japanese people.

Dr. Koji Eto, Department of Clinical Application

Dr. Eto's research group focuses on the basic research *in vitro* generation of blood cells, including hematopoietic stem cells (HSCs), platelets, or red blood cells from hiPS cells, aiming toward clinical applications. The objective is to develop a safe and stable blood supply for transfusion independently of blood donation and gene therapy. As an example, Dr. Eto described the need for platelets from hiPS cells. There is a constant demand for platelet supply because platelets cannot be stored frozen and have a short shelf life. From a safety point of view, anucleate cells (platelets, RBCs) can be gamma-irradiated to prevent teratoma formation. In the differentiation scheme there are two major roadblocks: (1) forming megakaryocytes from HSCs and (2) forming platelets from megakaryocytes (Fig. C.21).

While work on differentiating HSC to megakaryocytes is ongoing, progress has been made towards differentiating megakaryocytes into platelets. The group has shown that hiPS cells differentiate to platelets through "iPS-Sac" (Takayama et al. 2010) and more recently, using advanced *in vivo* imaging technology, that megakaryocytes generate platelets in various patterns including both through proplatelets and rapture mechanism (Nakamura et al. 2014).

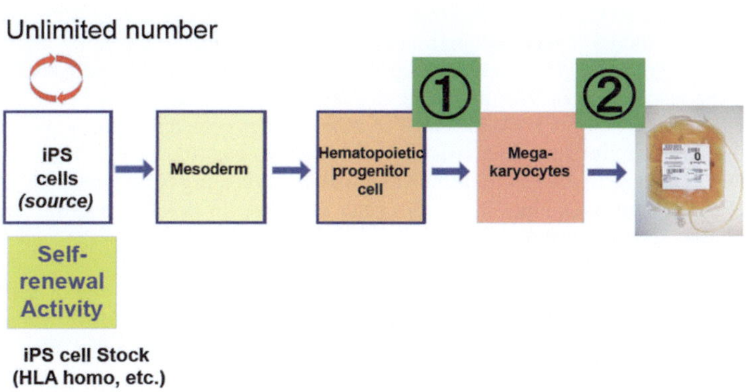

Fig. C.21 Dr. Eto's approach for platelet generation from hiPS cells (Unpublished; courtesy of Dr. K. Eto and colleagues)

A recent study in Dr. Eto's laboratory demonstrated an increase in the yield of platelet differentiation from hiPS cell-derived megakaryocytes using a two-dimensional flow culture system generated using sheetlike scaffold and medium circulation (Fig. C.22, Nakagawa et al. 2013). This biomimetic artificial blood vessel system, in which two flows in different directions with pressure and shear stress levels are applied, promoted platelet generation from megakaryocytes. These findings indicate that the two-dimensional flow culture system may be a feasible for *in vitro* generation of large numbers of platelets from hiPS cells, sufficient for transfusion therapy.

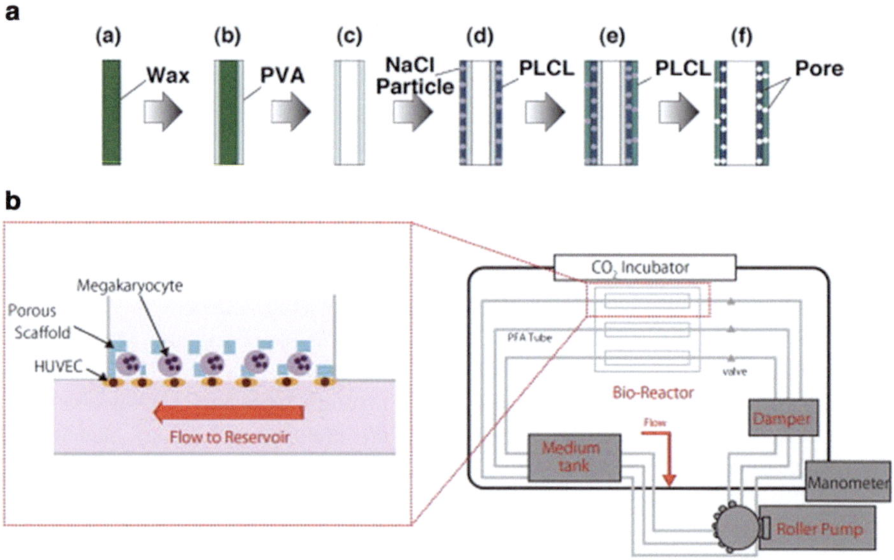

Fig. C.22 Fabrication process of a sheetlike scaffold and medium circulation model of bioreactor. (From Nakagawa 2013)
(*A*) (*a*) Produce a wax (toluene sulfonamide) model. (*b*) Dip the wax model into polyvinyl alcohol (PVA) solution and pull it upward with a dip coater (pull-up velocity: 1 mm/s). The dip coating of PVA onto the wax model is repeated six times, and each coating is followed by drying at room temperature for 30 min. (*c*) After an interval of 12 h, soak the PVA-coated wax model in acetone. Because of the insolubility of PVA and solubility of wax in acetone, the wax is selectively dissolved, leaving the structure of PVA. (*d*) The PVA model is then coated with the polymer and NaCl particle solution using the dip coater (velocity: 3 mm/s). To coat the poly-L-lactide-co-epsilon-caprolactone (PLCL) membrane as uniformly as possible, dip coating of polymer solution onto the PVA model is repeated three times in one direction and three times in the opposite direction by setting the model upside down. (*e*) Dip coating PLCL of the second layer in the same manner as (*d*). (*f*) Each coating is followed by drying at room temperature for 2 min. The PLCL-coated PVA models are then soaked in deionized (DI) water, followed by dissolution of PVA and the NaCl microparticles. After PVA and the NaCl microparticles are completely removed from the PLCL, and the salt-leached PLCL scaffold is dried for 24 h at room temperature. (*B*) Culture within the reactor system was performed at 37 °C in a CO_2 incubator. Circulation was achieved using a PFA tube connected to a roller pump to produce flow. The schematic at left shows the cross-sectional view of the bioreactor. HUVECs were preapplied for stable adhesion to the micropores in the scaffold 2 days before the MKs were applied

Translational Efforts

CiRA's main thrust is the development of hiPS cell technology for clinical use. The goals for the center's first 10 years are:

- Establish basic iPS cell technology and secure the associated intellectual property rights
- Build a stock of iPS cells for use in regenerative medicine.
- Carry out preclinical studies and work toward clinical studies
- Contribute to the development of therapeutic drugs using patient derived iPS cells.

Sources of Support

The fiscal budget implementation as of March 31, 2013 shows the following distribution of the ¥46 billion budget for CiRA:

- Other public research grants, 51 %
- FIRST grant (Cabinet Office), 28 %
- Grants-in-Aid for Scientific Research, 9 %
- Basic operating funds, 8 %
- Private sector grants, 3 %
- iPS Cell Research Fund (Donation), 1 %

Summary and Conclusions

CiRA is a leading institute in basic and translational research of hiPSCs. Led by Nobel Laureate Shinya Yamanaka, the institute's expertise extends from reprogramming to upscale and clinical utilization. The institute's activities are highly supported by the government and Kyoto University. The members are interacting with U.S. government (e.g., workshops and study sections) and academia on a regular basis.

References

Nakagawa, Y., Nakamura, S., Nakajima, M., Endo, H., Dohda, T., Takayama, N., Nakauchi, H., Arai, F., Fukuda, T., Eto, K.: Two differential flows in a bioreactor promoted platelet generation from human pluripotent stem cell-derived megakaryocytes. Exp. Hematol. **41**(8), 742–748 (2013). doi:10.1016/j.exphem.2013.04.007. [Epub ahead of print Apr 22, 2013]

Nakamura, S., Takayama, N., Hirata, S., Seo, H., Endo, H., Ochi, K., Fujita, K.I., Koike, T., Harimoto, K.I., Dohda, T., Watanabe, A., Okita, K., Takahashi, N., Sawaguchi, A., Yamanaka, S., Nakauchi, H., Nishimura, S., Eto, K. 2014. Expandable megakaryocyte cell lines enable clinically applicable generation of platelets from human induced pluripotent stem cells. Cell Stem Cell. doi:10.1016/j.stem.2014.01.011. [Epub ahead of print 13 Feb 2014]

Takayama, N., Nishimura, S., Nakamura, S., Shimizu, T., Ohnishi, R., Endo, H., Yamaguchi, T., Otsu, M., Nishimura, K., Nakanishi, M., Sawaguchi, A., Nagai, R., Takahashi, K., Yamanaka, S., Nakauchi, H., Eto, K.: Transient activation of c-MYC expression is critical for efficient platelet generation from human induced pluripotent stem cells. J. Exp. Med. **207**, 2817–2830 (2010)

Tanabe, K., Nakamura, M., Narita, M., Takahashi, K., Yamanaka, S.: Maturation, not initiation, is the major roadblock during reprogramming toward pluripotency from human fibroblasts. Proc. Natl. Acad. Sci. USA. **110**(30), 12172–12179 (2013). doi:10.1073/pnas.1310291110. (Epub ahead of print Jun 28, 2013)

http://www.cira.kyoto-u.ac.jp/j/pressrelease/pdf/CiRA_booklet_20130712.pdf

Kyoto University, Laboratory of Bioimaging and Cell Signaling

Site Address:	Yoshida-Konoe-cho, Sakyo-ku Kyoto Japan
Date Visited:	June 7, 2013
WTEC Attendees:	Sharon Gerecht, Owen McCarty, Parag Mallick (report author), Sean Hanlon, Hassan Ali
Host(s):	Dr. Kazuhiro Aoki Associate Professor, Laboratory of Bioimaging Graduate School of Biostudies Tel.: 81-75-753-9450 Fax: 81-75-753-4698 k-aoki@lif.kyoto-u.ac.jp
	Dr. Yasuhiro Inoue Associate Professor, Department of Biomechanics, Research Center for Nano Medical Engineering, Institute for Frontier Medical Sciences Tel.: 81-75-751-4125 Fax: 81-75-751-4125 inoue@frontier.kyoto-u.ac.jp
	Dr. Naoki Honda Associate Professor Imaging Platform for Spatio-Temporal Information Graduate School of Medicine Tel.: 81-75-753-4422 Fax: 81-75-753-4698 n-honda@sys.i.kyoto-u.ac.jp

Overview

The Graduate School of Biostudies was established in 1999, to engage in 21st Century Life Science as a comprehensive science. The school focuses on both research and education. The three goals of the graduate school are (1) provide education for pursuing the new biostudies at the world's top level, (2) train individuals to apply the new life sciences for the protection of the global environment and for human welfare, and (3) nurture individuals who can understand the various vital phenomena of the living organisms as a systemic function, and pursue these systemic functions.

Within the graduate school of biostudies is the Department of Molecular and System Biology, which is composed of the Laboratory of Neuroscience, the Laboratory of Immunology and Cell Biology and the Laboratory of Bioimaging and Cell Signaling directed by Prof. Michiyuki Matsuda. The Lab of Bioimaging and Cell Signalling is focused on visualizing the growth signal transduction cascades in living. In general, there is a significant focus on dynamic living systems based on multi-dimensional quantitative imaging and mathematical modeling. The program involves investigators from a number of programs (Graduate Schools of Medicine, Biostudies, and Informatics, Institute for Frontier Medical Sciences, Institute for Virus Research, and Imaging Platform for Spatio-Temporal Information). These investigators are conveniently co-located.

Functional Focus

Characterizing dynamic processes in cellular regulation.

Research and Development Activities

Kazuhiro Aoki

The focus of the Aoki group is on understanding cancer as a system, and using that information to help control cancer. They have a unique pipeline that integrates exceptional experiments to measure Kd *in vivo*, modeling and simulation to interrogate the EGFR-Ras-ERK pathway that is at the center of diverse cancers. The two major challenges they identified were the measurement of quantitative kinetic parameters and the establishment of a link between signaling and phenotype. Using fluorescence cross-correlation spectroscopy they were able to measure Kd values that differed from those measured *in vitro* because the Kds for any given interaction is affected by all other interactions for the two partners (as well as other factors such as crowding, local concentration, salt, etc.). Among the emergent behaviors they

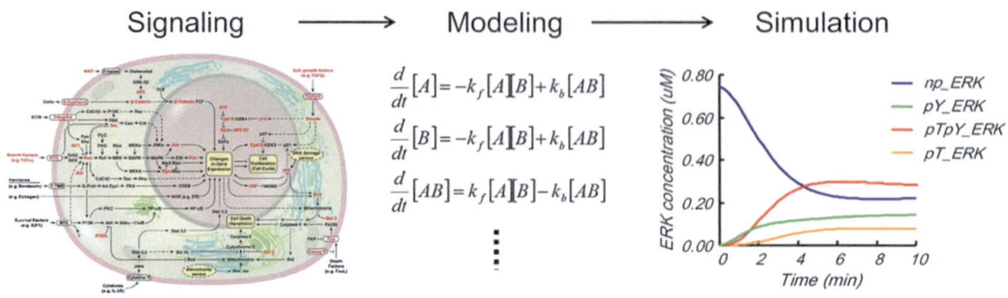

Fig. C.23 A schematic of quantitative biology for cell signaling (Courtesy of K. Aoki, Kyoto University)

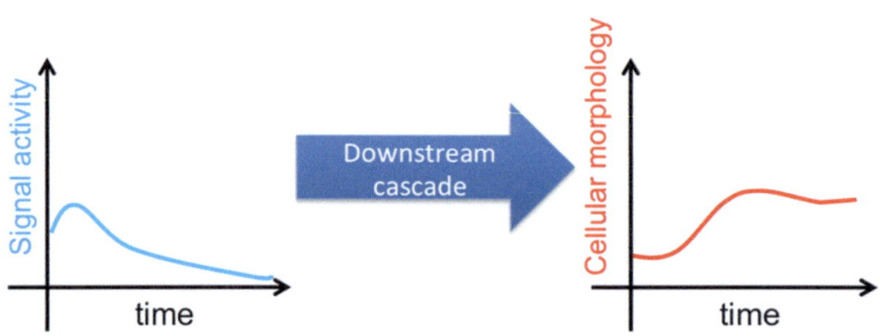

Fig. C.24 Transfer function of the downstream cascade (Courtesy of N. Honda, Kyoto University)

were able to observe were cell density-dependent proliferation rates that were driven by ERK signaling frequency rather than amplitude. To perform this study, they developed a FRET biosensor for ERK, which revealed complex dynamics (Fig. C.23).

Naoki Honda

The Honda group is focused on predictive biology. In particular, they seek to identify the dynamic control laws driving morphodynamics by the Rho family of small GTPases. The questions they are investigating include:

- How do the GTPases regulate cellular morphological changes?
- Does Rac1 regulate morphological change?
- Does Rac1 signal contain the information of morphological change?
- How is signal transmitted by the Rho GTPases to morphological changes via downstream cascade (Fig. C.24)?

Their approach is essentially to uncover the transfer function mapping signal activity to phenotype. They begin with detailed FRET imaging of Rac, followed by extensive imaging processing and quantification. They then pose the problem as a system identification problem and use the solutions for prediction.

Beyond this area, Dr. Honda is exploring a number of diverse topics in mapping the relationship between molecular behavior and phenotype.

Yasuhiro Inoue

Dr. Inoue is focused on the development of multiscale models to describe mechanobiology (Fig. C.25). In these studies, he focuses on the question: How do local forces and organization (molecule scale) ensure proper global organization (organ scale)? To answer this question, he models multi-cellular structures as aggregates of cells with formal cell-cell boundaries. These are specifically modeled as a network of polyhedrons with formal mechanical behavior. Simulations using the models demonstrated that the mechanical properties of environment (e.g., viscous versus elastic) significantly impact gross-structure.

Translational Efforts

The focus of the center is very much on basic studies with clinical relevance, rather than on translation. They noted that there was not much demand for translation.

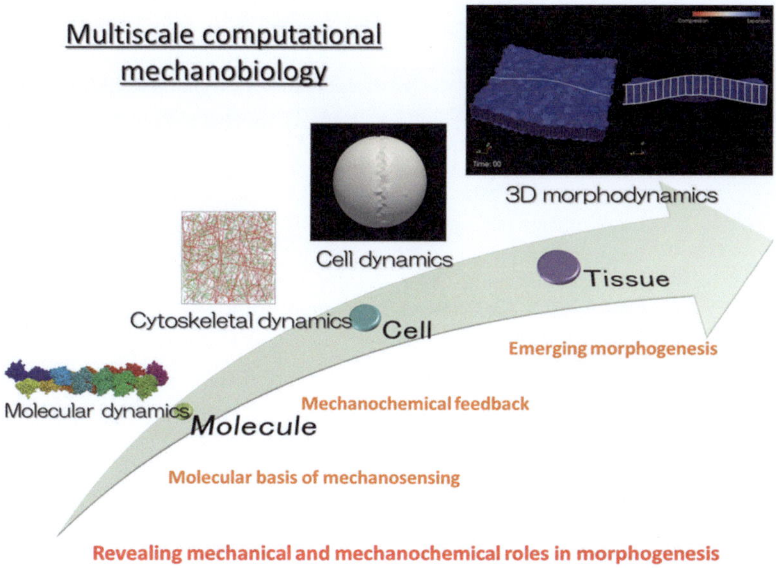

Fig. C.25 Multiscale computational mechanobiology of epithelial tissue morphogenesis (Courtesy of Y. Inoue, Kyoto University)

Sources of Support

Funding for the project comes primarily from the Ministry of Education, Culture, Sports, Science and Technology-Japan (MEXT). The structure allows for more flexibility as one does not need to justify one's project to the entire scientific community—only to the senior group leader. This structure allows for greater incubation of long-term projects.

Summary and Conclusions

Kyoto University is at the forefront of integrating novel physical sciences with oncology. They take a long view on many problems and take the time to do things extremely rigorously. Their focus on clinically important basic science is a notable strength.

Nanyang Technological University – School of Biological Sciences

Site Address:	SBS-01 s 45c 60 Nanyang Drive Singapore 637551 http://www.ntu.edu.sg/Pages/index.aspx
Date Visited:	June 4, 2013
WTEC Attendees:	Paul Janmey, Owen McCarty (report author), Cynthia Reinhart-King, Sharon Gerecht, Nastaran Kuhn, Larry Nagahara, Patricia Foland
Host(s):	Prof. Lars Nordenskiold Associate Dean (College of Science) Tel.: (65) 6592-7506 LarsNor@ntu.edu.sg
	Peter Preiser, Ph.D. Professor and acting Chair Director, CN Yan Scholars Programme Tel.: (65) 6316-2822 prpreiser@ntu.edu.sg

Overview

The mission of the Nanyang Technological University (NTU) School of Biological Sciences is to train students in life sciences in order to develop a talent pool of researchers for the life sciences industry. The School of Biological Sciences was established in 2002, and occupies 30,000 m2 total floor space comprised of state-of-the-art classrooms, 60 research laboratories and core instrumentation facilities within the NTU campus.

The School of Biological Sciences is a member of the College of Science, and has two main research divisions: Structural Biology and Biochemistry, led by Dr. Joe Toon, and Molecular Genetics and Cell Biology, led by Dr. Richard Sugrue. Dr. Peter Preiser is the Acting Chair of the School of Biological Sciences.

The School of Biological Sciences offers a Bachelor of Science degree, as well as a unique dual degree program that combines the Bachelor of Science with a Bachelor of Chinese Medicine offered in conjunction with the Beijing University of Chinese Medicine.

Research and Development Activities

Research within the School of Biological Sciences ranges across the fields of peptide chemistry, genetics, drug discovery, infectious disease, protein engineering and environmental biotechnology.

Dr. Julien Lescar

The most prevalent mosquito-borne viral pathogen that infects humans is Dengue virus, from genus Flavivirus in the family Flaviviridae. Dengue virus poses a public health threat to over 2.5 billion people worldwide, and is estimated to cause between 50 and 100 million infections per year. There are currently no known vaccine or antiviral therapies that are efficacious in treating or preventing Dengue. The Lescar group has partnered with the Novartis Institute for Tropical Disease to rationally design a methyltransferase (Mtase)-based small molecule inhibitor (compound 10*, Fig. C.26) that is selective for disease-related methyltransferases (Lim et al. 2011).

Dr. Peter Preiser

The most virulent form of human malaria is Plasmodium falciparum. This merozite utilizes multiple ligands, including members of the erythrocyte binding proteins/ligands (EBLs) and reticulocyte binding protein homologues (RBLs) families and to invade, replicate within and subsequently destroy red blood cells. The Preiser group is focused on defining the molecular interactions of EBLs or RBLs binding to red blood cell receptors. They have recently identified the eryothrocyte binding region of the P. falciparum reticulocyte binding protein homologue 1(PfRH1), and have demonstrated that antibodies designed to target this specific region inhibit invasion (Gao et al. 2008).

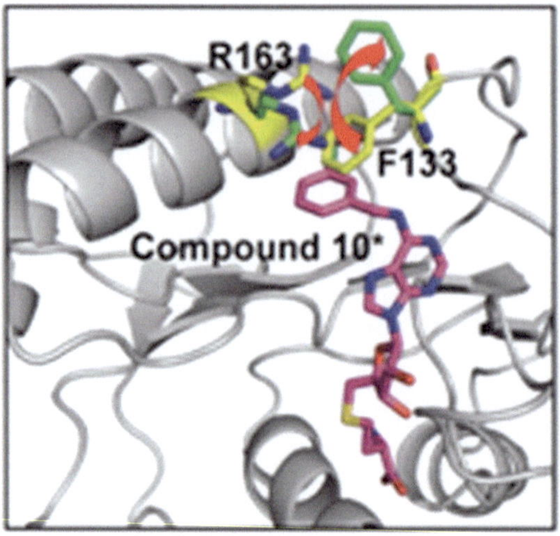

Fig. C.26 Co-crystal structures of DENV-3 MTase in complex with compound 10*. Compound-induced conformational change of amino acids. (Adapted from Lim et al. 2011)
Compound 10* is shown in stick presentation (*magenta*). Residues Phe-133 and Arg-163 in the SAM-MTase complex are colored in *yellow*; the same residues in the compound 10*-MTase complex are shown in *green*; two arrows (*red*) indicate the compound-induced conformational change

Dr. Lars Nordenskiöld

Dr. Nordenskiöld serves as the Associate Dean for the College of Science. His research group is focused on defining the biochemistry and biophysics of chromatin packaging within the cell nucleus. In particular, his group has two main aims: (1) use a combination of theoretical and experimental approaches to define the role of histone tail charges in chromatin folding, (2) define the physical biology of electrostatic and molecular interactions for DNA structure-function and dynamics in the context of chromatin. His group has recently developed a new course-grain model of nucleosome core particles (NCPs; Fig. C.27), which describes the role of Mg2+ and CoHex3+ ions in regulating NCP-NCP aggregation and stacking in a manner congruent with the experimental observed condensed phase of NCP (Fan et al. 2013).

Translational Efforts

The focus of translational science within the School of Biological Sciences is to address unmet clinical needs particular to Southeast Asia. For instance, translational research is focused on development of diagnostics and therapies for malaria, Dengue virus, and SARS.

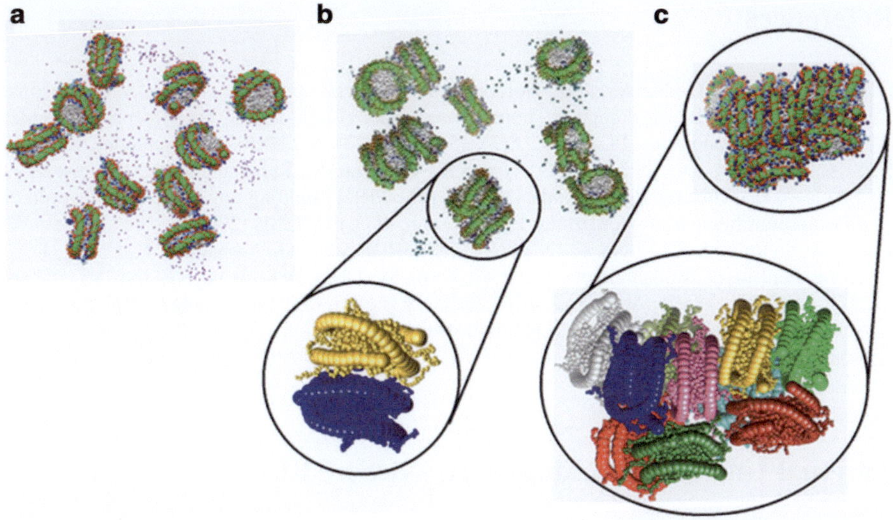

Fig. C.27 NCP-NCP interaction is defined by the valency of cation neutralizing NCP charge (Adapted from Fan et al. 2013)
Snapshots of the 10NCP/K+ (*A*), 10NCP/Mg2+ (*B*), and 10NCP/Co3+ (*C*) systems. Inserts below (*B*) and (*C*) illustrate close NCP-NCP interactions with the formation of stacked NCP pairs in the 10NCP/Mg2+ system and a dense NCP aggregate in the 10NCP/Co3+ system with combination of different types NCP-NCP contacts with stacking being dominant (note two NCP stacks, *green-yellow* and *pink-blue* forming a column). In the inserts, mobile cations and DNA phosphate groups are omitted for clarity

Sources of Support

Graduate student training is largely funded by the Singapore Ministry of Education. Emphasis is towards funding interdisciplinary research projects. Grant opportunities are also available through the Ministry of Defense.

Summary and Conclusions

The NTU School of Biological Sciences is designed to foster interdisciplinary training in life sciences in order to develop a talent pool of researchers for the life sciences industry, with an emphasis towards translational medicine in the areas of infectious disease.

References

Fan, Y., Korolev, N., Lyubartsev, A. P., Nordenskiold, L.: An advanced coarse-grained nucleosome core particle model for computer simulations of nucleosome-nucleosome interactions under varying ionic conditions. PLoS ONE **8**(2), e54228 (2013). doi:10.1371/journal.pone.0054228

Gao, X., Yeo, K.P., Aw, S.S., Kuss, C., Iyer, J.K., Genesan, S., Rajamanonmani, R., Lescar, J., Bozdech, Z., Preiser, P.R.: Antibodies targeting the PfRH1 binding domain inhibit invasion of Plasmodium falciparum merozoites. PLoS Pathog. **4**(7), e1000104 (2008)

Lim, S.P., Sonntag, L.S., Noble, C., Nilar, S.H., Ng, R.H., Zou, G., Monaghan, P., Chung, K.Y., Dong, H., Liu, B., Bodenreider, C., Lee, G., Ding, M., Chan, W.L., Wang, G., Jian, Y.L., Chao, A.T., Lescar, J., Yin, Z., Vedananda, T.R., Keller, T.H., Shi, P.Y.: Small molecule inhibitors that selectively block dengue virus methyltransferase. J. Biol. Chem. **286**(8), 6233–6240. doi:10.1074/jbc.M110.179184

National University of Singapore (NUS)-CBIS

Site Address:	Centre for Bioimaging Sciences (CBIS) Blk S1A, Level 2 14 Science Drive 4 Singapore http://mbi.nus.edu.sg/resources/nus-centre-for-bioimaging-sciences-cbis/
Date Visited:	June 3, 2013
WTEC Attendees:	Paul Janmey (report author), Owen McCarthy, Cynthia Reinhart-King, Sharon Gerecht, Nastaran Kuhn, Larry Nagahara, Patricia Foland
Host(s):	Prof. Paul Matsudaira Director Tel.: 65162723 dbshead@nus.edu.sg
	Thorsten Wohland, Ph.D. Associate Director chmwt@nus.edu.sg
	Dr. Dipanjan Bhattarcharya Research Fellow dbsdb@nus.edu.sg

Overview

The Centre for Bio Imaging Sciences (CBIS) was founded by faculty from the Biological Sciences, Chemistry, and Physics departments at the National University of Singapore as an interdisciplinary research center with ties to the Optical Bioengineering Laboratory in the NUS Division of Bioengineering, the medical imaging research facilities of the Singapore BioImaging Consortium, and the advanced confocal and super-resolution microscopes of the Mechanobiology Institute, Singapore. The mission of the Centre for Bio Imaging Sciences' research is focused on developing the science and technology of biological imaging by light and electron microscopy and the development of computational and microscopy-based methods and technologies.

Research and Development Activities

Research groups lead by 11 principal investigators study a wide range of problems in biology including molecular and cellular parasitology, developmental biology and genetics, stem cell biology, mechanobiology, cell motility and various aspects of structural biology.

A centerpiece of the CBIS is a set of four transmission electron microscopes housed in a climate controlled environment and adapted for biological samples. Among these instruments is a state-of-the-art Titan Krios transmission electron microscope (FEI, Netherlands) for visualizing native structures of biological samples to atomic resolution. In the future, the facility has plans for phase contrast electron microscopes and more sensitive and high-resolution CMOS cameras.

Development of light microscopy is also a core of the center, and techniques such as digital scanning laser sheet microscopy and fluorescence correlation spectroscopy are being pushed to new limits. One example of this strategy used single-particle tracking and fluorescence correlation spectroscopy to reveal the restricted dynamics of phosphatidyl serine in the inner leaflet of the cell plasma membrane and internal membranes (Kay et al. 2012). Light sheet microscopy has been used to image cells deep within the spinal structure of a zebrafish larva as shown in Fig. C.28.

An essential aspect of the CBNIS is its in-house facility for data storage, transfer, and processing, housed in a facility that is designed to adapt as technology improves. The current capacity allows for handling of TB (243 bits, close to the functional memory of the human brain) data sets produced by current imaging technologies. This computational core will be interconnected to the Mechanobiology Institute (MBI) data center using multiple high-speed 10 Gbps Ethernet circuits.

Translational Efforts

This center was established as an academic facility, and its mission is to advance technology that leads to scientific discovery rather than having a formal mission for applied science and translation. However, several of the research groups and the discoveries emerging from this facility naturally lead to potential applications that

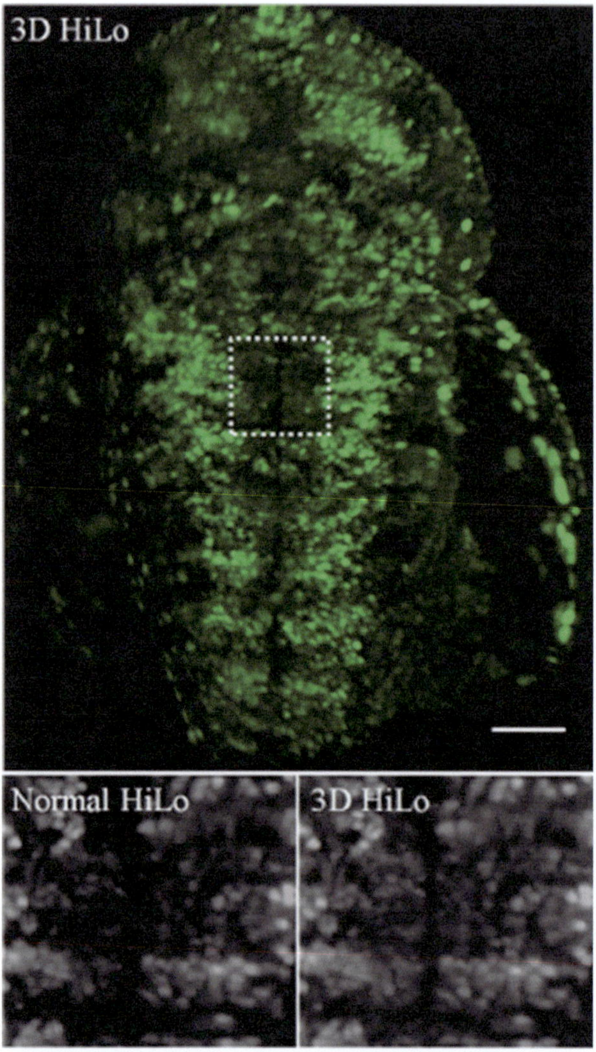

Fig. C.28 Deep tissue live embryo imaging with digital scanning laser sheet microscopy data processed with Hi-Lo filtering to detect EGFP-labeled cell nuclei within zebrafish spine (From Bhattacharya 2012) *Black* **and** *white* **images are magnifications of area outlined by** *dotted line*. **Scale bar = 100 μm**

are highlighted in some of the publications from this center. NUS-CBIS will become the Asian demo and training site for FEI (Hillsboro, Oregon; www.fei.com/). In addition CBIS will soon become a research resource for pharmaceutical companies to characterize and control the quality of biologicals from production and R&D. A JEOL 2200FS with phase contrast optics and CMOS camera will be delivered in November.

Sources of Support

Singapore Agency for Science, Technology and Research
National Research Foundation, Singapore

Summary and Conclusions

National University of Singapore (NUS)-CBIS has focused large-scale resources and expertise to build state of the art facilities in light and electron microscopy with the goal of driving discovery in biosciences in collaboration with multiple other institutions in Singapore with research interests in mechanobiology, bioengineering and other disciplines. Technical development of imaging technologies as well as creation of facilities and methods for data storage and analysis continue to be integral elements in this facility.

References

Anwar, M., Gupta, P.B. Palaniapan, R., Matsudaira, P.: Protein array patterning by diffusive gel stamping. PloS One **7**(10), e46382 (2012). doi:10.1371/journal.pone.0046382

Bhattacharya, D., Singh, V.R., Zhi, C., So, P.T.C., Matsudaira, P., Barbastathis, G.: Three dimensional HiLo-based structured illumination for a digital scanned laser sheet microscopy (DSLM) in thick tissue imaging. Opt. Expr. **20**, 27337–27347 (2012)

Kay, J.G., Koivusalo, M., Ma, X, Wohland, T., Grinstein, S.: Phosphatidylserine dynamics in cellular membranes. Mol. Biol. Cell. **23**, 2198–2212 (2012)

Lauterbach, T., Manna, M., Ruhnow, M., Wisantoso, Y., Wang, Y., Matysik, A., Oglecka, K., Mu, Y., Geifman-Shochat, S., Wohland, T., Kraut, R.: Weak glycolipid binding of a microdomain-tracer peptide correlates with aggregation and slow diffusion on cell membranes. PLoS One 7, e51222 (2012)

Mirsaidov, U.M., Zheng, H.M., Casana, Y., Matsudaira, P.: Imaging Protein Structure in Water at 2.7 nm Resolution by Transmission Electron Microscopy. Biophys. J. **102**, L15-L17 (2012)

Singh, A.P., Krieger, J.W., Buchholz, J., Charbon, E., Langowski, J., Wohland, T.: The performance of 2D array detectors for light sheet based fluorescence correlation spectroscopy. Opt. Expr. **21**, 8652–8668 (2013)

National University of Singapore (NUS)-Mechanobiology Institute

Site Address:	Mechanobiology Institute T-Lab, #10-01 5A Engineering Drive 1 Singapore 117411 http://mbi.nus.edu.sg/
Date Visited:	June 3, 2013
WTEC Attendees:	Paul Janmey, Owen McCarthy, Cynthia Reinhart-King (report author), Sharon Gerecht, Nastaran Kuhn, Larry Nagahara, Patricia Foland
Host(s):	Prof. Michael Sheetz, Ph.D. Director Mechanobiology Institute ms2001@columbia.edu Tel : 6601 1557
	G.V. Shivashankar Deputy Director dbsgvs@nus.edu.sg
	Ms. Cynthia Lee Sook Chen Administrative director mbileesc@nus.edu.sg
	Mr. Joshi Shashikant Senior Facility Manager mbisbj@nus.edu.sg

Overview

The Mechanobiology Institute (MBI) was created in 2009 with the goal of being the world's leading center in mechanobiology. It was established through funding from the National Research Foundation, the Ministry of Education, and the National University of Singapore to both advance mechanobiology as a discipline and to advance the overall research efforts in Singapore. To do so, they have recruited a dynamic, creative faculty from around the world who specialize in studies of mechanobiology at the molecular, cellular and tissue levels. Their multi-scale approaches and findings are being transferred into the classroom, to train the next generation of researchers, and will ultimately impact the clinic, influencing patient treatment.

Functional Focus

The MBI's focus is to integrate experiments with models to investigate the role of force in motility and morphogenesis at the molecular, cellular and tissue levels. They emphasize a multidisciplinary, multi-scale approach to understand mechanotransduction in a variety of physiological systems.

Research and Development Activities

The MBI is composed of 23 principal investigators; labs typically contain 5–8 people. They have implemented several different approaches to maximize intellectual stimulation and cross-disciplinary work and collaboration. As examples, they host visiting professors for varying amounts of time, labs are open format, and students are seated randomly. They recruit both engineering/physics students and biology students. They also host an annual Mechanobiology Conference and numerous joint international conferences with overseas institutes and organizations such as National Institute of Biological Sciences, Bangalore, Tata Institutes in Hyderabad, Bangalore and Kolkata, Weizmann Institute in Israel, and the Biophysical Society in the United States.

The facilities that are available are extensive. The microscope suites contain an array of modalities including TIRF, STORM, and SIM, which are in high use within the center. They also have fabrication facilities in-house and a cell culture core. To enhance the bench-to-classroom transition, the MBI has pioneered the creation of MBinfo, a website containing extensive information on mechanobiology (www.mechanobio.info/). It currently contains seven chapters authored and reviewed by experts in the field. It is wiki-based, allowing for others to add to and edit the current information. It is expected that this will the site to stay current and encourage engaged learning.

There are four main research focus groups: (1) Technology Innovation, (2) Molecular Mechanisms, (3) Cell-matrix and Cell-cell, and (4) Mechanotransduction. Research in these areas is extensive, covering molecular, cellular and tissue levels. Below are some highlights:

At the molecular level, Jie Yan's group is using transverse magnetic tweezers (Fig. C.29) to investigate the mechanism by which force drives structural changes in DNA (Zhang, Chen et al. 2013). Their work provides 3 possible mechanisms to explain the structural changes in torsion unconstrained DNA.

Michael Sheetz's group is interested in mechanisms of force generation, cell-cell and cell-matrix interactions. Their work in this area is extensive. As just one example, in recent work, they used magnetic tweezers, flexible pillar arrays and TIRF to show that alpha-actinin plays separate roles in nascent versus mature focal adhesions (Roca-Cusachs, del Rio et al. 2013). Their work indicates that it competes with talin in early focal adhesion formation but it can bind integrins and mediate mechanotransduction and adhesion maturation. Alpha-actinin may be a key linker in the force transmission between integrins and the cytoskeleton.

At the cellular level, G.V. Shivashankar's group has made a significant contribution to our knowledge of nuclear mechanics. In recent work, they used micropatterning and transcriptome analysis to investigate how cell shape affects gene expression patterns (Fig. C.30). The data indicate that shape and spreading can switch gene expression pathways from those that control cell-matrix interactions versus homeostasis (Jain et al. 2013).

At the tissue level, Chwee Teck Lim and Benoit Ladoux have used micro patterning and force measurements to investigate how individual cell-cell adhesions affect collective movements which regulate tissue organization (Fig. C.31, Vedula et al. 2012). They demonstrate a clear role for cell-cell adhesion in collective motions.

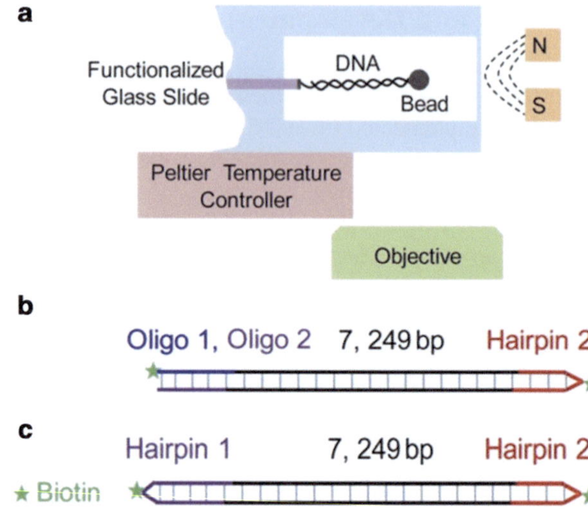

Fig. C.29 Work from Yan group investigating DNA structural changes in DNA in response to force (From Zhang et al. 2013). (*A*) Schematic of the transverse magnetic tweezers with temperature control. (*B*) End-opened DNA that allows peeling from the open end. (*C*) End-closed and torsion-unconstrained DNA (the same sequence as the end-opened DNA) that does not allow peeling topologically

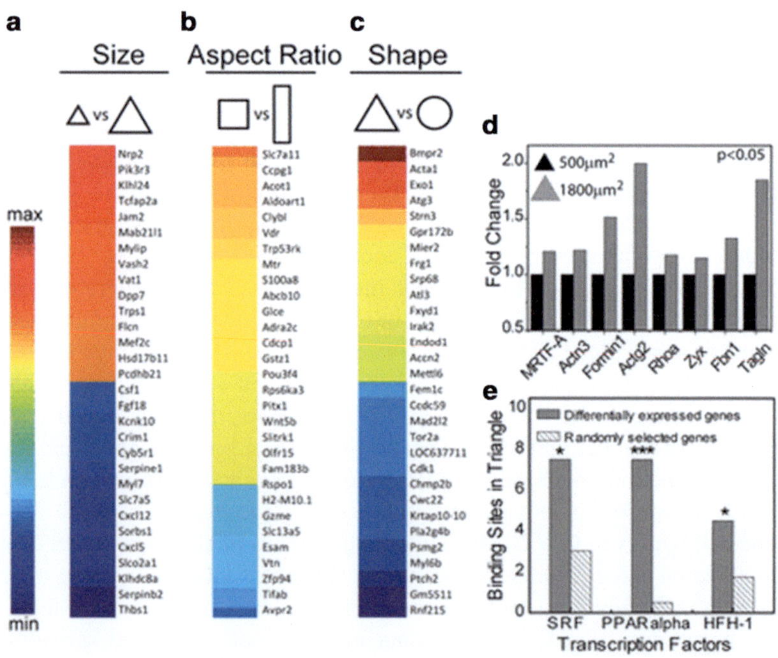

Fig. C.30 Geometric regulation of actin-related genes and TFs. (From Jain et al. 2013) (*A–C*) Color maps showing differentially regulated genes between cells of different geometries: size (*A*), AR (*B*), and shape (*C*). (*D*) Actin-related genes differentially expressed between cells cultured on smaller and larger triangular patterns. (*E*) TF showing significant binding sites in differentially regulated genes between circular and triangular cells of equal area (1,800 μm²)

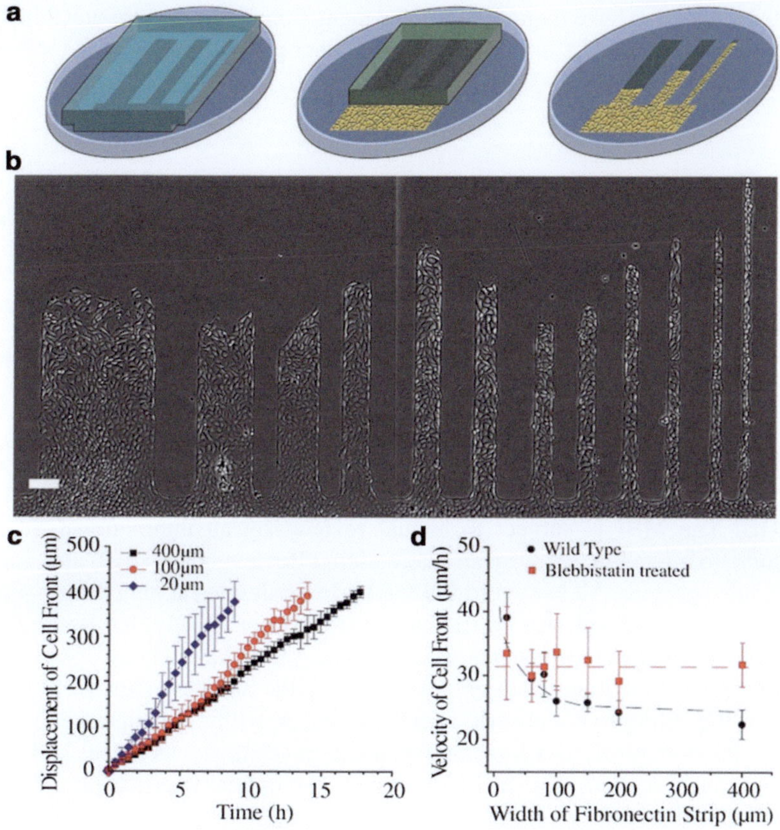

Fig. C.31 Migration of MDCK cell sheet on fibronectin strips of different widths (From Vedula et al. 2012)
(*A*) Schematic of the fibronectin stamped pattern with a block of PDMS (*gray*). Cells reach confluence in the reservoir (shown as a *yellow* area) and migrate into the strips when the PDMS block is lifted (as illustrated by the last step). (*B*) MDCK cell sheets migrating on fibronectin strips of different widths. (*C*) Average displacement of cell front over time in 400, 100, and 20-μm wide strips. (*D*) Velocity of cell front on strips of different widths for untreated (*black*) and blebbistatin-treated (*red*) MDCK cells. *Dashed lines* are a smooth fit to guide the eye. Scale bar = 100 μm)

Translational Efforts

In general, collaborations with clinicians are based on individual investigator interactions. Physicians generally have a high patient load and minimal time for research. There are multiple groups working on new devices to probe mechanotransduction and other applications. As an example, Chwee Teck Lim has a start-up, Clearbridge BioMedics, to detect and retrieve cancer cells from blood of cancer patients for further downstream analysis. It has won numerous local and international innovation awards.

Patents filed that have large MBI involvement include the MembraneChip™ (Jay Groves) and the MARC Chip (Evelyn Yim). Other patents filed are for methodologies and chips.

Sources of Support

Research funding for MBI investigators comes from the National Research Foundation, the Ministry of Education, and an HFSP grant. The original budget for the MIB was \$120 M USD over 10 years.

Summary and Conclusions

The MBI The MBI is subject to annual reviews by an international Scientific Advisory Board and in 2013, by an International Review board (IRP) appointed by the funding agencies. All have certified MBI to be positioned among the top research institutes of its kind and well on track to be (if it is not already) the leading center on mechanobiology. It integrates biology with physical sciences approaches to investigate mechanobiology at the molecular, cellular and tissue levels. It has demonstrated the ability to collaborate across the center, within Singapore and internationally. Its strength resides in some top research scientists, technical and support expertise, and excellent core facilities such as the optical microscope facility and fabrication resources.

References

Jain, N., Iyer, K. V., Kumar, A., Shivashankar, G.V.: Cell geometric constraints induce modular gene-expression patterns via redistribution of HDAC3 regulated by actomyosin contractility. Proc. Natl. Acad. Sci. USA. **110**(28), 11349–11354 (2013)

Roca-Cusachs, P., del Rio, A., Puklin-Faucher, E., Gauthier, N.C., Biais, N., Sheetz, M.P.: Integrin-dependent force transmission to the extracellular matrix by alpha-actinin triggers adhesion maturation. Proc. Natl. Acad. Sci. USA. **110**(15), E1361–E1370 (2013)

Vedula, S. R., Leong, M.C., Lai, T.L., Hersen, P., Kabla, A.J., Lim, C.T., Ladoux, B.: Emerging modes of collective cell migration induced by geometrical constraints. Proc. Natl. Acad. Sci. USA. **109**(32), 12974–12979 (2012)

Zhang, X., Chen, H., Le, S., Rouzina, I., Doyle, P.S., Yana, J.: Revealing the competition between peeled ssDNA, melting bubbles, and S-DNA during DNA overstretching by single-molecule calorimetry. Proc. Natl. Acad. Sci. USA. **110**(10), 3865–3870 (2013)

Osaka University – Department of Immunology and Cell Biology

Site Address:	2-2 Yamada-oka Suita Osaka 565-0871 Japan
Date Visited:	June 7, 2013
WTEC Attendees:	Paul Janmey, Daniel Fletcher (report author), Cynthia Reinhart-King, Nastaran Kuhn, Patricia Foland
Host(s):	Masaru Ishii, M.D., Ph.D. Professor and Chairman Department of Immunology and Cell Biology Osaka University Graduate School of Medicine mishii@icb.med.osaka-u.ac.jp

Overview

Dr. Masaru Ishii is Professor and Chairman of the Department of Immunology and Cell Biology in the Graduate School of Medicine at Osaka University. Dr. Ishii is also a Principal Investigator in the Immunology Frontier Research Center (iFReC) at Osaka University, which is a World Premier International Research Center established by the Japanese Ministry of Education, Culture, Sports, Science and Technology in 2007. Dr. Ishii is a leading expert in the development and use of intravital two-photon microscopy for *in vivo* imaging of bone tissues. The Immunology and Cell Biology group that Dr. Ishii leads is investigating the mechanisms that control bone homeostasis and how this homeostasis is disrupted in diseases such as rheumatoid arthritis. In particular, the group uses two-photon microscopy to identify normal and aberrant behavior of osteoclasts in response to chemokines and other signals. Their findings are pointing to novel regulators of motility that have implications for the study autoimmune diseases as well as cancer.

Research and Development Activities

A major focus of Dr. Ishii's laboratory is rheumatoid arthritis, an autoimmune disease that involves bone destruction caused by aberrant osteoclast activity. A straightforward way to monitor osteoclast activity would be image their movements and organization in live animals, but imaging of cellular behavior inside of bone tissue is difficult due to scattering of light by the bone. To address this problem, Dr. Ishii and his colleagues have addressed this issue with intravital two-photon microscopy. Two-photon fluorescence microscopy uses long wavelengths that penetrate deeper into tissue ($\approx$200 μm deep) than the shorter wavelengths of more conventional

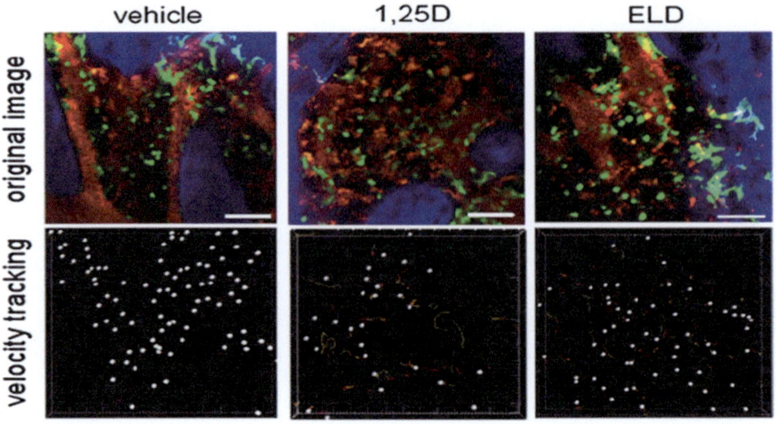

Fig. C.32 Intravital two-photon imaging of S1PR2-mediated control of migration of osteoclast precursor monocytes. (From Kikuta et al. 2013)
Imaging of the skull bone tissues of heterozygous CX3CR1-EGFP knockin mice showing CX3CR1-EGFP–positive cells (*green*), microvasculature containing Texas Red-conjugated 70 kDa dextran (*red*), and bone surface (*blue*). The study revealed that the active form of vitamin D affects migration of osteoclast precursors to S1P. (Scale bars = 50 μm)

one-photon fluorescence microscopy. This allows fluorescently-labeled cells and molecules to be imaged in live animals that have had the tissue of interest surgically exposed.

Dr. Ishii's laboratory combines *in vitro* and *in vivo* studies to identify factors that influence osteoclast activity. In one important advance for understanding the underlying mechanism of rheumatoid arthritis, Dr. Ishii discovered that osteoclasts *in vitro* will exhibit directed motility toward sphigosine-1-phosphate (S1P), a signaling sphingolipid. His recent work has expanded on that finding to develop methods for studying and controlling osteoclast chemotaxis and bone resorption activity *in vivo*. This work takes full advantage of the ability of two-photon imaging to track single cells over time, allowing study of their movement speeds, paths, and morphology. For example, Dr. Ishii and his colleagues recently showed that the hormonally active form of Vitamin D reduces bone resorption by modulating S1P-mediated motility (Fig. C.32, Kikuta et al. 2013).

In a significant advance for the field, Dr. Ishii has been able to directly image the bone resorption process with two-photon microscopy, using a novel probe for osteoclast secretions. Analysis of osteoclast movements has revealed that there are two distinct states of the cell, a static resorbing state and a moving non-resorbing state. The osteoclast is observed to switch between these states as it moves within the bone tissue and modifies the bone. Ongoing work has identified a role for cell cycle in the behavior of osteoclasts that may be a general mechanism modulating cell motility in rheumatoid arthritis and other situation.

Translational Efforts

There are direct translational implications of an improved understanding of osteoclast movement and activity at the single cell level. Dr. Ishii's work is leading to a new model for immune regulation of osteoclast function that identifies opportunities for therapeutic intervention. As described above, intravital imaging has revealed the mechanism of vitamin D-based reduction of bone resorption, and new pharmaceutical targets are being tested now. Importantly, the molecular mechanisms that have been found to modulate osteoclast activity may also be important in cancers such as colon cancer.

Sources of Support

Dr. Ishii's work is supported by the Funding Program for the Core Research for Evolutional Science and Technology, established by the Japan Agency of Science and Technology, and is also supported by several Grants-in-Aid from the Japanese Ministry of Education, Culture, Sports, Science and Technology. Dr. Ishii's work is also partly supported by the FIRST Program from the Japan Science and Technology Agency and the Research Grants from the International Human Frontier Science Program.

Summary and Conclusions

Intravital two-photon imaging is a powerful technology that can reveal single-cell behavior in live animals. Dr. Ishii has developed and applied imaging technologies to study osteoclast motility, demonstrating the value of two-photon imaging. Continued advancement of two-photon microscopy to enable "5D imaging" (space, time, and wavelength) will increase the information that can be obtained on cells in their natural environments and will be important for studies in immunology as well as cancer.

References

Ishii, M., Egen, J.G., Klauschen, F., Meier-Schellersheim, M., Saeki, Y., Vacher, J., Proia, R.L., Germain, R.N.: Sphingosine-1-phosphate mobilizes osteoclast precursors and regulates bone homeostasis. Nature. **458**(7237), 524–528 (2009)

Ishii, M., Fujimori, S., Kaneko, T., Kikuta, J.: Dynamic live imaging of bone: opening a new era with 'bone histodynametry.' J. Bone. Miner. Metab. (Epub ahead of print, Apr 2, 2013) (2013)

Kikuta, J., Kawamura, S., Okiji, F., Shirazaki, M., Sakai, S., Saito, H., Ishii, M.: Sphingosine-1-phosphate-mediated osteoclast precursor monocyte migration is a critical point of control in antibone-resorptive action of active vitamin D. Proc. Natl. Acad. Sci. USA. **110**(17), 7009–7013 (2013)

Peking University

Site Address:	5 Yiheyuan Road Haidian District Beijing 100871 China
Date Visited:	June 3, 2013
WTEC Attendees:	Dan Fletcher (report author), Parag Mallick, Sean Hanlon, Hassan Ali
Host(s):	Dr. Qiushi Ren Distinguished Professor and Chairman Department of Biomedical Engineering Tel.: +86-10-6276-7113 renqsh@coe.pku.edu.cn
	Dr. Jeff Jianzhong Xi Department of Biomedical Engineeering Tel.: +86-10-6276-7113 jzxi@pku.edu.cn xi@coe.pku.edu.cn
	Dr. Fan Bai Biodynamic Optimal Imaging Center, Peking University Tel. :86-10-6275.6164 chinabaifan@hotmail.com
	Dr. Zhifei Dai Department of Biomedical Engineering Tel.: +86-10-82529377 zhifei.dai@pku.edu.cn
	Dr. Hongbin Han Department of Radiology, Peking University Third Hospital hanhongbin@bjmu.edu.cn
	Dr. Zuhong Lu Department of Biomedical Engineering Tel.: +86-10-62756160 zuhonglu@coe.pku.edu.cn
	Prof. Chao Tang Director, Center for Theoretical Biology, University of Peking Professor, University of California San Francisco School of Pharmacy chao.tang@ucsf.edu
	Dr. Jun Jie Wang Department of Radiation Oncology, Cancer Center, Peking University Third Hospital doctorwangjunjie@yahoo.com.cn
	Dr. Yugang Wang Institute of Basic Medical Sciences ei16-beijing@pku.edu.cn

Overview

Peking University has built an active scientific community that is pursuing innovative research on cancer diagnosis and treatment. Combining the expertise of bioengineers, physicists, and physicians, Peking University is a model of collaboration between physical sciences and oncology that is producing exciting new technologies and ideas. This collaboration is the result of several factors that have increased government support for collaborative research in the biomedical sciences: (1) long term strategic science and technology goals that include cancer and cardiovascular disease, (2) an effort to overhaul healthcare for rural areas, and (3) increasing focus on higher education and associated research and development. Increased funding for biomedical sciences has fostered the growth of biomedical engineering at Peking University, as well as collaborating research areas including cell biology, genetics, and physical sciences. Academic researchers at Peking are joined in their studies by physicians at associated hospitals who are highly engaged despite full-time clinical schedules. The opportunity to form strong and productive collaborations that combine engineering, physical sciences, and medicine has produced not only new technologies but also the opportunity to test many technologies in the clinic with physicians. The close link between the engineering laboratory and the clinic has driven innovation and will continue to be an important advantage and asset of Peking University.

A hub of activity at the interface between physical sciences and oncology in China is the Department of Biomedical Engineering at Peking University. Established in 2005, the Department of Biomedical Engineering has grown rapidly and now includes 18 core faculty and additional adjunct faculty and staff. Collaborating groups include the Department of Physics, Biodynamics Optical Imaging Center, and the Peking University Medical School. The Department of Bioengineering also has a joint program with the Department of Bioengineering at Georgia Tech that offers students and faculty further research and education opportunities. The APHELION team visited with a group of researchers from Peking University organized by Dr. Qiushi Ren of the Department of Biomedical Engineering to discuss research on cancer diagnosis and treatment. Below is a summary of highlights from the meeting.

Research and Development Activities

Early detection and treatment of cancer relies on advanced imaging and on high throughput screening technologies. In both of these topic areas, researchers at Peking University are making important contributions.

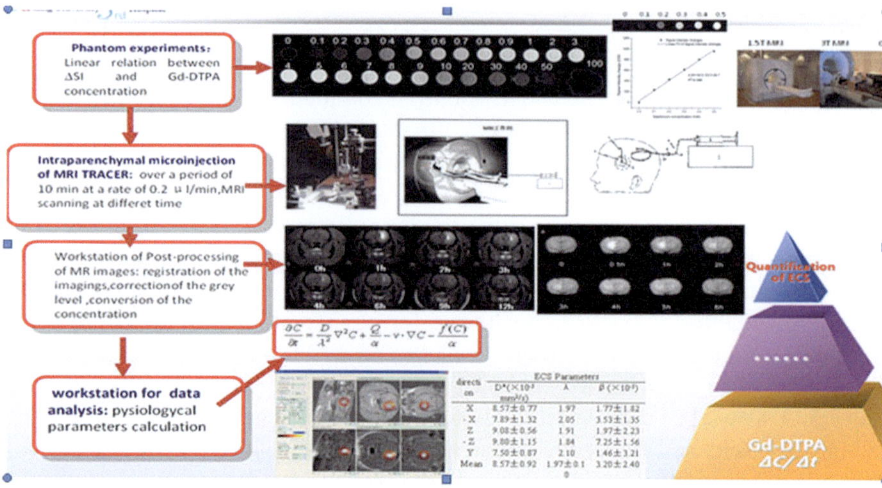

Fig. C.33 Procedures for measuring brain ECS by using MRI developed by Dr. Hongbin Han (Courtesy of H.B. Han, based on work reported in Han et al. 2011, 2012, and Han 2012)

Imaging

Dr. Qiushi Ren described advancements in multi-modality molecular imaging system (CT-PET-SPECT-FMT) that take advantage of each imaging technique to enhance data collection from individual patients. This combination of information can help physicians to make better diagnoses and design better therapies that use new technologies such as Dr. Jun Jie Wang's Axesse Linac, which is for radiosurgery and radiotherapy applications. Dr. Ren also described research in the Department of Biomedical Engineering on gold rods as theranostics, 3D photoacoustic imaging, *in vivo* confocal reflectance imaging, and flexible endoscope for hyperspectral imaging.

Dr. Hongbin Han described a new method for using magnetic resonance imaging to study diffusion, such as that of a drug or other fluid, in the extracellular space in the brain (Fig. C.33). This work will aid in the design of better drug delivery to the brain.

Screening Technologies

Dr. Zuhong Lu described the clinical potential of personal genome analysis and the impact that low cost DNA microarrays and low cost DNA sequencing could have. Dr. Lu explained recent applications of high throughput sequencing to myeloid leukemia and to lung cancer, and he also presented recent uses of microfluidics for cell sorting. Dr. Jeff Jianzhong Xi presented a high throughput siRNA screen platform based on self-assembly cell microarray based on siRNA array (SAMCell) (Fig. C.34). Dr. Xi is

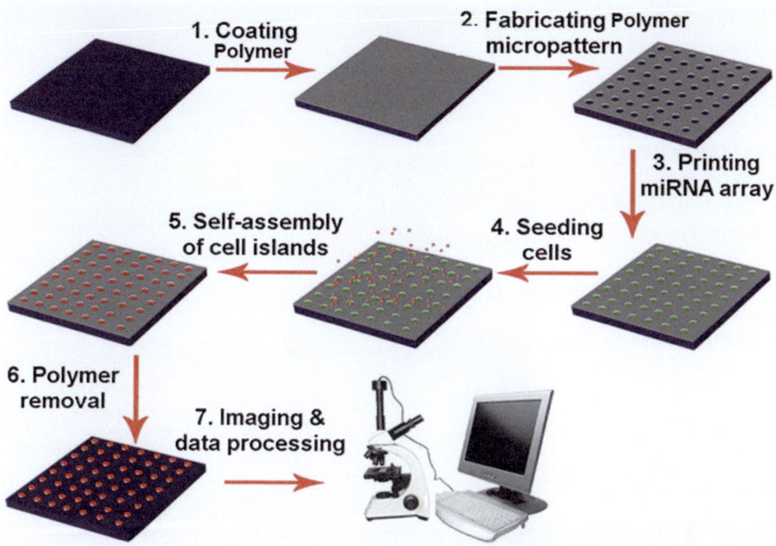

Fig. C.34 Method for screening migratory miRNA known as a Self-Assembly Cell Microarray (SAMCell) developed by Dr. Jeff Jianzhong Xi (From Zhang et al. 2011)

extending his screening of RNA to include micro RNA (miRNA), which play important roles in transcriptional and post-transcriptional gene regulation.

Dr. Fan Bai discussed progress in understanding the role of circulating tumor cells in cancer progression. He described a system for studying circulating tumor cells (CTCs) at the single cell level by first using a commercial system for a first round of CTC selection, following that with custom automated imaging, and then applying single cell sequencing to the final collection of CTCs. This approach has the potential to reveal new properties, similarities, and differences in CTCs, and the methods that Dr. Bai outlined have exciting clinical promise.

Dr. Yugang Wang presented new data on cancer stem cells (CSCs) and non-stem cancer cells (NSCCs). He showed that CSCs are more radioresistant than NSCCs when treated with γ-radiation (Fig. C.35). He also showed how CSCs and NSCCs can interconvert and presented a model to describe how that interconversion might work.

Translational Efforts

The projects described above have great translational potential, and active collaborations with physicians are facilitating clinical testing. Of particular promise is the novel imaging methods under development that will enable multiplexing of molecular imaging modalities, the circulating tumor cell analysis at the single cell level,

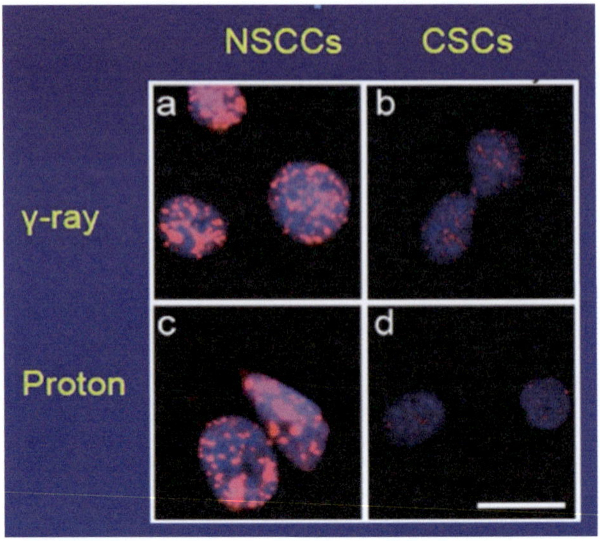

Fig. C.35 Comparison of the response of cancer stem cells (CSCs) and non-stem cancer cells (NSCCs) to radiotherapy and proton/carbon ions. CSCs are more radio-resistant than NSCCs, and proton/carbon ions are more efficient at eliminating CSCs than γ-rays (Courtesy of Dr. Hongbin Han)

and the ability to target and control cancer stem cells. The engagement of physicians with these projects will be a key to their continued success.

Sources of Support

Various sources are supporting the research projects described here, though the primary source is the Chinese government through different funding agencies.

Summary and Conclusions

Peking University is clearly a leading research institution working at the interface between physical sciences and oncology. The Department of Biomedical Engineering has grown rapidly and established strong connections with other departments and nearby hospitals. As demonstrated by the exciting projects currently under way, these collaborations will continue to advance novel methods for imaging and new technologies for high throughput screening.

References

Han, H., Xia, Z., Chen, H., Hou, C., Li, W.: Simple diffusion delivery via brain interstitial route for the treatment of cerebral ischemia. Sci. China Life Sci. **54**, 235-239 (2011). doi:10.1007/s11427-011-4141-6

Han, H.: Method for measuring the physiological parameters of brain interstitial fluid and brain extracellular space. Patent China ZL201020237182.1, US 2012/0238865 AI (2012)

Han, H.B., Li, K., Yan, J.H., Zhu, K., Fu, Y.: An *in vivo* study with an MRI tracer method reveals the biophysical properties of interstitial fluid in the rat brain. Sci. China Life Sci. **55**(9), 782–787 (2012). doi:10.1007/s11427-012-4361-4

Zhang, H., Hao, Y., Yang, J., Zhou, Y., Li, J., Yin, S., Sun, C., Ma, M., Huang, Y., Xi, J.J.: Genome-wide functional screening of miR-23b as a pleiotropic modulator suppressing cancer metastasis. Nat. Commun. **2**, 554 (2011)

RIKEN Center for Developmental Biology –Morishita Lab

Site Address:	2-2-3 Minatojima-minamimachi, Chuo-ku Kobe, Hyogo 650-0047 Japan
Date Visited:	June 7, 2013
WTEC Attendees:	Paul Janmey, Daniel Fletcher, Cynthia Reinhart-King (report author), Nastaran Kuhn, Patricia Foland
Host(s):	Dr. Yoshikiro Morishita Research Unit Leader morishita@cdb.riken.jp
	Dr. Tatsuo Shibata Research Unit Leader Laboratory for Physical Biology tatsuoshibata@cdb.riken.jp
	Dr. Yoshiki Sasai Deputy director +81-78-306-1841 yoshikisasai@cdb.riken.jp http://www.cdb.riken/jp/sasai/index-e.html
	Dr. Urs Frey Team Leader Quantitative Biology Center (QBiC) ufrey@riken.jp
	Dr. Douglas Sipp Manager, Office for research Communications Leader, Science Policy and Ethics Studies Unit Tel.: +81-90-4641-4990 sipp@cdb.riken.jp

Overview

The Center for Developmental Biology (CDB) at Riken has core strengths in developmental biology, regeneration, and stem cell biology. According to its 2012 Annual Report (www.cdb.riken.jp/en/ 01_about/0105_annual02.html), the organization's focus going forward is to strengthen its quantitative approaches and collaborations within the RIKEN system and around the world. They have entered into several new collaborative agreements with institutes in Barcelona, Spain and they have hired many international investigators. The CDB offers an array of core facilities including Animal Resources and Genetic Engineering, Electron Microscopy, Bio-Imaging, Genomics, and Proteomics.

Morishita's Lab was started in January 2012. He has a background in engineering, applied math and mathematical biology. His lab uses theoretical and experimental approaches to understand morphogenesis.

Functional Focus

The overall focus of the efforts in Yoshikiro Morishita's lab is to develop a general model for limb morphogenesis based on microscopic observations and tissue deformations.

Research and Development Activities

Tissue morphogenesis is a multi-scale, spatially and temporally changing process. Morishita's lab is using tissue-level data to link cellular and molecular behaviors to changes in shape. Given a specific tissue shape, there are infinite ways to deform that tissue to generate that shape. Therefore, they use a Bayesian statistical model to find the likelihood that a certain deformation created a specific shape. This computational work is motivated by experimental work, mapping the movements of cells before and after deformations, obtaining information about local deformations, tissue growth rate and the direction of anisotropy. They can incorporate temporally and variably changing conditions including morphogen gradients (Fig. C.36 and Hironaka and Morishita 2012). To perform this work, they use model systems that optically transparent and contain a minimal number of cells, like the *Drosophila* wing (Hironaka, Iwasa and Morishita 2012). His lab has already demonstrated the high predictive ability of his model.

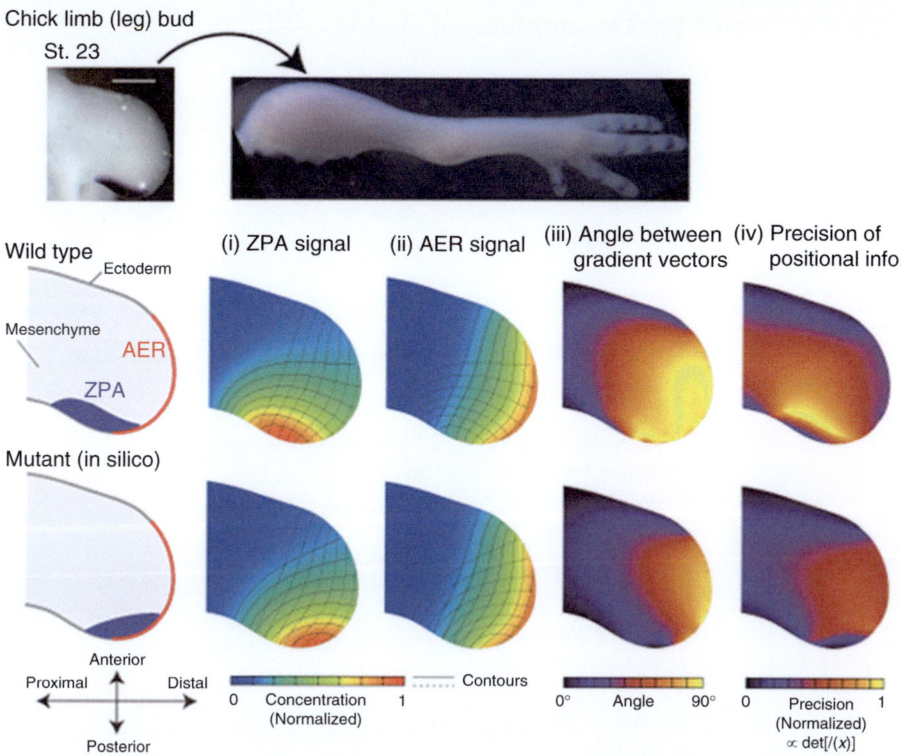

Fig. C.36 Optimal arrangement of morphogen source (Shh expression region) to maximize the precision of positional information in vertebrate limb bud (From http://www.cdb.riken.jp/en/02_research/0207_strategic04.html)

Summary and Conclusions

There is demonstrated power in combining the strength of the experimental strengths available at the CDB with the computational approaches within the Morishita lab to provide new insights into developmental systems. Having been trained at the intersection of computational biology and applied math and now being at the CDB provides an excellent platform to work at the intersection of math and biology.

References

Hironaka, K., Morishita, Y.: Encoding and decoding of positional information in morphogen-dependent patterning. Curr. Opin. Genet. Dev. **22**(6), 553–561 (2012)

Hironaka, K., Iwasa, Y., Morishita, Y.: Multiple feedback loops achieve robust localization of wingless expression in *Drosophila notum* development. J. Theor. Biol. **292**, 8–29 (2012)

RIKEN Center for Developmental Biology –Shibata Lab

Site Address:	2-2-3 Minatojima-minamimachi, Chuo-ku Kobe, Hyogo 650-0047 Japan CDB RIKEN Kobe since 2000
Date Visited:	June 7, 2013
WTEC Attendees:	Paul Janmey (report author), Daniel Fletcher, Cynthia Reinhart-King, Nastaran Kuhn, Patricia Foland
Host(s):	Dr. Tatsuo Shibata Research Unit Leader Laboratory for Physical Biology tatsuoshibata@cdb.riken.jp
	Dr. Yoshiki Sasai Deputy director Tel.:+81-78-306-1841 yoshikisasai@cdb.riken.jp http://www.cdb.riken/jp/sasai/index-e.html
	Dr. Yoshikiro Morishita Research Unit Leader morishita@cdb.riken.jp
	Dr. Urs Frey Team Leader Quantitative Biology Center (QBiC) ufrey@riken.jp
	Dr. Douglas Sipp Manager, Office for research Communications Leader, Science Policy and Ethics Studies Unit Tel.: +81-90-4641-4990 sipp@cdb.riken.jp

Overview

The Shibata Laboratory in Physical Biology is constituted to develop use concepts and methodology from physical and mathematical sciences to enhance the study and elucidation current emerging questions in biology, with a current emphasis on determining mechanisms of morphogenesis and information processing in cells and tissues that are involved in development and regeneration.

Research and Development Activities

Mathematical analysis of time and space-dependent changes in signaling and morphogenetic markers during cell movement and development is essential for understanding how these dynamic processes are controlled *in vivo*. The Shibata laboratory works closely with experimentalists within RIKEN and elsewhere to analyze experimental data related to morphogenesis and tissue development. The first topic is to analyze the controlled synthesis and degradation of polyphosphoinositide lipids that are involved in regulation of cell polarity and motility in amoeboid cells (Arai, Shibata et al. 2010, Shibata et al. 2012) as shown schematically in Fig. C.37.

As a second line of the study, in a collaboration with the Sasai laboratory at RIKEN, quantitative modeling by the Shibata group has led to a model by which dorsal-ventral (DV) patterning in the amphibian embryo depends on facilitated degradation of Chordin and the expression of the Chordin-proteinase inhibitor Sizzled, as depicted in Fig. C.38 (Inomata et al. 2013).

Translational Efforts

The mathematical modeling undertaken in this laboratory is fundamental research. Its utility in analysis complex data sets has potential for applied work in cell biology testing contexts.

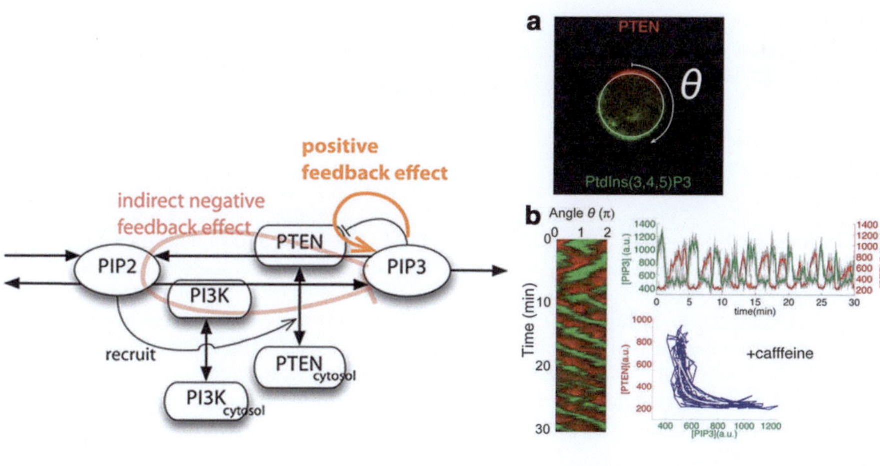

Fig. C.37 Diagram of the feedback loops controlling enzymatic production (*left panel*) that leads to self-organization of PtdIns lipids controlling cell polarity and migration (*right panel*) (From Shibata et al. 2012)

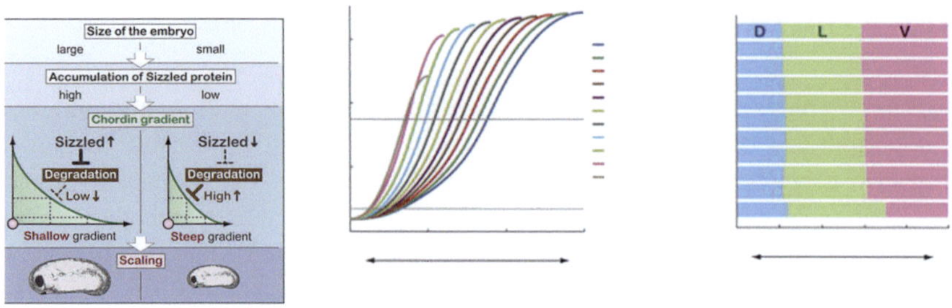

Fig. C.38 Schematic diagram of how gradients in the protein chordin lead to differences in the size of embryonic tissue (*left*) and the mathematical result of how BMP gradient changes depending the embryo size (*right*) (From Shibata et al. 2012)

Sources of Support

Ministry of Education, Culture, Sports, Science, and Technology Program, Japan
Japan Science and Technology Agency

Summary and Conclusions

As biological data sets become larger, more comprehensive, and more precise, mathematical modeling of temporal and spatial changes in biochemical parameters as undertaken by this group are increasingly becoming essential elements for understanding complex biological phenomena such as tissue development and morphogenesis.

References

Arai, Y., Shibata, T. Matsuoka, S. Sato, M.J., Yanagida, T., Ueda, M.: Self-organization of the phosphatidylinositol lipids signaling system for random cell migration. Proc. Natl. Acad. Sci. USA. **107**, 12399–12404 (2010)

Inomata, H., Shibata, T., Haraguchi, T., Sasai, Y.: Scaling of dorsal-ventral patterning by embryo size-dependent degradation of Spemann's organizer signals. Cell **153**, 1296–1311 (2013)

Shibata, T., Nishikawa, M., Matsuoka, S., Ueda, M.: Modeling the self-organized phosphatidylinositol lipid signaling system in chemotactic cells using quantitative image analysis. J. Cell Sci. **125**, 5138–5150 (2012)

RIKEN Center for Developmental Biology –The Sasai Lab

Site Address:	2-2-3 Minatojima-minamimachi, Chuo-ku Kobe, Hyogo 650-0047 Japan RIKEN Kobe since 2000
Date Visited:	June 7, 2013
WTEC Attendees:	Paul Janmey (report author), Daniel Fletcher, Cynthia Reinhart-King, Nastaran Kuhn, Patricia Foland
Host(s):	Dr. Yoshiki Sasai Deputy director Tel.: +81-78-306-1841 yoshikisasai@cdb.riken.jp http://www.cdb.riken/jp/sasai/index-e.html
	Dr. Tatsuo Shibata Research Unit Leader Laboratory for Physical Biology tatsuoshibata@cdb.riken.jp
	Dr. Yoshikiro Morishita Research Unit Leader morishita@cdb.riken.jp
	Dr. Urs Frey Team Leader Quantitative Biology Center (QBiC) ufrey@riken.jp
	Dr. Douglas Sipp Manager, Office for Research Communications Leader, Science Policy and Ethics Studies Unit Tel.: +81-90-4641-4990 sipp@cdb.riken.jp

Overview

The Sasai laboratory investigates molecular and cellular mechanisms underlying the pattern formation and morphogenesis of neuronal tissues in several contexts. Current emphases include elucidation of the mechanisms by which emergent properties arise in biological systems and in applying genetic and cell biologic knowledge to guide stem cell development toward restoration of retinal tissues. In addition to heading the laboratory of Organogenesis and Neurogenesis Prof. Sasai is also the deputy director of the RIKEN Center for Developmental Biology, a large multi-investigator institution with approximately 30 research groups. The mission of the

CDB is basic research, but a significant fraction of core directors and laboratory heads have medical training, and a significant amount of the research results have clear relevance for biomedicine and bioengineering applications.

Research and Development Activities

This summary is adapted in part from www.cdb.riken.jp/sasai/index-e.html.

One focus involves "Induction and Early Patterning of Vertebrate Neural Tissues." This work aims at revealing the molecular and cellular bases of the initial step of neurogenesis, and early ectodermal patterning following the primary induction. The lab has isolated a number of regulatory genes essential for neural determination and neural crest development. For this work, the primary experimental system is *Xenopus*.

A second focus is entitled "Secreted Patterning Signals that Provide Positional Information in CNS Development." This work identifies new patterning signals that give positional information to differentiating neural cells using both chick and mouse models.

Focus 3 is entitled "Establishment of *in vitro* Neural Differentiation Systems Using ES Cells." This work uses mouse embryonic stem (ES) cells to study molecular mechanisms of neural induction and specification. This project also aims to develop feeder-free neural induction systems to understand signals that define the positional identity of cells in the embryonic nervous systems.

The fourth focus is entitled "Application of *in vitro* Produced Neural and Sensory Cells to Regenerative Medicine." The goal of this work is to establish the technical bases of the production of neurons to be used in regenerative medicine for diseases. This study is in close collaboration with Kyoto University Hospital.

A current emphasis is on issues of emergent properties, i.e., how self-organization of biological systems occurs through mechanisms involving reaction-diffusion processes, convection, and other phenomena commonly considered in physics and engineering sciences and increasingly so in biological science (Sasai 2013). This work requires extensive collaborations among multiple experimental groups together with theoreticians (Inomata et al. 2013; Sasai 2013) (Fig. C.39).

One of the most remarkable achievements of this large, ambitious, and outstandingly productive laboratory is the integration of genetics, physical modeling, stem cell biology, and physiology to recapitulate many features of retinal development in the eye starting from a culture of stem cells and making use of the wealth of knowledge acquired from studies of eye development *in vivo*. Detailed steps in differentiation and self-organization of these cells are shown in Fig. C.40 (Eiraku and Sasai 2012). The technique is invaluable for basic researchers and may also lead to new treatments for people whose vision deteriorates.

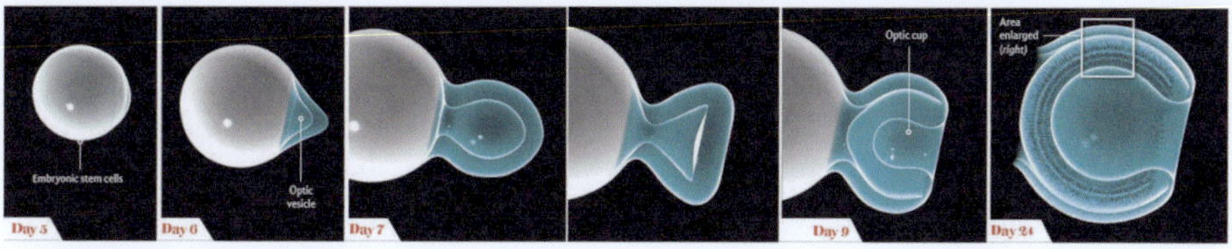

Fig. C.39 Growing a retina in culture from embryonic stem cells recapitulates development of the eye. (From Sasai 2012)
As the illustrations show, embryonic stem cells aggregate and begin to form the very early optic vesicle after about 5 days of being mixed with molecules called growth factors. The vesicle balloons out by day 7, and a few days later the structure collapses to form the optic cup, which by day 24 has delineated all the layers of the retina

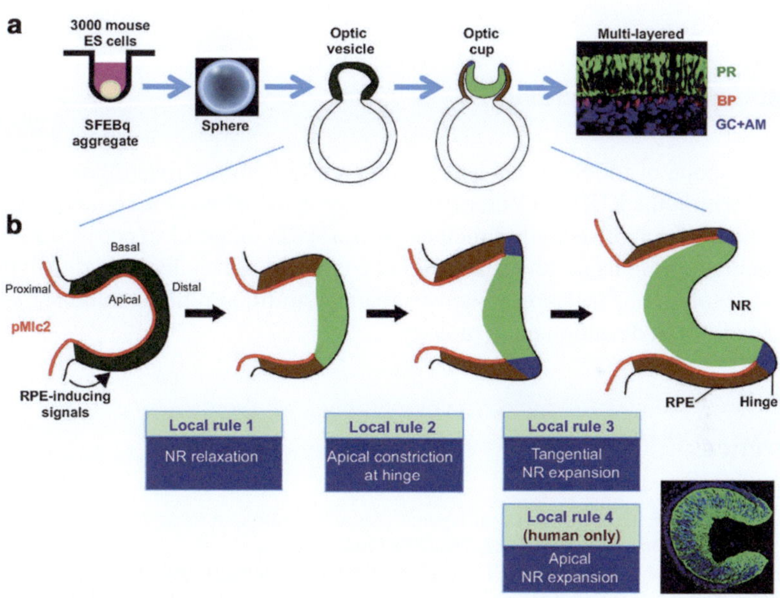

Fig. C.40 Self-organization of optic cup *in vitro* (From Sasai et al. 2012)
(*A*) Self-formation of optic cup stratified from an embryonic stem (ES) cell aggregate. Photomicrograph shows a cross-section of mouse ES cell-derived stratified neural retina. PR, recoverin+photoreceptors; BP, Chx10+ bipolar cells; GC, ganglion cells (Pax6+); AM, amacrine cells (Pax6+). (*B*) Relaxation-expansion model for self-driven optic cup morphogenesis. Three local rules are common for mouse ES cell and human ES cell cultures, whereas the fourth local rule is specific to large and thick human retinal epithelium. The lower right image shows a cross-section of human ES cell-derived optic cup. (NR, neural retina; pMlc2, phospho-myosin light chain 2 (pMlc2); RPE, retinal pigment epithelium)

Translational Efforts

Much of the basic science is directed at fundamental questions of tissue formation and development. The fourth focus specifically applies methods to develop neuronal and sensory cells for use in regenerative medicine. Controlled growth and differentiation of stem cells toward retinal development demonstrate one example for potential application of these methods for regenerative medicine.

Sources of Support

Ministry of Education, Culture, Sports, Science, and Technology Program, Japan Leading Project for Realization of Regenerative Medicine

Summary and Conclusions

The large scope of the Sasai lab and its integration with multiple other groups and facilities within the RIKEN Center for Developmental Biology serve as a model for engagement of scientists and clinicians with a wide range of expertise to address fundamental problems in development that cannot be solved by a single lab or discipline in isolation. The steps toward recreating stratified retinal structures *in vitro* are a landmark in developmental biology.

References

Eiraku, M., Sasai, Y.: Self-formation of layered neural structures in three-dimensional culture of ES cells. Curr. Opin. Neurobiol. **22**, 768–777 (2012)

Inomata, H., Shibata, T., Haraguchi, T., Sasai, Y.: Scaling of dorsal-ventral patterning by embryo size-dependent degradation of Spemann's organizer signals. Cell. **153**, 1296–1311 (2013)

Sasai, Y.: Grow your own eye: biologists have coaxed cells to form a retina, a step toward growing replacement organs outside the body. Sci. Am. **307**, 44–49 (2012)

Sasai, Y.: Cytosystems dynamics in self-organization of tissue architecture. Nature. **493**, 318–326 (2013)

Sasai, Y., Eiraku, M., Suga, H.: *In vitro* organogenesis in three dimensions: self-organising stem cells. Development. **139**, 4111–4121 (2012)

RIKEN QBiC

Site Address:	Minatojima-minamimachi 2-2-3 Chuo-ku, Kobe, Hyogo Japan 650-0047 http://www.qbic.riken.jp/english/research/outline/lab-21.html
Date Visited:	June 7, 2013
WTEC Attendees:	Paul Janmey, Daniel Fletcher (report author), Cynthia Reinhart-King, Nastaran Kuhn, Patricia Foland
Host(s):	Dr. Urs Frey Team Leader, Frey Initiative Research Unit Tel.: +81-78-306-3458 Fax: +81-78-306-3459 ufrey@riken.jp
	Dr. Yoshiki Sasai Deputy Director Tel.: +81-78-306-1841 yoshikisasai@cdb.riken.jp http://www.cdb.riken/jp/sasai/index-e.html
	Dr. Tatsuo Shibata Research Unit Leader Laboratory for Physical Biology tatsuoshibata@cdb.riken.jp
	Dr. Yoshikiro Morishita Research Unit Leader morishita@cdb.riken.jp
	Dr. Douglas Sipp Manager, Office for Research Communications Leader, Science Policy and Ethics Studies Unit Tel.: +81-90-4641-4990 sipp@cdb.riken.jp

Overview

Dr. Urs Frey is head of the Frey Initiative Research Unit at the RIKEN Quantitative Biology Center (QBiC) in Kobe. With a background in circuit design for microsensors and microfabrication, Dr. Frey and his group at RIKEN are developing novel CMOS-based sensors for biology with applications in systems biology and systems neuroscience. The group is pioneering the design and use of microelectrode array technology that can simultaneously collect electrical signals from multiple positions, providing a method for tracking the propagation of signals and studying collective electrical behavior of groups of cells. This research has great potential to advance understanding of systems level behavior of cells and tissues.

Research and Development Activities

The Frey Initiative Research Unit is involved in the development of bioelectronics for neuroscience, integrated circuit design for sensor interfaces, and nanofabrication of biosensors. The novel microelectrode array devices produced by the group are being applied to study systems biology and systems neuroscience of acute brain slices, dissociated neurons, retinal cells, and cardiomyocytes. Additional research in the laboratory is aimed at developing and applying carbon nanotube sensors for neural interfacing and creating label-free sensors for biomolecule detection from extended-gate field-effect transistors.

The bioelectric sensors developed by the Frey group have applications to a broad set of biological systems to better understand how they electrically communicate. This is of particular importance for groups of neurons but is also relevant for other cell types that respond to electrical signals, such as muscle cells. Development of bioelectric sensors involves three major areas of research: (1) Design, (2) Fabrication, and (3) Testing, which are the topics of ongoing research in the Frey group.

Design

CMOS technology is well developed, widely available, and amenable to custom design. Of note, custom fabrication can be accomplished commercially, through CMOS foundries. By adapting CMOS design techniques to the needs of bioelectronics, highly integrated sensors can be developed.

Fabrication

While the CMOS electronics can be fabricated in a commercial foundry, the sensor itself and the steps needed to make the system biocompatible are not available from foundries. This is accomplished in-house at RIKEN or collaborator facilities by, for example, adding a passivation layer that allows the sensor to operate in aqueous environments.

Testing

Microelectrode arrays designed and fabricated to be compatible with cells enable recording of electrical signals with spatially high density. The devices are being tested for use in several application areas. For example, propagation of a cardiomyocyte contraction wave (Fig. C.41) and axonal action potentials (Bakkum 2013) can be observed with a two-dimensional microelectrode array.

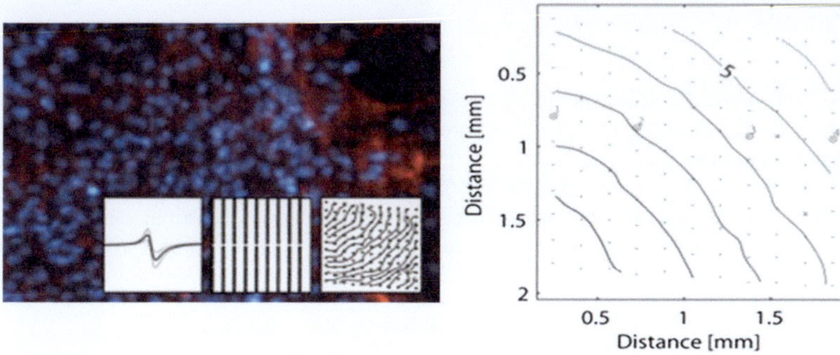

Fig. C.41 Microelectrode array analysis of cardiomyocyte activity. Cultured cardiomyocytes exhibit propagating waves of electrical activity associated with contraction. Use of the microelectrode array device enables analysis of electrical activity over long distances (Courtesy of U. Frey)

Translational Efforts

Current research on bioelectronics sensors is focused on the collection of basic data relevant to systems biology and systems neuroscience. Future applications of the microelectrode arrays could include active electrical triggering of neurons or other cells, rather than sensing, to control activity. Applications of this technology in the brain, eye, heart, or other tissue faces challenges with long-term interfacing with the tissue, but it is an interesting and important direction for further research.

Sources of Support

Support for the Frey group is provided by RIKEN and through grants from the Japanese Government.

Summary and Conclusions

The development of bioelectronic sensors that leverage the multiplexing capabilities of CMOS technology represents an exciting and important direction for systems biology and systems neuroscience. While currently focused on basic science questions rather than clinical problems, further development of microelectrode arrays could have useful implications for the control of multi-cellular systems.

References

Bakkum, D. J., Frey, U., Radivojevic, M., Russell, T.L., Muller, J., Fiscella, M., Takahashi, H., Hierlemann, A.: Tracking axonal action potential propagation on a high-density microelectrode array across hundreds of sites. Nat. Commun. 4, 2181 (2013)

Fiscella, M., Farrow, K., Jones, I.L., Jäckel, D., Müller, J., Frey, U., Bakkum, D.J., Hantz, P., Roska, B., Hierlemann, A.: Recording from defined populations of retinal ganglion cells using a high-density CMOS-integrated microelectrode array with real-time switchable electrode selection. J. Neurosci. Meth. **211**(1), 103–113 (2012)

The Institute of Physics, Chinese Academy of Sciences

Site Address:	P.O. Box 603 Beijing 100190 China
Date Visited:	June 4, 2013
WTEC Attendees:	Dan Fletcher, Parag Mallick (report author) Sean Hanlon, Hassan Ali
Host(s):	Dr. Liyu Liu Principal Investigator Microfluidic Biophysics Laboratory, Institute of Physics, CAS liu@iphy.ac.cn
	Dr. Yue Yang Department of Radiation Oncology, Peking University Cancer Hospital, Beijing zlyangyue@bjmu.edu.cn
	Dr. Jingfeng Chen Department of Radiation Oncology, Peking University Cancer Hospital, Beijing chengjinfeng123@163.com
	Dr. Guangying Zhu Department of Radiation Oncology, Peking University Cancer Hospital, Beijing zgypu@yahoo.com.cn
	Dr. Jing Huang Department of Immunology, Peking University Health Science Center huangjing82@bjmu.edu.cn

Overview

The Institute of Physics (IOP), Chinese Academy of Sciences (CAS) was established in 1950 through merging of the National Research Institute of Physics, Academia Sinica (established in 1928) and the Institute of Physics, National Academy of Peiping. Soon after the founding of the People's Republic of China it was known as the Institute of Applied Physics, Chinese Academy of Sciences, then in October 1958 the present name was adopted. The Institute is a multidisciplinary institution engaged in research on basic and applied physics, concentrating mainly on condensed matter physics, optical physics, atomic and molecular physics, plasma physics and theoretical physics. The Center for Condensed Matter Physics, CAS was set up in July 1996, based on the three State Key Labs, three Open Labs of the Academy (Lab. of Optical Physics, Beijing Lab. of Electron Microscopy and Beijing Lab. of Vacuum Physics), and certain other research groups of the Institute. Recently, there has been an increased focus on the physical aspects of biology.

Research and Development Activities

We met with Dr. Jing Huang whose lab focuses on tumor immunology and tumor metastasis. Jing Huang is working on 3D nano biochips for characterizing the molecular mechanism of primary hepatocellular carcinoma and colorectal cancer liver metastases. Among the key questions being interrogated are about how tumor infiltrating lymphocytes are affected by CD8+ T-cell depletion. The key questions they are asking include: Why do liver metastases occur in colorectal cancer? What is the biological mechanism of metastasis? Can metastases be inhibited? In addition they are hoping to use biochips to identify novel compounds that differentiate between metastatic and non-metastatic cells (Fig. C.42).

We next met with Dr. Liyu Liu, who is familiar with the Physical Sciences Oncology Center (PS-OC) initiative, having trained with Bob Austin (Princeton PS-OC). The research in the Liu group focuses on the use of microfabricated devices for mimicking 3D environments (Fig. C.43). Their belief is very strong that cancer metastasis is the major origin of cancer deaths. With the help of the advanced machinery, they are able to reconstruct cancer metastasis landscapes *in vitro*. They are working to track and study cell dynamic responses in the controlled microenvironment

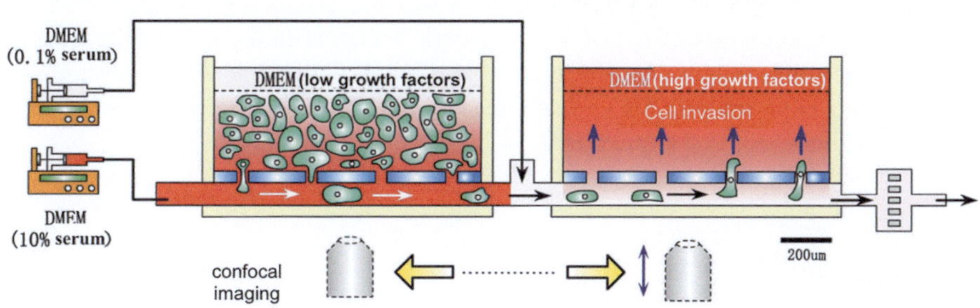

Fig. C.42 Metastasis model under development in Dr. Liu's lab. (Courtesy of L. Liu, Peking University)

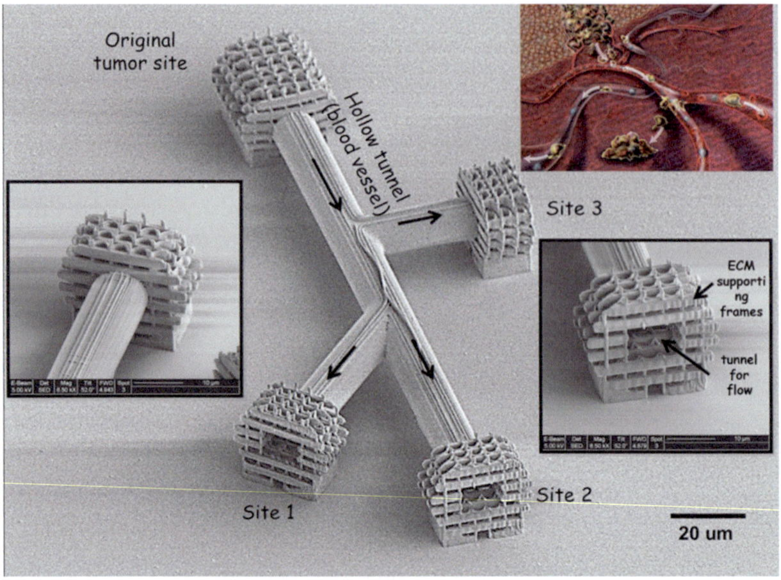

Fig. C.43 The 3D microstructure as an early approach to construct *in vitro* multiple-site metastasis model (Courtesy of L. Liu, Peking University)

so as to understand the origins of metastasis. Notably, there is generally the perception that 3D environments lead to emergent phenomena.

Extending beyond their initial model, they are now moving in fabricating structures that are mimetic of vasculature.

As a network, IOP, Peking University Health Science Center, and Peking University Cancer Hospital work closely on cancer problems. These units represent the highest research levels in physics, cancer biology, and cancer clinics. Benefiting from the integration of the resources within the network, the collaborations allow their researchers to study a broad range of cancer questions. However, they currently focus on a specific question: how to construct and use an *in vitro* 3D model to understand the origin and development of cancer metastasis and its application in clinics. The superior advantage of their network is that, with the help of the Cancer Hospital, their researchers can access to a great quantity of clinical samples and use them in their experiments. This advantage will greatly help the validation of their findings in future clinical applications.

Translational Efforts

Dr. Liu's lab is now developing customized, microscopic live-cancer cell incubation systems, which are available for various kinds of upright, inverted, and microscopes. These devices facilitate long-term observation and analysis of cell phenotypic and behavioral changes. They have been turned into a small range of commercial products.

Sources of Support

- National Basic Research Program of China ($800 K USD; 5 year program)
- CAS career startup ($800 K USD)
- One thousand talent project for recruiting overseas talent ($500 K USD)
- CAS key research support ($150 K USD)

Summary and Conclusions

The Institute of Physics is an eminent center for physics research. Their foray into biosciences in quite new, but well supported.

References

Liu, L., Sun, B., Aw Yong, K., Getzenberg, R., Stone, H., Austin, R.: Probing the invasiveness of prostate cancer cells in a 3D microfabricated landscape. Proc. Natl. Acad. Sci. USA. **108**, 6853–6856 (2011)

The University of Hong Kong, Li Ka Shang Faculty of Medicine

Site Address:	G/F, William MW Mongo Block 21 Sassoon Road Pokfulam Hong Kong http://www.med.hku.hk/v1/
Date Visited:	June 5, 2013
WTEC Attendees:	Sharon Gerecht, Owen McCarty (report author), Parag Mallick, Sean Hanlon, Hassan Ali

Host(s):	Prof. George Sai-Wah Tsao Department of Anatomy Li Ka Shing Faculty of Medicine Tel.: (852) 2819 9227 gswtsao@hku.hk
	Dr. Kwan Man Department of Surgery Li Ka Shing Faculty of Medicine Tel.: (852) 2819 9646 kwanman@hku.hk
	Prof. Wilson Y. P. Ching Department of Anatomy Li Ka Shing Faculty of Medicine Tel.: (852) 2819 9434 ypching@hku.hk
	Dr. Michael Zhang

Overview

The mission of the University of Hong Kong Centre for Cancer Research is four fold:

- identification of risk factors that contribute to cancer development
- identification of biomarkers for early detection of cancer
- development of vaccines and chemoprevention strategies for cancer prevention
- development of a comprehensive understanding of the biological drivers of cancer

The Centre for Cancer Research is designed to integrate clinical, basic and translational research programs in order to reduce mortality and increase the well-being of those diagnosed with cancer. The research initiatives for the Cancer Centre are focused in the areas of (1) cancer stem cells, (2) genetics, genomics, and epigenetic, (3) molecular targeted therapies, (4) cancer cell signaling, (5) inflammation and cancer, and (6) cancer imaging and detection.

Dr. George Tsao serves as the Director of the Centre for Cancer Research. He has established a world-class imaging center at the University of Hong Kong that serves as a core for the Centre for Cancer Research faculty, fellows and graduate students.

Research and Development Activities

Dr. Wilson Ching

The Rho family of small GTP binding proteins, including members RhoA, Cdc42, and Rac1, are master organizers of the actin cytoskeleton and direct cell spreading, motility and growth (Hall 1998). For the past two decades, these small GTPases have been investigated for separable as well as interdependent functions in filopodia and lamellipodia formation, membrane ruffling and stress fiber formation (Vega and Ridley 2008). The Rho GTPase family members Cdc42 and Rac promote the autocatalytic activation of the p21 activated kinases, or PAKs, a family of serine/threonine protein kinases that represent the best characterized Rho GTPase effectors. PAKs have been shown to localize to actin-rich adhesions and to the leading edge of migrating cells to coordinate actin cytoskeletal dynamics and cell motility (Berven and Crouch 2000).

The Ching group has focused on understanding how the dysregulation of the Rho GTPase signaling pathways, and in particular PAK, contribute to the pathogenesis of carcinogenesis. His group has recently shown that overexpression of a novel activator of PAK4, namely the CDK5 Kinase-Associated Protein CDK5RAP3, promotes human hepatocellular carcinoma (HCC) metastasis through PAK4 activation (Mak et al. 2011). Moreover, this study provided the first evidence that CDK5RAP3 is overexpressed in human HCCs. Future work is required to determine whether CDK5RAP3 not only activates PAK4 but also stabilizes PAK4, making it a potent regulator of PAK4 activity.

Dr. Kwan Man

Hepatocellular carcinoma (HCC) is one of the leading causes of cancer death in Asia due in part to endemic chronic hepatitis B virus infection. There are an estimated 300 million individuals infected with HBV in Asia, representing about 75 % of the world's chronic HBV carriers (Llovet, Burroughs, and Bruix 2003). The high incidence of HCC and the low organ donation rate in Asia represent a serious challenge with regards to liver transplantation for HCC patients. The Man group is focused on the development of therapeutic strategies for liver graft injury and cancer recurrence after liver transplantation. Utilizing a rat liver transplantation model, they have demonstrated the role that hepatic stellate cell (HSCs) activation plays an important role in small-for-size fatty liver graft injury (Cheng et al. 2010), and defined the molecular signature linked to acute phase injury and tumor invasiveness in small-for-size liver grafts (Man et al. 2010).

Dr. George Tsao

The Tsao laboratory is focused on understanding the link between Epstein–Barr virus (EBV) infection with human malignancies (Tsao et al. 2012). His group has identified several of the key molecular mechanisms by which EBV infection induces human epithelial malignancies including gastric and nasopharyngeal carcinomas. In particular, his group has demonstrated a role for cyclin D1 in the persistence of EBV infection in premalignant nasopharyngeal carcinomas (Tsang et al. 2010).

Translational Efforts

The identification of molecular biomarkers to guide the clinical decisions around liver transplantation addresses an important unmet clinical need in Asia and worldwide. The University of Hong Kong Centre for Cancer Research is ideally situated to make a major impact in the development of surgical and therapeutic strategies for liver transplantation.

Sources of Support

The research within the University of Hong Kong Centre for Cancer Research is supported in part by the clinical mission of the Faculty of Medicine.

Summary and Conclusions

The University of Hong Kong Centre for Cancer Research is poised to be the leader in the development of therapeutic strategies for liver transplantation. The recent development of a state-of-the-art imaging core facility provides the foundation for the integration of clinical, basic and translational research programs.

References

Berven, L.A., Crouch, M.F.: Cellular function of $p70^{S6K}$: a role in regulating cell motility. Immunol. Cell. Biol. **78**(4), 447–451 (2000)

Cheng, Q., Ng, K.T., Fan, S.T., Lim, Z.X., Guo, D.Y., Liu, X.B., Liu, Y., Poon, R.T., Lo, C.M., Man, K.: Distinct mechanism of small-for-size fatty liver graft injury – W nt4 signaling activates hepatic stellate cells. Am. J. Transplant. **10**(5), 1178–1188 (2010)

Hall, A.: Rho GTPases and the actin cytoskeleton. Science. **279**(5350), 509–514 (1998). doi:10.1126/science.279.5350.509

Llovet, J.M., Burroughs, A., Bruix, J.: Hepatocellular carcinoma. Lancet. **362**(9399), 1907–1917 (2003)

Mak, G.W., Chan, M.M., Leong, V.Y., Lee, J.M., Yau, T.O., Ng, I.O., Ching, Y.P.: Overexpression of a novel activator of PAK_4, the CDK5 kinase-associated protein CDK_5RAP_3, promotes hepatocellular carcinoma metastasis. Cancer Res. **71**(8), 2949–2958 (2011)

Man, K., Shih, K.C., Ng, K.T., Xiao, J.W., Guo, D.Y., Sun, C.K., Lim, Z.X., Cheng, Q., Liu, Y., Fan, S.T., Lo, C.M.: Molecular signature linked to acute phase injury and tumor invasiveness in small-for-size liver grafts. Ann. Surg. **251**(6), 1154–1161 (2010)

Tsang, C.M., Zhang, G., Seto, E., Takada, K., Deng, W., Yip, Y.L., Man, C., Hau, P.M., Chen, H., Cao, Y., Lo, K.W., Middeldorp, J. M., Cheung, A.L., Tsao, S.W.: Epstein-Barr virus infection in immortalized nasopharyngeal epithelial cells: Regulation of infection and phenotypic characterization. Int. J. Cancer. **127**(7), 1570–1583 (2010)

Tsao, S.W., Tsang, C.M., Pang, P.S., Zhang, G., Chen, H., Lo, K.W.: The biology of EBV infection in human epithelial cells. Semin. Cancer Biol. **22**(2), 137–143 (2012)

Vega, F.M., Ridley, A.J.: Rho GTPases in cancer cell biology. FEBS Lett. **582**(14), 2093–2101 (2008)

Appendix D. Site Visit Reports – United Kingdom

Site visit reports are arranged in alphabetical order by organization name.

P. Janmey et al. (eds.), *Physical Sciences and Engineering Advances in Life Sciences and Oncology*, Science Policy Reports,
DOI 10.1007/978-3-319-17930-8

Beatson Institute for Cancer Research

Site Address:	Garscube Estate Switchback Road, Bearsden Glasgow G61 1BD United Kingdom http://www.beatson.gla.ac.uk/	CANCER RESEARCH UK | BEATSON INSTITUTE
Date Visited:	October 3, 2013	
WTEC Attendees:	Paul Janmey (report author), Daniel Fletcher	
Host(s):	Dr. Laura Machesky Professor of Cell Biology Beatson Institute for Cancer Research l.machesky@beatson.gla.ac.uk	
	Dr. Robert Insall Group Leader, Beatson Institute for Cancer Research Professor of Genetics and Cell Biology, University of Glasgow r.insall@beatson.gla.ac.uk	
	Dr. Huabing Yin Senior Lecturer of Biomedical Engineering University of Glasgow Huabing.Yin@glasgow.ac.uk	

Overview

The Beatson Institute is a research facility with the mission to understand the mechanisms that regulate cancer cell proliferation, survival, and dissemination. Identifying the relevant molecules and signal transduction pathways that control cancer cell function can identify targets for novel cancer therapies. The Institute has close ties to the University of Glasgow's basic and clinical cancer research groups.

Research and Development Activities

In addition to the numerous studies to identify specific proteins involved in cancer biology, some of the work at the Beatson Institute also involves biophysical aspects of cell biology—especially cytoskeletal and motile changes that are associated with malignant cells.

Most studies related to physical aspects of cell biology involve defining the molecular mechanisms that drive cell motility and invasion through the extracellular matrix, usually involving actin binding proteins and their links to integrins. For example, the Machesky lab studies how actin polymerization regulators affect the formation of podosomes, adhesive cell protrusions that can direct matrix invasion that are also involved in mechanosensing (Schachtner et al. 2013).

Similarly, recent work from the Insall lab using the model organism *Dictyostelium discoideum* identified effects of several of the subunits of the conserved actin-nucleation SCAR/WAVE complex that control initiation of actin assembly (Davidson et al. 2013). More directly, physical studies have been initiated by the Yin lab, which studies how the mechanical properties of cells depend on the topography and chemical structure of their substrates (McPhee et al. 2010).

Sources of Support

The Beatson Institute is funded primarily by Cancer Research UK.

Summary and Conclusions

A focus on the mechanisms that control cytoskeletal assembly in cellular processes related to cancer cell migration are beginning to lead some laboratories within the research program of the Beatson Institute in directions related to defining the physical basis of cell motility and tumor proliferation.

References

Davidson, A.J., Ura, S., Thomason, P.A., Kalna, G., Insall, R.H.: Abi is required for modulation and stability but not localization or activation of the SCAR/WAVE complex. Euk. Cell. **12**, 1509–1516 (2013)

McPhee, G., Dalby, M.J., Riehle, M., Yin, H.: Can common adhesion molecules and microtopography affect cellular elasticity? A combined atomic force microscopy and optical study. Med. Biol. Eng. Comp. **48**, 1043–1053 (2010)

Schachtner, H., Calaminus, S.D.J., Thomas, S.G., Machesky, L.M.: Podosomes in adhesion, migration, mechanosensing and matrix remodeling. Cytoskeleton. **70**, 572–589 (2013)

Edinburgh University, School of Mathematics

Site Address:	Mayfield Road Edinburgh Scotland EH9 3JZ http://www.maths.ed.ac.uk/
Date Visited:	October 3, 2013
WTEC Attendees:	Paul Janmey, Daniel Fletcher (report author)
Host(s):	Prof. Tibor Antal School of Mathematics University of Edinburgh Mayfield Road Edinburgh, Scotland EH9 3JZ Tibor.Antal@ed.ac.uk

Overview

The University of Edinburgh, founded in 1583, is one of the oldest universities in the world and ranked in the top 20 universities by multiple surveys. It has a long history of training and research in physics and mathematics. Researchers in the School of Mathematics are applying analytical and computational techniques to study physical and evolutionary aspects of cancer.

Research and Development Activities

Prof. Tibor Antal served as our host for a visit to the School of Mathematics at the University of Edinburgh. With training in theoretical physics and research experience in physics, mathematics, and evolutionary dynamics, Prof. Antal applies probability theory and computational techniques to questions in biology.

Prof. Antal's recent work has advanced understanding of the role of stochasticity and genetic diversity in cancer. In one study, Prof. Antal and colleagues developed a mathematical model of tumor progression that relates the number of driver mutations to tumor size and growth rate (Bozic 2010). The team found that the stochastic nature of the mutations leads to the heterogeneity in tumor size and development time that is observed in patients, and they revealed that typical mutations in somatic cells confer only very slight selective advantage.

The accumulation of driver mutations not only affects the rate of tumor growth but also its spatial organization. In recent work, Prof. Antal developed a model that captures both the time evolution and spatial organization of solid tumor cells that acquire driver mutations that speed their growth and distort the initial spherical tumor shape (Antal 2013).

Models that capture the evolutionary dynamics of cancer growth have the potential to improve therapies. Prof. Antal and colleagues showed that cancer cell resistance to a single chemotherapy can be overcome by combination therapy designed using a mathematical model of cancer evolutionary dynamics (Bozic 2013).

Prof. Antal's work demonstrates that international collaborations that link clinicians with experimentalists and mathematicians can be highly productive over many years. Funding sources that promoted local collaboration would help to encourage the creation of similarly productive and dynamic groups at the University of Edinburgh.

Summary and Conclusions

The application of stochastic modeling and evolutionary theory to cancer has demonstrated its value in understanding tumor heterogeneity and designing treatment plans that increase the effectiveness of chemotherapy. It is likely that this type of modeling will become a mainstay of cancer treatment in the future.

References

Antal, T., Krapivsky, P.L., Nowak, M.A.: Spatial evolution of tumors with successive driver mutations. arXiv:1308.1564v1 (2013)

Bozic, I., Antal, T., Ohtsuki, H., Carter, H., Kim, D., Chen, S., Karchin, R., Kinzler, K.W., Vogelstein, B., Nowak, M.A.: Accumulation of driver and passenger mutations during tumor progression. Proc. Natl. Acad. Sci. USA. **107**(43), 18545–18550 (2010). doi:10.1073/pnas.1010978107

Bozic, I., Reiter, J.G., Allen, B., Antal, T., Chatterjee, K., Shah, P., Moon, Y.S., Yaqubie, A., Kelly, N., Le, D.T., Lipson, E.J., Chapman, P.B., Diaz, L.A., Jr., Vogelstein, B., Nowak, M.A. Evolutionary dynamics of cancer in response to targeted combination therapy. eLife. **2**, e00747. (2013). doi:10.7554/eLife.00747

Yachida, S., Jones, S., Bozic, I., Antal, T., Leary, R., Fu, B., Kamiyama, M., Hruban, R., H. Eshleman, J.R., Nowak, M.A., Velculescu, V.E., Kinzler, K.W., Vogelstein, B., Iacobuzio-Donahue, C.A.: Distant metastasis occurs late during the genetic evolution of pancreatic cancer. Nature. **467**, 1114 (2010)

Imperial College London, Institute of Biomedical Engineering

Site Address:	South Kensington Campus Royal School of Mines Prince Consort Road London SW7 2AZ United Kingdom http://www3.imperial.ac.uk/	Imperial College London
Date Visited:	October 2, 2013	
WTEC Attendees:	Paul Janmey (report author), Daniel Fletcher	
Host(s):	Dr. Harry Lamble Research Development Director Institute of Biomedical Engineering South Kensington Campus h.lamble@imperial.ac.uk	
	Prof. Peter Weinberg Professor in Cardiovascular Mechanics Faculty of Engineering, Department of Bioengineering p.weinberg@imperial.ac.uk	
	Dr. Sylvain Ladame Joint Lecturer in Biosensor Development Faculty of Engineering, Department of Bioengineering South Kensington Campus s.ladame@imperial.ac.uk	
	Prof. Eric Aboagye Campus Director of the CRUK-EPSRC-MRC-NIHR Comprehensive Cancer Imaging Centre Head of the MRC-CSC Molecular Therapy Group eric.aboagye@imperial.ac.uk	

(continued)

	Dr. Fang Xie Lecturer in Materials Faculty of Engineering, Department of Materials Royal School of Mines, South Kensington Campus f.xie@imperial.ac.uk
	Prof. Martyn Boutelle Faculty of Engineering Royal School of Mines, South Kensington Campus Department of Bioengineering, Professor of Biomedical Sensors Engineering m.boutelle@imperial.ac.uk
	Prof. James Moore, Jr. The Bagrit & RAEng Chair in Medical Device Design Faculty of Engineering, Department of Bioengineering Royal School of Mines, South Kensington Campus james.moore.jr@imperial.ac.uk
	Dr. Jürgen Pozimski Reader in Accelerator Physics Faculty of Natural Sciences, Department of Physics South Kensington Campus j.pozimski@imperial.ac.uk
	Dr. Armando Del Rio Hernandez Senior Lecturer Faculty of Engineering, Department of Bioengineering South Kensington Campus a.del-rio-hernandez@imperial.ac.uk
	Prof. Bob Brown Head of Division of Cancer, Chair in Translational Oncology Faculty of Medicine, Department of Surgery & Cancer Institute of Reproductive and Developmental Biology Hammersmith Campus b.brown@imperial.ac.uk
	Dr. Jennifer Podesta CRUK Centre Administrator Faculty of Medicine, Department of Surgery & Cancer Hammersmith Campus j.podesta@imperial.ac.uk
	Dr. Mengxing Tang Senior Lecturer Faculty of Engineering, Department of Bioengineering Royal School of Mines, South Kensington Campus mengxing.tang@imperial.ac.uk
	Prof. Charles Coombes Professor of Medical Oncology Faculty of Medicine, Department of Surgery & Cancer Hammersmith Campus c.coombes@imperial.ac.uk
	Prof. Margaret Dallman Dean of the Faculty of Natural Sciences and Professor of Immunology Faculty of Natural Sciences, Dept of Immunology m.dallman@imperial.ac.uk

Overview

A Network of Excellence in Cancer Engineering, directed by Prof. Peter Weinberg, has been established within the Imperial College London to promote collaborations among biomedical scientists, including clinicians and physical scientists/engineers. This network arose from a joint initiative between the Imperial Cancer Research UK (CRUK) Centre and the Faculty of Engineering at Imperial College London (ICL). The Cancer Engineering Network of Excellence is one of eight networks within the Institute of Biomedical Engineering, each of which is devoted to a specific area of clinical need.

Research and Development Activities

The scale of the Institute of Biomedical Engineering allows the research within the network to span a large range of topics. A few examples illustrate the broad range of activities in this center and the degree to which basic science and engineering has been directed to clinical problems.

One project in the Ladame lab, develops new reagents and techniques for identifying cancer markers such as DNA released from tumor cells, using, for example, substrates that alter fluorescence as they bind the cancer-related target (Najah et al. 2013). Other projects such as those from the Aboagye lab involve testing of microbubbles and physical methods to detect tumor boundaries such as those in the brain that are not visualized by conventional imaging techniques and to improve methods for analyzing heterogeneity within tumors and surrounding tissue (Gallo et al. 2013, Willaime et al. 2013). New nanomaterials and spectroscopic methods are also being developed in the Xie lab, such as metal enhanced fluorescence for use in biotechnology and diagnostics (Darvill et al. 2013).

Access to bioengineering and mechanical engineering expertise has also fostered the development of new instruments and methods that allows researchers in the Moore lab to study lymphatic vessel structure (Rahbar et al. 2012) and integrate that information into models of lymphatic pumping (Jamalian et al. 2013).

Diagnostic methods based in part on discoveries in bioanalytical sensors are being developed in the Boutelle lab to detect changes in glucose turnover and other aspects of metabolism in intact tissue. The need for such diagnostic devices is motivated by clinical issues in neuropathology as well as cancer and other diseases. Examples of this technology are methods to measure the dynamics of glucose levels (Rogers et al. 2013) and methods to quantify ATP release from tissues (Patel et al. 2011).

Some of the projects are very large-scale and long-term, including participation in the development of a new particle accelerator for medical applications. This facility was designed to deliver protons and carbon ions for charged particle therapy more effectively than existing technology based on cyclotrons (Peach et al. 2013).

Other radiation therapy projects involve the development of novel methods based on theoretical work from the Pozimski group to focus radiation beams more effectively than current methods allow (Pozimski and Meusel 2005).

A new facility lead by Prof. del Río Hernández within the Department of Bioengineering will use high resolution imaging at cell and molecular levels to study how forces impact the structures of specific proteins and alter their functions within cells that potentially contribute to the phenotype of malignant cells and tissues (del Rio et al. 2009).

From an educational perspective, every student is required to have at least two advisors in different disciplines. This approach, which appears to be appreciated by both students and faculty, not only enhances the learning experience but acts as a bridge between different laboratories. A Master's program in biotechnology and entrepreneurship has also been established with a significant interest in issues related to cancer research and treatment.

Translational Efforts

The Moore Lab is employing their models of lymphatic system pumping to design better treatment techniques and implantable devices for secondary lymphedema, which typically develops after cancer surgery involving lymph node removal. Researchers in the Boutelle lab have translated their research devices into clinical use as an integrated instrument for the automatic detection of a range of secondary insults (electrical, physical and chemical) in the injured human brain, and are planning to extend this work to monitoring of human tissue for cancer growth.

Sources of Support

Sources of support include the U.S. NIH and Cancer Research UK. In addition, two Ph.D. studentships have been funded jointly by Scottish Power and the Faculty of Engineering and four M. Res. studentships have been funded by the Wellcome Trust Institutional Strategic Support Fund.

Summary and Conclusions

The Institute of Biomedical Engineering and its connections to departments in both the Faculty of Engineering and clinical cancer research laboratories is a leading center in the United Kingdom for bringing together scientific teams from different disciplines to address important problems in biomedicine. The large number of scientists affiliated with the Institute and the tradition at the Imperial College London for multidisciplinary approaches to address large-scale projects have created an environment and atmosphere where scientists at all stages of their careers can engage for both research purposes and student training.

References

Darvill, D., Centeno, A., Xie, F.: Plasmonic fluorescence enhancement by metal nanostructures: shaping the future of bionanotechnology. Phys. Chem. Chem. Phys. **15**, 15709–15726 (2013)

del Río, A., Perez-Jimenez, R., Liu, R., Roca-Cusachs, P., Fernandez, J.M., Sheetz, M.P.: Stretching single talin rod molecules activates vinculin binding. Science. **323**, 638–641 (2009)

Gallo, J., Long, N.J., Aboagye, E.O.: Magnetic nanoparticles as contrast agents in the diagnosis and treatment of cancer. Chem. Soc. Rev. **42**, 7816–7833 (2013)

Jamalian, S., Bertram, C.D., Richardson, W.J., Moore, J.E. Jr.: Parameter sensitivity analysis of a lumped-parameter model of a chain of lymphangions in series. AJP Heart Circul. Physiol. **305**(12), H1709-717 (2013). doi:10.1152/ajpheart.00403.2013

Najah, M., Mayot, E., Mahendra-Wijaya, I.P., Griffiths, A.D., Ladame, S., Drevelle, A.: New glycosidase substrates for droplet-based microfluidic screening. Anal. Chem. **85**, 9807–9814 (2013)

Patel, B.A., Rogers, M., Wieder, T., O'Hare, D., Boutelle, M.G.: ATP microelectrode biosensor for stable long-term *in vitro* monitoring from gastrointestinal tissue. Biosens. Bioelect. **26**, 2890–2896 (2011)

Peach, K.J., Aslaninejad, M., Barlow, R.J., Beard, C.D., Bliss, N., Cobb, J.H., Easton, M.J., Edgecock, T.R., Fenning, R., Gardner, I.S.K., Hill, M.A., Owen, H.L., Johnstone, C.J., Jones, B. Jones, Jones, T., Kelliher, D.J., Khan, Machida, S., McIntosh, P.A., Pattalwar, S., Pasternak, J., Pozimski, J., Prior, C.R., Rochford, J., Rogers, C.T., Seviour, R., Sheehy, S.L., Smith, S.L., Strachan, J., Tygier, S., Vojnovic, B., Wilson, P., Witte, H., Yokoi, T.: Conceptual design of a nonscaling fixed field alternating gradient accelerator for protons and carbon ions for charged particle therapy. Phys. Rev. Spec. Top. Accel. Beams. **16**, e30101, 34 pp. (2013). 10.1103/PhysRevSTAB.16.030101

Pozimski, J., Meusel, O.: Space charge lenses for particle beams. Rev. Sci. Inst. **76**, 063308. (2005). http://dx.doi.org/10.1063/1.1904203

Rahbar, E., Weimer, J., Gibbs, H., Yeh, A.T., Bertram, C.D., Davis, M.J., Davis MJ, Hill, M.A., Zawieja, D.C., Moore, J.E., Jr.: Passive pressure-diameter relationship and structural composition of rat mesenteric lymphangions. Lymph. Res. Biol. **10**, 152–163 (2012). doi:10.1089/lrb.2011.0015

Rogers, M.L., Feuerstein, D., Leong, C.L., Takagaki, M., Niu, X.Z., Graf, R., Boutelle, M.G.: Continuous online microdialysis using microfluidic sensors: Dynamic neurometabolic changes during spreading depolarization. ACS Chem. Neurosci. **4**, 799–807 (2013). doi:10.1021/cn400047x

Willaime, J.M.Y., Turkheimer, F.E., Kenny, L.M., Aboagye, E.O.: Quantification of intra-tumour cell proliferation heterogeneity using imaging descriptors of 18 F fluorothymidine-positron emission tomography. Phys. Med. Biol. **58**, 187–203 (2013)

Oxford University, Center for Mathematical Biology, Mathematical Institute

Site Address:	Andrew Wiles Building Radcliffe Observatory Quarter Woodstock Road Oxford OX2 6GG United Kingdom http://www.maths.ox.ac.uk/groups/mathematical-biology	UNIVERSITY OF OXFORD
Date Visited:	October 2, 2013	
WTEC Attendees:	Paul Janmey, Daniel Fletcher (report author)	
Host(s):	Prof. Philip Maini, Professor of Mathematical Biology Wolfson Centre for Mathematical Biology, Mathematical Institute Andrew Wiles Building Radcliffe Observatory Quarter, Woodstock Road Oxford OX2 6GG United Kingdom maini@maths.ox.ac.uk	
	Prof. Alain Goriely, Professor of Mathematical Modelling, Director of the Oxford Centre for Collaborative Applied Mathematics (OCCAM) Andrew Wiles Building Radcliffe Observatory Quarter, Woodstock Road Oxford OX2 6GG United Kingdom goriely@maths.ox.ac.uk	
	Prof. Helen Byrne, University Lecturer in Computational Biology Andrew Wiles Building Radcliffe Observatory Quarter, Woodstock Road Oxford OX2 6GG United Kingdombyrneh@maths.ox.ac.uk	
	Prof. David Gavaghan, Professor of Computational Biology Department of Computer Science Wolfson Building, Parks Road Oxford OX1 3QD England david.gavaghan@cs.ox.ac.uk	

Overview

The University of Oxford has a long history as a center of excellence in mathematics. The Wolfson Centre for Mathematical Biology (WCMB), part of the Mathematical Institute, is the home for mathematics applied to biological processes, including development and cancer. The WCMB is led by Prof. Philip Maini and involves approximately 10 faculty and affiliates, as well as numerous research fellows, postdoctoral researchers, and graduate students. Research includes the development and application of analytical and computational techniques applied to a diverse set of complex biological problems. The WCMB also plays an active role in the suite of interdisciplinary doctoral training centers within Oxford.. These programs provide educational and research opportunities for students from a broad range of backgrounds to gain the mathematical skills to make important contributions to the biological sciences.

Research and Development Activities

Research in the WCMB aims to build realistic models of biological processes that advance understanding of fundamental mechanisms. The work includes topics at the intersection of physical sciences, life sciences, and oncology, such as tumor growth, chemotaxis, gene regulatory networks, and pattern formation.

Philip Maini and colleagues recently addressed the link between tumor development and changes in metabolism of cancers (McGillen et al. 2013). In particular, they developed a general reaction-diffusion model to study the Warburg Effect, in which tumors switch to using cytosolic glycolysis rather than oxidative phosphorylation for glucose breakdown. The non-linear reaction diffusion model enabled prediction of the acid-mediated behavior of tumor invasion, including its speed and shape.

Alain Goriely applies mathematical modeling to address the role of large deflection mechanics in biology. Large deflections occur during growth and remodeling and can significantly affect the behavior of dynamic systems. For example, Goriely and colleagues developed a model of cellular blebs that showed the importance of both geometrical and biochemical constraints in the determination of bleb size (Woolley 2013).

The response of tumors to chemotherapy is being studied by Helen Byrne, who has used mathematical modeling based on drug pharmacokinetics to predict the potential of combination therapies. Specifically, Byrne and colleagues found that the use of a platinum-based drug together with inhibition of Bcl-xL could delay the onset of drug resistance in ovarian cancers (Jain 2014).

David Gavaghan has developed a generalized framework for computational physiology and biology called "Chaste" (Cancer, Heart And Soft Tissue Environment; Mirams 2013). This open source C++ library will facilitate the development and testing of cell-based simulations by providing a common framework, including meshes and solvers for ordinary and partial differential equations, that can continue to grow and evolve.

Translation Efforts

The WCMB has close collaborative links with the pharmaceutical company Roche and with the Moffitt Cancer Centre.

Sources of Support

Biotechnology and Biological Sciences Research Council (BBSRC) of the United Kingdom and the U.S. National Cancer Institute (NIH).

Summary and Conclusions

Oxford University and the Wolfson Centre for Mathematical Biology are rich environments for advancing mathematics applied to life sciences. The students involved in the training programs serve as a kind of "glue" that links the diverse mathematical, physical, and biological topics necessary to solve complex problems such as cancer.

References

Jain, H.V., Richardson, A., Meyer-Hermann, M., Byrne, H.M.: Exploiting the synergy between carboplatin and ABT-737 in the treatment of ovarian carcinomas. PLoS One. **9**(1), e81582 (2014)

McGillen, J.B., Gaffney, E.A., Martin, N.K., Maini, P.K.: A general reaction-diffusion model of acidity in cancer invasion. J. Math. Biol. (2013) 10.1007/s00285-013-0665-7, [Epub 28 March 2013]

Mirams, G.R., Arthurs, C.J., Bernabeu, M.O., Bordas, R., Cooper, J., Corrias, A., et al.: Chaste: an open source C++ library for computational physiology and Biology. PLoS Comput. Biol. **9**(3), e1002970 (2013). doi:10.1371/journal.pcbi.1002970

Woolley, T.E., Gaffney, E.A., Oliver, J.M., Baker, R.E., Waters, S.L., Goriely, A.: Cellular blebs: pressure-driven, axisymmetric, membrane protrusions. Biomech. Model Mechanobiol. (2013) [Epub 16 Jul 2013]

The Institute of Cancer Research

Site Address:	237 Fulham Road London SW3 6JB England http://www.icr.ac.uk/	ICR The Institute of Cancer Research
Date Visited:	October 2, 2013	
WTEC Attendees:	Paul Janmey, Daniel Fletcher (report author)	
Host(s):	Dr. Chris Bakal Chester Beatty Laboratories, Division of Cancer Biology The Institute of Cancer Research London SW3 6JB cbakal@icr.ac.uk	

Overview

The Institute of Cancer Research is a leading academic research institute dedicated to advancing cancer treatment through basic research. The ICR specializes in identifying and isolated genes related to cancer and in finding new drugs that specifically target cancer cells. Advanced technologies and computational methods are used for both basic scientific discovery and applied drug screening. Close connection with The Royal Marsden, a nearby cancer hospital, motivates research at ICR and provides a link to clinical applications of basic research.

Research and Development Activities

Dr. Chris Bakal hosted our visit to the Chester Beatty Laboratories of the ICR in London. Dr. Bakal is a biochemist, cell biologist, and computational biologist who runs the Dynamical Cell Systems Team at the ICR. His group uses high-throughput functional genomics, image-based screens, and computational analysis methods to investigate signal transduction in multicellular systems and uncover aberrations in cancer, with a particular interest in breast cancer and pancreatic cancer.

Recently, Dr. Bakal and his team combined RNAi screens with high-throughput microscopy to identify key regulators of cell shape (Yin et al. 2013). Using *Drosophila* hemocytes as a model system, Dr. Bakal developed computational methods to automatically characterize cell shape from microscopy images and identified five discrete morphologies of hemocytes. An RNAi screen was then used to determine whether variations in gene expression would cause cell shape to smoothly transition from one morphology to another, or whether cell shape was restricted to discrete morphologies. Dr. Bakal found that cell shape was restricted to discrete morphologies, and that changes in gene expression could decrease the shape heterogeneity of the hemocyte population, causing enrichment of rounded and elongated cells. The genes responsible for this include the tumor suppressor PTEN and serve as shape regulators in metastatic melanoma cells.

The image-based screens developed by Dr. Bakal and his team are powerful tools for understanding signal transduction networks (Evans 2013). They can be used together with RNAi to explore the effect of oxidative stress (Garcia 2012) and can be extended to investigate the role of such factors as substrate stiffness, cell densities, media composition, and dynamic changes in cellular environments.

While technology for carrying out high-throughput screens has advanced, a growing challenge for image-based screens is the need for data handling support and methods for rapidly analyzing large data sets. Further developments in "big data" methods and hardware will help to expand the accessibility of the approaches advanced by Dr. Bakal and colleagues.

Summary and Conclusions

The Institute of Cancer Research is a highly interdisciplinary environment that combines basic research with drug discovery to pursue new cancer therapies. The use of high-throughput screens by the Bakal group is a promising direction for further research.

References

Evans, L., Sailem, H., Vargas, P.P., Bakal, C.: Inferring signalling networks from images. J. Microsc. **252**(1), 1–7 (2013)

Garcia, M.A., Alvarez, M.S., Sailem, H., Bousgouni, V., Sero, J., Bakal, C.: Differential RNAi screening provides insights into the rewiring of signalling networks during oxidative stress. Mol. Biosyst. **8**(10), 2605–2613 (2012)

Yin, Z., Sadok, A., Sailem, H., McCarthy, A., Xia, X., Li, F., Garcia, M.A., Evans, L., Barr, A.R., Perrimon, N., Marshall C.J., Wong, S.T.C., Bakal, C.: A screen for morphological complexity identifies regulators of switch-like transitions between discrete cell shapes. Nat. Cell Biol. **5**(7), 860–871 (2013)

The University of Sheffield (Virtual Site Visit Report)[2]

Site Address:	The University of Sheffield Western Bank Sheffield S10 2TN United Kingdom
Date Visited:	September 20, 2013
WTEC Attendees:	Owen McCarty (report author)
Host(s):	Prof. Graham Leggett Department of Chemistry Tel.: +44-(0)114-22-29556 Fax: +44-(0)114-22-29436 graham.leggett@sheffield.ac.uk

Overview

Winner of the UK University of the Year in 2011, the University of Sheffield is home to over 25,000 students from 117 countries. The more than 1,200 academic staff members work within the 50 academic departments and 82 research centers and institutes that make up the University.

Dr. Graham Leggett, a Professor of Nanoscale Analytical Chemistry at the University of Sheffield, serves as the Network Chair for the UK Engineering and Physical Sciences Research Council (EPSRC) Grand Challenge Network entitled "Understanding the Physics of Life." The network consists of a steering group of 11 faculty members from the University of Cambridge, Leeds, Oxford, Bristol, Sheffield, York, Edinburgh, Durham and Liverpool within the UK. The goal of network is to develop a comprehensive understanding of biology through an integrative approach across length and time scales—from molecules to systems.

The Physics of Life network was launched in April 2013, and has organized a series of Plenary Events on topics including The Living Cell, Synthetic Biology, and Multicellularity. Moreover, a series of focused workshops are planned on the topics of coupling of nuclei and cells, molecular machines, coupling lower level systems to their environment, quantum mechanics and life, and diversity, ergodicity and optimization in evolution. The development of early stage investigators will be facilitated through summer schools planned for 2014 and 2015. New collaborative partnerships will be supported through funds for travel between laboratories to generate proof-of-concept work.

[2]This report was prepared based on the observations of one member of the WTEC panel, not the full panel.

Research and Development Activities

Building on the fundamentals of physical chemistry and analytical chemistry, the Leggett group is focused on defining the physical properties of molecular surfaces at the nanometer scale. The goal is to develop a fundamental understanding of how nanoproperties regulate biological and physical processes.

The main research interests of the Leggett group are in the areas of nanofabrication and nanotribology. Their work in nanofabrication relies on the use of a photochemical platform to design selective molecular transformations in nanometer-scale regions on surfaces. In particular, they utilize near-field optical probes to initiate selective chemical transformations, resulting in self-assembled monolayers with a resolution of 9 nm. Examples of nanofabricated structures are illustrated in Fig. D.1.

Nanotribology is the study of sliding contacts between materials. The Leggett group utilizes atomic force microscopy (AFM) techniques to study nanometer-scale contacts between molecular materials. Their work has established a direct link between solution-phase thermodynamics and hydrogen bonding in nanoscale molecular contacts (Busuttil et al. 2012).

Translational Efforts

The characterization of nanoscale properties is required for the development of organic thin films and monolayers for lubrication of miniaturized devices, such as microelectromechanical systems. Moreover, development of approaches to characterize molecular nanostructures, such as friction force microscopy, is utilized for mapping variations in chemical structure and composition of surfaces.

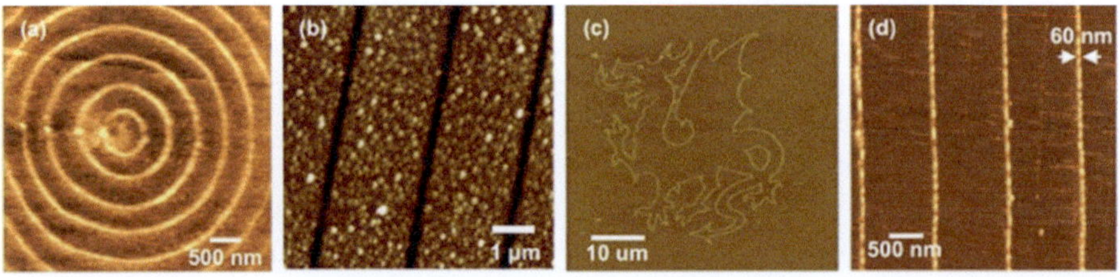

Fig. D.1 Nanofabricated structures from the work of Shuqing Sun and colleagues. (From Sun et al. 2006)
(*a*) Concentric rings of carboxylic acid groups fabricated on silicon surfaces by photochemical dehalogenation of a benzylchloride functionalized silane. (Sun et al. 2006); (*b*) 100 nm trenches fabricated in aluminum by the use of scanning near-field photolithography of alkylphosphonate SAMs on aluminum oxide, followed by wet etching through the SAM resist. (Sun and Leggett 2007); (*c*) Dragon structure formed by photochemical modification of a SAM of aryl azide functionalized phosphonic acids on aluminum oxide. (El Zubir et al. 2013); (*d*) 60 nm gold nanowires formed by near-field sintering of trilayers of thiol-stabilized gold nanoparticles on a silicon substrate

Sources of Support

The research within the University of Sheffield is supported by a variety of UK Research Councils, including the Biotechnology and Biological Sciences Research Council, Engineering and Physical Sciences Research Council, and the Medical Research Council.

Summary and Conclusions

The University of Sheffield is a leader in both education and research in the UK. The Department of Chemistry has notable strengths in chemical biology, polymer and surface chemistry, spectroscopy and synthetic chemistry research. The Physics of Life network builds on these interdisciplinary research clusters in order to develop an understanding from single molecules to systems in biology.

References

Busuttil, K., Nikogeorgos, N., Zhang, Z., Geoghegan, M., Hunter, C.A., Leggett, G.J.: The mechanics of nanometre-scale molecular contacts. Faraday Disc. **156**, 325 (2012)

El Zubir, O., Barlow, I., ul Haq, E., Tajuddin, H., Williams, N.H., Leggett, G.J.: Generic methods for micrometer- and nanometer-scale surface derivatization based on photochemical coupling of primary amines to monolayers of aryl azides on gold and aluminum oxide surfaces. Langmuir. **29**, 1083 (2013)

Sun, S., Leggett, G. J.: Micrometer and nanometer scale photopatterning of self-assembled monolayers of phosphonic acids on aluminum oxide. Nano. Lett. **7**, 3753–3758 (2007)

Sun, S., Montague, M., Critchley, K., Chen, M.S., Dressick, W.J., Evans, S.D., Leggett, G.J.: Fabrication of biological nanostructures by scanning near-field photolithography of chloromethylphenylsiloxane monolayers. Nano Lett. **6**, 29–33 (2006)

Sun, S., Mendes, P., Critchley, K., Diegoli, S., Hanwell, M., Evans, S.D., Leggett, G.J., Preece, J.A., Richardson, T.A.: Fabrication of gold micro- and nanostructures by photolithographic exposure of thiol-stabilized gold nanoparticles. Nano Lett. **6**, 345–349 (2006)

University of Dundee, College of Life Sciences

Site Address:	Dow Street Dundee DD1 5EH United Kingdom http://www.lifesci.dundee.ac.uk/
Date Visited:	October 4, 2013
WTEC Attendees:	Paul Janmey, Daniel Fletcher (report author)
Host(s):	Professor Timothy Newman Biophysics & SULSA Research & Systems Biology Division of Cell and Developmental Biology & Division of Computational Biology College of Life Sciences University of Dundee Dundee, United Kingdom T.Newman@dundee.ac.uk
	Professor Alastair Thompson University of Dundee, Nethergate, Dundee DD1 4HN United Kingdom a.m.thompson@dundee.ac.uk
	Professor Alastair Munro Radiation Oncology Tayside Cancer Centre Ninewells Hospital & Medical School Dundee DD1 9SY United Kingdom a.j.munro@dundee.ac.uk
	Dr. Andrew South Assistant Professor, UC Department of Pediatrics Senior Lecturer Division of Cancer Research, Medical Research Institute Jacqui Wood Cancer Centre, University of Dundee James Arrott Drive Ninewells Hospital And Medical School Dundee DD1 9SY United Kingdom A.P.South@dundee.ac.uk
	Professor Mark Chaplain Ivory Chair of Applied Mathematics Division of Mathematics University of Dundee Dundee DD1 4HN United Kingdom m.a.j.chaplain@dundee.ac.uk

	Professor Julian Blow Head of The Centre for Gene Regulation and Expression College of Life Sciences University of Dundee Dundee DD1 5EH United Kingdom j.j.blow@dundee.ac.uk
	Professor Inke Näthke Epithelial Biology and Cancer Cell and Developmental Biology Division College of Life Sciences University of Dundee Dundee DD1 5EH United Kingdom i.s.nathke@dundee.ac.uk
	Professor Edik Rafailov Photonics and Nanoscience Group Harris Building University of Dundee, Nethergate Dundee DD1 4HN United Kingdom E.U.Rafailov@dundee.ac.uk
	Professor Mike Ferguson Molecular Parasitology and Dean of Research College of Life Sciences University of Dundee Dundee DD1 5EH United Kingdom m.a.j.ferguson@dundee.ac.uk
	Professor Sandy Cochran Biophysical Science and Engineering, Team Leader Medical Ultrasound at the Institute for Medical Science and Technology University of Dundee Wilson House, 1 Wurzburg Loan Dundee Medipark Dundee DD2 1FD S.Cochran@dundee.ac.uk

Overview

The University of Dundee has established a vibrant community of physicists, biologists, physicians, engineers, and mathematicians who are developing novel approaches to address biological problems, with an emphasis on their clinical implications. Under the leadership of Prof. Tim Newman, the College of Life Sciences at the University of Dundee has embraced computational approaches to biology and formed strong connections with the clinical centers through the Public Health

System and Cancer Research UK Dundee. Research teams resulting from active interaction among those at the interface of life sciences, physical sciences, and oncology are attracting funding and producing exciting research advances.

Research and Development Activities

Researchers at the University of Dundee are pursuing a broad range of research applying physical science approaches to cancer and development. The work can be broadly grouped into biological modeling, tissue engineering, and advanced imaging. This report captures only a fraction of the excellent work that was presented during our brief visit.

To understand the robustness of gene regulatory networks through modeling, Tim Newman is applying the concept of buffered qualitative stability (BQS), which states that robust feedback loops cannot have more than two elements. Newman and co-workers are developing the theoretical framework to address robustness and feedback in gene regulatory networks involved in normal and cancer cells. Intriguingly, normal cells appear to follow the rule of having no feedback beyond two elements, while the gene regulatory networks of cancer cells do, suggesting that this increased feedback could be involved in the loss of regulation in cancer. Additional work by the Newman group is examining the fluctuations of genetic feedback loops (Grima 2012) and rare event statistics of metastasis.

Inke Näthke and Andrew South are advancing tissue biology and engineering through the development of *in vitro* tissue systems. Ultrasound imaging has revealed stiffness differences associated with cancer of the digestive tract, and Näthke is exploring cell and tissue polarity in the small intestine and its connection with tissue development and cancer. In a recent study of tissue organoids, Näthke and colleagues showed that tissue polarity is one of the earliest properties lost during cancer progression, while cell polarity is retained until later in the progression process (Fatehullah 2013). Andrew South has developed a skin primary cell culture system based on dissociation of cells from tissue and selective expansion to get specific cell types. The system is providing new insight into the role of the extracellular matrix proteins in tumor behavior, such as that of Type VII collagen (Dayal 2013).

The Institute for Medical Science and Technology (IMSaT) at the University of Dundee is a hub for development and application of novel technologies. Mike MacDonald is developing new light sheet tomography techniques for *in situ* imaging (Yang 2013), and Sandy Cochran is pioneering new high resolution ultrasound methods for elastography that permit the measurement of mechanical properties *in situ* (Munirama 2013).

Ninewells Hospital in Dundee provides a critical connection to real-world problems in cancer that help to guide research and development of new technologies in Dundee. Luc Bidaut, a physician, heads Imaging and Technology at the University of Dundee School of Medicine, where he is advancing the use of multi-modality imaging and image guided medicine. Alistair Munro, a radiation oncologist also based at Ninewells, is an expert in radiotherapy and is pursuing modeling to enable better design of radiation therapy doses to avoid "bystander" effects.

Summary and Conclusions

The impressive breadth and concentration of scientists, engineers, and physicians who collaborate are making the University of Dundee a leader in applying physical sciences to oncology research and clinical treatment. Training of students and post-doctoral fellows at this interface is a major goal and challenge, and the University of Dundee could be a model of interdisciplinary research and education for others to follow.

References

Dayal, J.H., Cole, C.L., Pourreyron, C., Watt, S.A., Lim, Y.Z., Salas-Alanis, J.C., Murrell, D.F., McGrath, J.A., Stieger, B., Jahoda, C., Leigh, I.M., South, A.P.: Type VII collagen regulates tumour expression of organic anion transporting polypeptide OATP1B3, promotes front to rear polarity and increases structural organisation in 3D spheroid cultures of recessive dystrophic epidermolysis bullosa tumour keratinocytes. J. Cell. Sci. (2013). doi: 10.1242/jcs.128454. [Epub 19 Dec 2013]

Fatehullah, A., Appleton, P.L., Näthke, I.S.: Cell and tissue polarity in the intestinal tract during tumourigenesis: cells still know the right way up, but tissue organization is lost. Philos. Trans. R Soc. Lond. B Biol. Sci. **368**(1629), 20130014 (2013)

Grima, R., Schmidt, D., Newman, T. J.: Steady-state fluctuations of a genetic feedback loop: an exact solution. J. Chem. Phys. **137**, 035104 (2012)

Munirama, S., Joy, J., Eisma, R., Corner, G., Cochran, S., McLeod, G.: Images in anesthesiology: shear wave elastography: novel technology for ultrasound-guided regional anesthesia. Anesthesiology. **119**(3), 698 (2013)

Yang, Z., Downie, H., Rozbicki, E., Dupuy, L.X., MacDonald, M.P.: Light Sheet Tomography (LST) for *in situ* imaging of plant roots. Opt. Expr. 21(14), 16239–16247 (2013)

Appendix E. Site Visit Reports – Brazil

Site visit reports are arranged in alphabetical order by organization name.

P. Janmey et al. (eds.), *Physical Sciences and Engineering Advances in Life Sciences and Oncology*, Science Policy Reports,
DOI 10.1007/978-3-319-17930-8

Brazilian National Cancer Institute (INCA; Virtual Site Visit Report)[3]

Site Address:	Brazilian National Cancer Institute (INCA) Rua André Cavalcant Rio de Janeiro, RJ Brazil
Date Visited:	September 24, 2013
WTEC Attendees:	Owen McCarty (report author), Nastaran Kuhn
Host(s):	Dr. José Andrés Morgado-Díaz Divisão de Biologia Celular, Centro de Pesquisas jmorgado@inca.gov.br
	Dr. Luiz Felipe Ribeiro Pinto Departamento de Bioquímica Universidade do Estado do Rio de Janeiro lfrpinto@inca.gov.br

Overview

Founded in 1937, the mission of the Brazilian National Cancer Institute (INCA) is to lead a country-wide policy for cancer control in Brazil. To achieve this goal, the efforts of INCA are focused in the following areas:

- Prevention
- Diagnosis
- Treatment
- Supportive and Palliative Care
- Education and Research
- Cancer Registries

The research mission at INCA is focused in four areas: cellular biology, pharmacology, genetics, and immunology. The education mission of INCA is to train professionals to meet the needs of the SUS throughout Brazil.

Research and Development Activities

Dr. José Andrés Morgado-Díaz

The incidence of colorectal cancer has dramatically increased over the past decade in Brazil (Jemal et al. 2011) The Morgado-Diaz lab is focused on the development and characterization of inhibitors to interfere with the process of malignant

[3]This report was prepared based on the observations of one member of the WTEC panel accompanied by a sponsor representative, not the full panel.

transformation. For instance, the Diaz group has recently characterized the anticancer activity of two N-glycan biosynthesis inhibitors (swainsonine and tunicamycin) in an *in vitro* model of colorectral cancer colony formation, migration and invasion (de Freitas Junior et al. 2012). Moreover, they demonstrated that the combination of swainsonine with chemotherapy drugs, such as cisplatin or irinotecan, enhanced the toxicity of swainsonine in HCT-116 cells as compared to cisplatin or irinotecan alone. As aberrant glycosylation is one of the hallmarks of epithelial tumor progression, the goal of this line of work is to provide a rationale for the development of N-glycan biosynthesis inhibitors for the treatment of colorectal cancer.

Dr. Luiz Felipe Ribeiro Pinto

Dr. Pinto serves as the Head of Education for INCA. His laboratory group is focused on the identification of genetic mutations in carcinomas, in order to provide a platform for predicting patient response to therapy. As an example, they have recently characterized the mutations and polymorphisms in human epidermal growth factor receptors, EGFR and HER2, in esophageal squamous cell carcinomas (Gonzaga et al. 2012). The results demonstrated that only a small percentage of patient samples (<10 %) had mutations in the EGFR pathway or overexpression of either EGFR or HER2, suggesting that this pathway is only altered in a minority of in esophageal squamous cell carcinoma patients. This work suggests that HER receptor-targeted therapies would only be efficacious in a minority of this patient group.

Translational Efforts

The characterization of the aberrant signaling pathways present in tumor samples provides rationale for the utility and development of targeted therapies for patients with cancer.

Sources of Support

INCA is a technical branch of the Federal Government, and is under the direct administration of the Ministry of Health. INCA is primarily funded by the Integrated Public Health System (SUS).

Summary and Conclusions

Brazil faces a unique challenge as the disease burden in the population shifts from infectious to chronic diseases, such as heart disease, stroke, and cancer. Brazil has experienced a marked increase in the rates of prostate and breast cancer in the more industrialized regions, while cervical cancer is still major threat in regions of greater poverty. The majority of cancers are diagnosed at a late and often incurable stage. The mission of INCA is to train the next generation of Brazilian physicians and scientists to meet this challenge.

References

de-Freitas-Junior, J.C., Bastos, L.G., Freire-Neto, C.A., Rocher, B.D., Abdelhay, E.S., Morgado-Díaz, J.A.: N-glycan biosynthesis inhibitors induce *in vitro* anticancer activity in colorectal cancer cells. J. Cell Biochem. **113**(9), 2957–2966 (2012)

Gonzaga, I.M., Soares-Lima, S.C., de Santos, P.T., Blanco, T.C, de Reis, B.S., Quintella, D.C., de Oliveira, I.M., de Faria, P.A., Kruel, C.D., Andreollo, N.A., de Simão, T.A., Pinto, L.F.: Alterations in epidermal growth factor receptors 1 and 2 in esophageal squamous cell carcinomas. BMC Cancer. **12**, 569 (2012)

Jemal, A., Bray, F., Center, M.M., Ferlay, J., Ward, E., Forman, D.: Global cancer statistics. CA Cancer J. Clin. **61**(2), 69–90 (2011)

Federal University of Rio de Janeiro (UFRJ; Virtual Site Visit Report)[4]

Site Address:	Av. Erasmo Braga Rio de Janeiro, RJ Brazil
Date Visited:	September 25, 2013
WTEC Attendees:	Owen McCarty (report author), Nastaran Kuhn
Host(s):	Dr. Jerson Lima da Silva Institute of Medical Biochemistry Tel.: 55-21-25626756 Fax: 55-21-38814155 jerson@bioqmed.ufrj.br
	Dr. Robson de Queiroz Monteiro Institute of Medical Biochemistry Te.l: +55 212562 6782 Fax: +55 212270 8647 robsonqm@bioqmed.ufrj.br

[4] This report was prepared based on the observations of one member of the WTEC panel accompanied by a sponsor representative, not the full panel.

Overview

As the largest public university in Brazil, the Universidade Federal do Rio de Janeiro (UFRJ) was founded in 1920, UFRJ is home to over 50,000 students and nearly 10,000 faculty and staff. Spread over 3 campuses, UFRJ consists of schools of medicine and health sciences, engineering and technology, natural and mathematical sciences, law and economics, language and fine arts, and philosophy and humanities.

Research and Development Activities

Dr. Robson de Queiroz Monteiro

Dr. Monteiro heads the Laboratory for Thrombosis and Cancer with the UFRJ Institute of Medical Biochemistry. His research mission is to define the role that blood coagulation proteins and cells play in cancer metastasis, as well as to identify the mechanisms underlying the prothrombotic phenotype associated with malignant neoplasms. His group's research efforts are focused in four areas: (1) functional characterization, spanning from *in vitro* to *in vivo* studies, of novel antithrombotic therapies; (2) elucidation of the role of crosstalk between blood coagulation enzymes and cell surface protease-activated receptors in cancer metastasis; (3) evaluation of the antitumor/antiangiogenic efficacy of novel inhibitors of blood coagulation factors; (4) characterization of the role of microvesicles (fragments of plasma membranes, often termed exosomes or microparticles) in the prothrombotic phenotype of malignant neoplasms.

In a recent study designed to test the hypothesis that inhibition of the procoagulant action of tumor cells would inhibit tumor growth and metastasis, the Monteiro group demonstrated that the tick salivary protein, ixolaris, which inhibits tissue factor (Carneiro-Lobo et al. 2012), reduced both the primary tumor growth and the metastatic potential in a murine model of melanoma (Fig. E.1, de Oliveira et al. 2012). Moreover, they demonstrated that ixolaris inhibited of tumor angiogenesis. This study was designed to provide rationale for the development of inhibitors of the tissue factor pathway for the treatment of aggressive malignancy.

Dr. Jerson Lima da Silva

Trained in medicine and biophysics at UFRJ, Dr. Silva serves as the Scientific Director of the Rio de Janeiro State Funding Agency and the Coordinator for the National Institute of Science and Technology for Structural Biology and Bioimaging (INBEB) at UFRJ.

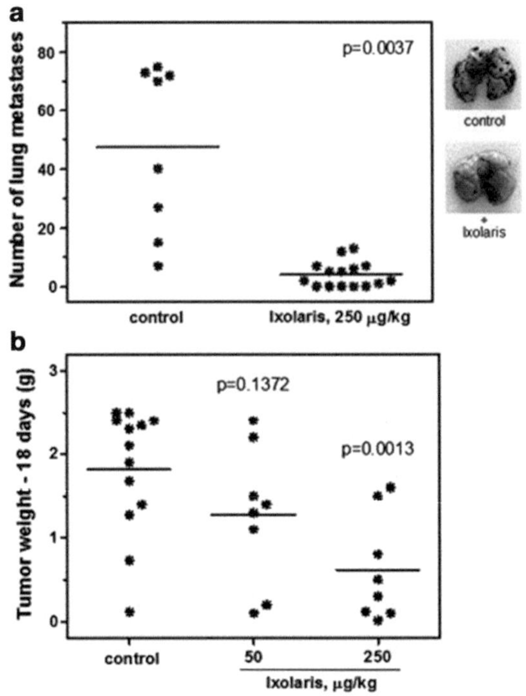

Fig. E.1 Ixolaris inhibits the establishment of B16F10 melanoma cells *in vivo*. (Adapted from de Oliveira et al. 2012)
(*A*) B16F10 cells (2.5 × 105) were injected intravenously into C57BL/6 mice after previous injection of ixolaris (250 μg/kg) or PBS. After 15 days, tumor nodules on harvested lungs were counted. Lines represent the mean of the number of pulmonary tumor nodules observed in the animals from each group. Representative lungs of each group are shown. (*B*) B16F10 cells (3.5 × 105) were injected subcutaneously into C57BL/6 mice. Daily treatment with ixolaris was initiated 3 days after tumor cell inoculation. Control animals were treated with an equivalent volume of PBS. After 18 days of tumor cell inoculation, the animals were sacrificed and the tumors were removed and weighed. Lines represent the mean of tumor weight in each group

The INBEB is comprised of faculty from over 20 research groups from over 20 institutions representing all 7 Brazilian states. The mission of the INBEB is to create and consolidate a technical and scientific infrastructure that facilitates the study of the structure of biological systems (from the macromolecular level to the whole-organism level) while making use of the most advanced analytical techniques and the highest possible resolution images. The research initiatives underway at the INBEB are five-fold: (1) the study of macromolecules involved in infectious diseases, neurodegenerative illnesses, and cancer; (2) the study of viruses, such as Dengue fever, yellow fever, and others; (3) the study of complex structures found in protozoan parasites that are the agents responsible for causing relevant illnesses such as Leishmaniasis, Chagas disease, malaria, and toxoplasmosis; (4) monitoring the evolution of viral and protozoan parasite infections in small experimental animals and their behavior in animals undergoing experimental chemotherapy; and (5) the study of the *in vivo* behavior of stem cells in order to analyze their biodistribution, localization, and function as cellular therapies for degenerative diseases.

Dr. Silva's research interest is in the study of the fundamental mechanisms of protein folding, and the development of an understanding of the role of protein misfolding and aggregation in neurodegenerative diseases and cancer (Silva et al. 2013). As an example, his group is working to understand how the aggregation of p53, a nuclear phosphoprotein which acts as a tumor suppressor by regulating cell-cycle, may contribute to the loss of p53's tumor suppressor function (Ano Bom et al. 2012). This work spans from investigation at the protein level to the cell, tissue, and organism level in order to understand the role of p53 misfolding in cancer (Fig. E.2).

Translational Efforts

UFRJ is nationally recognized as a center for excellence in innovative entrepreneurship. As an example, the format of the INBEB has been designed to foster translation of imaging approaches for diagnostic and therapeutic utility in diseases ranging from cancer to neurodegenerative diseases.

Sources of Support

UFRJ is supported as part of the mission of the Brazilian Ministry of Education by both the state and federal government.

Summary and Conclusions

Brazil invests over 5 % of its GDP on education, with a goal of increasing this towards 7 %. Brazil has a rich history in leading the world's research efforts in infectious diseases, such as Chagas disease. UFRJ is in a strong position for the training and education of the next generation of researchers which utilize physical biology approaches to study complex diseases.

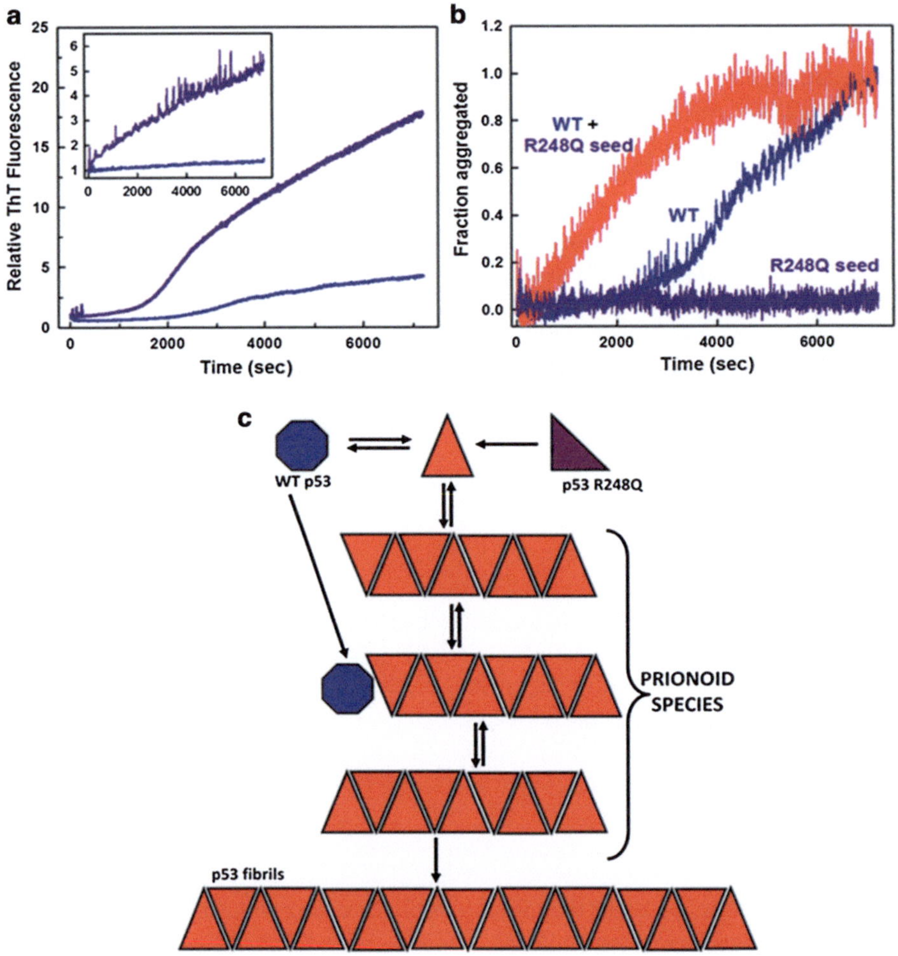

Fig. E.2 Dominant-negative phenomenon and gain-of-function prion-like effect. (Adapted from Ano Bom et al. 2012)
(*A*) Aggregation kinetics of WT (*blue*) or R248Q p53C (*purple*) at 37 °C at pH 7.2 or 5.0 (inset). (*B*) Aggregation of WT p53 at 10 μM without a seed (*blue*) or seeded with aggregated R248Q (*red*); R248Q was seeded alone at 2 μM as a control (*purple*). Aggregation of WT p53 or R248Q was monitored by thioflavin T binding at 37 °C for 30 min. (*C*), Scheme showing conversion of native WT p53 (*blue*) or R248Q (*purple*) into a misfolded species (*red triangle*) that will further aggregate. The prionoid species are oligomers and protofibrils that bind anti-oligomer antibody

References

Ano Bom, A.P., Rangel, L.P., Costa, D.C., de Oliveira, G.A., Sanches, D., Braga, C.A., Gava, L.M., Ramos, C.H., Cepeda, A.O., Stumbo, A.C., De Moura Gallo, C.V., Cordeiro, Y., Silva, J.L.: Mutant p53 aggregates into prion-like amyloid oligomers and fibrils: implications for cancer. J. Biol. Chem. **287**(33), 28152–28162 (2012)

Carneiro-Lobo, T.C., Schaffner, F., Disse, J., Ostergaard, H., Francischetti, I.M., Monteiro, R.Q., Ruf, W.: The tick-derived inhibitor Ixolaris prevents tissue factor signaling on tumor cells. J. Thromb. Haemost. **10**(9), 1849–1858 (2012)

de Oliveira, A.S., Lima, L.G., Mariano-Oliveira, A., Machado, D.E., Nasciutti, L.E., Andersen, J.F., Petersen, L.C., Francischetti, I.M., Monteiro, R.Q.: Inhibition of tissue factor by ixolaris reduces primary tumor growth and experimental metastasis in a murine model of melanoma. Thromb. Res. **130**(3), e163–e170 (2012). doi:10.1016/j.thromres.2012.05.021

Silva, J.L., Rangel, L.P., Costa, D.C., Cordeiro, Y., De Moura Gallo, C.V.: Expanding the prion concept to cancer biology: dominant-negative effect of aggregates of mutant p53 tumor suppressor. Biosci. Rep. **33**(4), 593–603 (2013)

Appendix F. Recent Conferences

The vitality of this field in Europe and Asia is also evident in the number of innovative conferences on the physics of cancer. A sample of these conference and workshops is listed below:

P. Janmey et al. (eds.), *Physical Sciences and Engineering Advances in Life Sciences and Oncology*, Science Policy Reports,
DOI 10.1007/978-3-319-17930-8

Physics of Cancer

European Science Foundation Exploratory Workshop; Physical and Engineering Sciences and Medical Sciences
Convened by Stefano Zapperi and Caterina La Porta
13–15 September 2012 – Varenna, Italy

International Conference on Translational Research in Radio-Oncology and CERN's Physics for Health in Europe

European Organization for Nuclear Research
10–14 February 2014 – Geneva, Switzerland

Physics of Living Matter

University of Cambridge
19–20 September 2013 – Cambridge, United Kingdom

Computational Physics Methods for Cancer

Centre Européen de Calcul Atomique et Moléculaire
27–29 June 2012 – Lausanne, Switzerland

PhysCell 2012

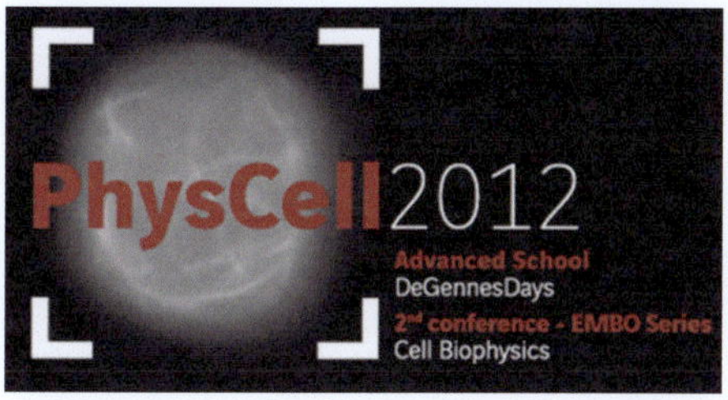

Advanced School on Cellular Biophysics; Conference Physics of Cell: From Soft to Living Matter
Fondation Pierre-Gilles de Gennes and the European Molecular Biology Organization
2–8 September 2012 – Hyères, France

4th Annual Symposium: Physics of Cancer

International Meeting of the German Society for Cell Biology
24–27 September 2013 – Leipzig, Germany

WTEC Publications

WTEC Books

Stem Cell Engineering: A WTEC Global Assessment. Robert M. Nerem, Jeanne Loring, Todd C. McDevitt, Sean P. Palecek, David V. Shaffer, Peter W. Zandstra (Ed.) Springer, 2013.

Convergence of Knowledge, Technology, and Society: Beyond Convergence of Nano-Bio-Info-Cognitive Technologies. Mihail C. Roco, William S. Bainbridge, Bruce Tonn, George Whitesides (Ed.) Springer, 2013.

Nanotechnology Research Directions for Societal Needs in 2020: Retrospective and Outlook. Mihail C. Roco, Chad Mirkin, and Mark Hersam (Ed.) Springer, 2011.

International Assessment of Research and Development in Simulation-Based Engineering and Science. S. C. Glotzer (Ed.) Imperial College Press, 2011

International Assessment of Research and Development in Catalysis by Nanostructured Materials. R. Davis (Ed.) Imperial College Press, 2011

Brain-Computer Interfaces: An International Assessment of Research and Development Trends. Ted Berger (Ed.) Springer, 2008.

Robotics: State of the Art and Future Challenges. George Bekey (Ed.) Imperial College Press, 2008.

Micromanufacturing: International Research and Development. Kori Ehmann (Ed.) Springer, 2007.

Systems Biology: International Research and Development. Marvin Cassman (Ed.) Springer, 2007.

Nanotechnology: Societal Implications. Mihail Roco and William Bainbridge (Eds.) Springer, 2006. Two volumes.

Biosensing: International Research and Development. J. Shultz (Ed.) Springer, 2006.

Spin Electronics. D.D. Awschalom et al. (Eds.) Kluwer Academic Publishers, 2004.

P. Janmey et al. (eds.), *Physical Sciences and Engineering Advances in Life Sciences and Oncology*, Science Policy Reports,
DOI 10.1007/978-3-319-17930-8

Converging Technologies for Improving Human Performance: Nanotechnology, Biotechnology, Information Technology and Cognitive Science. Mihail Roco and William Brainbridge (Eds.) Kluwer Academic Publishers, 2004.
Tissue Engineering Research. Larry McIntire (Ed.) Academic Press, 2003.
Applying Molecular and Materials Modeling. Phillip Westmoreland (Ed.) Kluwer Academic Publishers, 2002
Societal Implications of Nanoscience and Nanotechnology. Mihail Roco and William Brainbridge (Eds.) Kluwer Academic Publishers, 2001.
Nanostructure Science and Technology: R&D Status and Trends in Nanoparticles, Nanostructured Materials and Nanodevices. R.S. Siegel, E. Hu, and M.C. Roco (Eds.) Kluwer Academic Publishers, 2000.

Selected WTEC Panel Reports

Assessment of Physical Sciences and Engineering Advances in Life Sciences and Oncology (APHELION) in Europe. P. Janmey (Ed.) (8/2012).
International Assessment of Research and Development in Human-Robot Interaction (HRI) M. Veloso (Ed.) (5/2012).
European Research and Development in Mobility Technology for People with Disabilities. D. Reinkensmeyer (Ed.) (8/2011).
International Assessment of Research and Development in Rapid Vaccine Manufacturing. J. Bielitzki (Ed.) (7/2011).
International Assessment of Research and Development in Flexible Hybrid Electronics. A. Dodabalapur (Ed.) (7/2010).
Research and Development in Carbon Nanotube Manufacturing and Applications. P. C. Eklund (Ed.) (6/2007).
High-End Computing Research and Development in Japan. A. Trivelpiece (Ed.) (12/2004).
Additive/Subtractive Manufacturing Research and Development in Europe. J.L. Beaman (Ed.) (11/2004).
Microsystems Research in Japan. R. T. Howe (Ed.) (9/2003).
Environmentally Benign Manufacturing. T. Gutowski and C. Murphy (Eds.) (4/2001).
Wireless Technologies and Information Networks. A. Ephremides (Ed.) (7/2000).

Selected Workshop Reports Published by WTEC

NNI Supplement to the President's 2013 Budget (2/2012).
Regional, State, and Local Initiatives in Nanotechnology (2/2011).
NanoEHS Series: Capstone: Risk Management Methods & Ethical, Legal, and Societal Implications of Nanotechnology (3/2010).

Defense Nanotechnology Research and Development Program: Report to Congress (12/2009).
International assessment of R&D in Stem Cells for Regenerative Medicine and Tissue Engineering (4/2008).
Manufacturing at the Nanoscale (2007).
Building Electronic Function into Nanoscale Molecular Architectures (6/2007).
Infrastructure Needs of Systems Biology (5/2007).
X-Rays and Neutrons: Essential Tools for Nanoscience Research (6/2005).
Sensors for Environmental Observatories (12/2004).
Nanotechnology in Space Exploration (8/2004).
Nanoscience Research for Energy Needs (3/2004).
Nanoelectronics, Nanophotonics, and Nanomagnetics (2/2004).

Selected Staff Research Papers

Publish or Patent: Bibliometric Evidence for Empirical Trade-offs in National Funding Strategies. *Journal of the American Society for Information Science and Technology.* Vol. 63(3): 498-511. R.D. Shelton and L. Leydesdorff (2012).
The Race for World Leadership of Science and Technology: Status and Forecasts. *Science Focus* Vol. 5, No. 1, pp. 1-9 (Feb. 2010) in Chinese. Also, Proceedings of the 12th International Conference on Scientometrics and Informetrics, pp. 369-380, Rio de Janeiro, July, 2009. R. D. Shelton and P. Foland.
Relations Between National Research Investment Input and Publication Output: Application to an American Paradox, 9th International Conference on S&T Indicators, Leuven, Sept., 2006 and Scientometrics. Vol. 74 No. 2, 191-205, Feb., 2008. R.D. Shelton.

All WTEC reports are available on the Web at http://www.wtec.org.
Webcasts of recent workshops are available at http://www.tvworldwide.com.

MIX
Papier aus verantwortungsvollen Quellen
Paper from responsible sources
FSC® C105338

If you have any concerns about our products,
you can contact us on
ProductSafety@springernature.com

In case Publisher is established outside the EU,
the EU authorized representative is:
Springer Nature Customer Service Center GmbH
Europaplatz 3, 69115 Heidelberg, Germany

Printed by Libri Plureos GmbH
in Hamburg, Germany